Optimization of Logistics and Supply Chain Systems

Theory and Practice

Turkay YILDIZ

DEDICATION

To my parents...

Table of Contents

CHAPTER 7. A CLOSED-LOOP SUPPLY CHAIN SYSTEM: A MULTI-ECHELON, MULTI-PERIOD, MULTI-PRODUCT SYSTEM 173

CHAPTER 8. OPTIMAL SCHEDULING OF DISTRIBUTIONS TO DEMANDING NODES BASED ON PREFERENCES 253

PART III – LOGISTICS IMPACT ASSESSMENT OF TRADE POLICIES BY CGE MODELING 277

CHAPTER 9. AN OVERVIEW OF THE COMPUTABLE GENERAL EQUILIBRIUM (CGE) MODELING 277

CHAPTER 10. THE STANDARD CGE MODEL WITH GAMS CODE 279

Preface

This book is the collection of my own studies in the optimization of logistics and supply chain systems, targeted to a broad readership. In this book, first few chapters of the optimization concepts come out of the materials I used to refer while doing my research. I brought together some of these materials to form a guidance material on the fundamentals of the optimization concepts along with my own studies in the application of optimization methods.

For a macro level logistics and supply chain systems optimization perspective, I added several chapters that are the results of my studies, which use the Global Trade Analysis Project (GTAP and GDyn) simulations. The *static* and *dynamic* computable general equilibrium (CGE) concept is also the essential part of the optimization study of logistics and supply chain systems. Therefore, the fundamentals of the static and dynamic (CGE) models are introduced. As the demand for logistics depends mostly on the volume of trade and trade patterns, international trade affects the transport and logistics, as it might generate a higher or lower demand for transport and logistics services, which are calculated by various approaches of optimization studies.

This book consists of six parts and twenty chapters. **The first part** of the book, which includes three chapters, is about introduction to optimization with typical base problems and algorithms for solving these problems. **The second part** of this book includes five my own researches in the application of optimization methods. **The third part** of the book shortly introduces you to the general concepts of the computable general equilibrium models (CGE) and presents you the fundamentals of a CGE model. In each chapter of **the fourth part**, short articles that include five simulations based on various scenarios are presented. **The fifth part** of the book briefly introduces you to the basic concepts of the computable general equilibrium models (CGE) and then, presents you the fundamentals of dynamic general equilibrium models. In each chapter of **the sixth part**, two short articles that simulate various scenarios are presented.

All the chapters in this book are independent of each other. I hope you will find this book informative, beneficial and appropriate for your needs.

Turkay Yildiz, M.A., Ph.D.

ii

Part I

Chapter 1. The Optimization Concept

Optimization is already a part of our life. You might not be aware of it. You are already doing optimization in your life. For example, you wake up early in the morning and try to catch a bus or train on time. Your objective is to catch the bus/train on time while properly using your limited time early in the morning. You go to shopping and decide what to purchase within your budget constraint. Here, you try to maximize your satisfaction within your budget and in a time constraint. You choose the shortest and accessible paths to arrive from point A to B. You try to allocate more time for your leisure activities while considering allocating more time for studying and doing your home works.

There are so many examples in our lives that we try to optimize while maximizing/minimizing our objectives with many other constraints as such time, cost, etc.

Optimization also plays important role in business organizations. Indeed, systematically applying optimization techniques in business processes can help organizations save millions of dollars by helping in reducing costs and eliminating redundant activities. In return, these savings might help in investing in areas that are most relevant and might make the organization more competitive.

Depending on the number of variables, optimization problems range from a very simple, short definition of a problem to a more complex longer definition. Retrieving the best solutions for these kind of problems highly depend on the capabilities of the algorithms and thus the computer power. Thanks to the recent advancement of computers with powerful CPUs and larger memories, many complex problems could be solved within seconds.

This chapter introduces the fundamentals of optimization. In subsections, types of optimization problems, optimization methods for solving optimization problems and the concept of complexity of algorithms and problems are introduced.

1.1 Introduction

Optimization is an old subject that has shown a resurgence since the advent of computers and whose methods are applied in many fields: economics, management, planning, logistics, robotics, optimal design, engineering, signal processing, etc. (Allaire and Craig 2007). Optimization is a mature area because of the extensive research that has been conducted over the last 60 years (Arora 2007). The optimization is a broad topic that touches on the calculus of variations, operations research (optimization of management processes or decision) and optimal control

(Allaire and Craig 2007). Optimization is an important tool in the decision science and in the analysis of physical systems (Nocedal and Wright, 1999). Optimization is everywhere, from business to engineering design, planning your vacation to your daily routine (Yang 2008).

Optimization is very important and relevant to virtually every discipline (Rangaiah 2010). Professional organizations must maximize profit and minimize cost (Yang 2008). The technical design is to maximize the performance of any product designed while minimizing costs at the same time (Yang 2008). The study of logistics systems typically involves several phases (Yalaoui *et al.* 2013):

- The objective of the first phase is the acquisition of information, which is to identify the parameters that affect the system. This could be profit, time, potential energy, or the quantity or combination of quantities that can be represented by a single number (Nocedal and Wright, 1999).
- The second, very important, phase is the development of a model to integrate all the parameters identified.
- The third phase focuses on the evaluation and analysis of the performance of the system using the original model. This performance is understood by its quantitative and qualitative aspects. However, the model can be large and difficult to use. A phase of further simplification can then reduce the order of the system, while maintaining the characteristics and properties of the original model.
- Finally, an optimization phase is regularly used to improve system performance.

Any optimization problem has three basic ingredients (Arora 2007):

- The optimization variables, also called the design variables.
- Cost function, also called the objective function.
- The constraints expressed as equalities or inequalities.

The variables for the problem can be continuous or discrete. Depending on the types of variables and functions, one might obtain a continuous variable, discrete variable, differentiable and non-differentiable problems (Arora 2007).

Depicting an optimization problem in a mathematical formulation is a critical step in the process of solving the problem (Arora 2007). The decision consists of the following steps (Talbi 2009):
- **Formulating the problem:** In this first step, a decision problem is identified. Then, an initial problem statement is made.

- **Modeling the problem:** In this important step, a mathematical model is constructed for the problem.
- **Optimizing the problem:** Once the problem is modeled, the solving procedure generates a good solution to the problem. The solution may be optimal or suboptimal.
- **Implementing a solution:** The solution is tested in practice by the decision maker and is implemented if it is acceptable.

1.2 Types of optimization problems

Many types of problems have been addressed and many different types of algorithms have been studied in the literature (Arora 2007). Many methodologies have been used in various practical applications and the range of applications is constantly increasing (Arora 2007). Optimization problems can be divided into the following general categories depending on the type of decision variables, function(s), objectives and constraints (Diwekar 2008):

- **Linear programming (LP):** The objective function and constraints are linear. The decision variables are scalar and continuous.
- **Nonlinear programming (NLP):** The function and / or objective constraints are not linear. The decision variables are scalar and continuous.
- **Integer Programming (IP):** The decision variables are scalar and integers.
- **Mixed integer linear programming (MILP):** The objective function and constraints are linear. The decision variables are scalar; some of them are integers, while others are continuous variables.
- **Mixed integer nonlinear programming (MINLP):** Integer linear programming and continuous decision variables are involved.
- **Discrete optimization:** Problems involving discrete variables (integer). These problems are IP, MILP, and includes MINLPs.
- **Optimal Control:** The decision variables are vectors.
- **Stochastic programming or stochastic optimization:** It is also called optimization under uncertainty. In these problems, the objective function and / or constraints have uncertain variables (random). Often involves the above categories as subcategories.
- **Multi-objective optimization:** Problems involving more than one objective. Often involves the above categories as subcategories.

It is important to note that the term "programming" in this context means planning activities that consume resources and / or meet the requirements as expressed in the constraints, not the other meaning – i.e. coding computer programs (Chen *et al.* 2011).

A linear programming problem (LP) is a class of mathematical programming problem, a problem of constrained optimization, in which we seek to find a set of values for continuous variables that maximizes or minimizes a linear objective function z, while satisfying a set of linear constraints (a system of simultaneous linear equations and / or inequalities) (Chen *et al.* 2011).

Linear programming is a powerful mathematical modeling technique, which is widely used, in business planning, technical design, the oil industry, and many other optimization applications (Yang 2008). Linear programming (LP) problems involve linear objective function and linear constraints (Diwekar 2008). The basic idea of linear programming is to find the maximum or minimum under linear constraints (Yang 2008).

An integer problem (linear) programming (IP) is a linear programming problem, in which at least one of the variables is restricted to integer values (Chen *et al.* 2011).

In the last two decades there has been an increasing use of a problem in other term – **mixed integer programming (MIP)** – for LPs with integer restrictions on part or all of the variables (Chen *et al.* 2011).

A **combinatorial optimization problem (COP)** is a discrete optimization problem in which we seek to find a solution within a finite set of solutions that maximizes or minimizes an objective function (Chen *et al.* 2011). This type of problem usually arises in the selection of a finite set of mutually exclusive alternatives (Chen *et al.* 2011).

In **Nonlinear (NLP) programming problems**, either the objective function, constraints, or both the objective and constraints are non-linear (Diwekar 2008).

Discrete optimization problems can be classified as **integer programming (IP)** problems, the **mixed linear integer programming (MILP)**, and **mixed integer non-linear programming (MINLP)** problems (Diwekar 2008).

Optimization under uncertainty: optimization under uncertainty refers to that branch of optimization where there are uncertainties in the data or model, and is popularly known as **stochastic programming** or **stochastic optimization problems** (Diwekar 2008). In this terminology, stochastic randomness and programming are related to mathematical programming techniques such as LP, NLP, IP, MILP, and MINLP (Diwekar 2008). The simplest approach to deal with uncertainty is to estimate the average value of each parameter and solve a deterministic problem (Talbi 2009).

Multi-objective optimization: Suppose one tries to improve product performance while trying to minimize the cost at the same time. In this case, he or she is dealing with a multi-objective optimization problem. Many new approaches are needed to solve the multi-objective optimization (Yang 2014).

Dynamic optimization: In dynamic optimization problems, the objective function is deterministic at some point, but varies over time (Talbi 2009).

The main issues to solve dynamic optimization problems are (Talbi 2009):

- Detect the change in the environment where it occurs. For most problems of real life, change is smooth rather than radical.
- Respond to the changing environment to follow the new global optimal solution. The objective is to dynamically track change of the optimal.

Robust Optimization: To solve the problem, one needs to consider the fact that a solution must be acceptable with respect to slight variations in the values of the decision variables (Talbi 2009). Robust optimization can be regarded as a specific type of problem with uncertainties (Talbi 2009).

In this class of problems, the solution implementation should be insensitive to small variations in the design parameters (Talbi 2009). This variation can be caused by production tolerances or drift in parameters during operation (Talbi 2009).

1.3 Optimization Methods

There are three major classes of algorithms to solve problems of integer programming (Pop 2012):

- **Exact algorithms** are guaranteed to find the optimal solution, but it can take an exponential number of iterations. In practice, they are usually only applicable to small instances, due to long operating times caused by the complexity. They include: branch-and-bound, branch-and-cut, cutting plane, and dynamic programming algorithms.
- **Approximation algorithms** provide a sub-optimal solution in polynomial time, with a bounding constraint on the degree of sub-optimality. Unlike heuristics, which generally provide pretty good solutions within a reasonable time, approximation algorithms provide provable solution quality and provable execution limits (Talbi 2009).
- **Heuristic algorithms** provide an optimum solution, but make no guarantee as to its quality. Heuristic algorithms use trial and error, learning and adapting to solve problems (Yang 2010). Although the running time is not guaranteed to be polynomial, empirical evidence suggests that some of these algorithms quickly find a good solution. These algorithms are particularly suited for large instances of optimization problems. Unlike exact methods, meta-heuristics are

used to tackle cases of large-scale problems by providing satisfactory solutions within a reasonable time (Talbi 2009). Metaheuristics received increasing popularity over the last 20 years. Their use in many applications shows their effectiveness in solving large, complex problems (Talbi 2009). Modern metaheuristics algorithms are almost guaranteed to work well for a wide range of optimization problems difficult (Yang 2010).

Evolutionary computation techniques have attracted increasing attention in recent years to solve complex optimization problems (Sarker *et al.* 2002). They are more robust than the traditional methods (Sarker *et al.* 2002). Evolutionary computation techniques can cope with complex problems of better optimization of conventional optimization techniques (Sarker *et al.* 2002).

Finally, it should be noted that there is "no free lunch" in the optimization (Yang 2010). It has been proved by Wolpert and Macready in 1997 that if the algorithm A is better than algorithm B for some problems, then B to outperform A over other problems (Yang 2010). That is to say, a universally efficient algorithm does not exist (Yang 2010).

1.4 Complexity of algorithms and problems

The first step in the study of a combinatorial problem is whether the problem is "easy" or "hard" (Pop 2012). This classification is all about the complexity theory.

The efficiency of the algorithm is often measured by the computational complexity. In the literature, this complexity is also called Kolmogorov complexity. For a given problem size n, the complexity is noted using the **Big-O** notation as $O(n^2)$ or $O(n \log n)$ (Yang 2008).

$O(n)$ denotes a linear complexity while $O(n^2)$ has a quadratic complexity (Yang 2008). That is, if n is doubled, the time of the linear complexity will double, but will quadruple the quadratic complexity (Yang 2008).

Complexity of algorithms (Talbi 2009): An algorithm needs two important resources to solve a problem: time and space. The time complexity of the algorithm is the number of steps required to solve a problem of size n. Complexity is usually defined as the worst-case analysis. The purpose of determining the computational complexity of an algorithm is not to obtain a number of correct but not an asymptotic bound on the number of steps. Big-O notation makes use of the asymptotic analysis. It is one of the most popular notations in the analysis of algorithms.

Complexity of the problems (Talbi 2009): The complexity of a problem is equivalent to the complexity of the best algorithm solving this problem. One problem is soluble (or easy) if it is for a polynomial time algorithm to solve it. One problem is insoluble (or difficult) if no polynomial time algorithm exists to solve the problem. The theory of the complexity of issues deals with decision problems. A decision problem always has a "yes" or "no".

An important aspect of the theory of computation is to classify problems in complexity classes. **A complexity class** is the set of all problems that can be solved using a given amount of computing resources. There are two important classes of problems: P and NP (Talbi 2009).

Therefore, the P represents the family of problems where a known polynomial-time algorithm exists to solve the problem. Problems in class P are then relatively easy to solve (Talbi 2009).

A problem is called non-deterministic polynomial (NP) if its solution can be guessed and evaluated in polynomial time (hence, non-deterministic) (Yang 2008).

Some problems of the class P: Some classical problems in class P are minimum spanning tree, shortest path problems, maximum matching bipartite, and programming linear continuous models (Talbi 2009).

The question of whether P = NP is one of the most important open questions due to the large impact, the answer would have on the theory of computational complexity. Obviously, for every problem in P we solve a non-deterministic algorithm (Talbi 2009).

If a polynomial deterministic algorithm exists to solve an NP-complete problem, then all problems in class NP can be solved in polynomial time (Talbi 2009).

No known algorithms exist to solve NP-hard problems, however, approximate solutions or heuristic solutions are possible (Yang 2008). Thus, heuristics and metaheuristics are very promising in obtaining approximate solutions or suboptimal / nearly optimal solutions (Yang 2008). Many popular academic problems are NP-hard (Talbi 2009):

- Sequencing problems and planning such as flow-shop, job-shop planning, or open shop scheduling.
- Assignment and location problems such as quadratic assignment problem (QAP), generalized assignment problem (GAP), the facility location, and p-edian problem.
- Problems such as data clustering, graph partitioning and graph coloring.
- Routing and covering problems such as vehicle routing problem (VRP), set covering problem (SCP), the problem of Steiner tree and covering tour problem (CTP).
- Knapsack and packing/cutting problems, and so on.

References

Allaire, Gregoire; Craig, Alan (Translated by). Numerical Analysis and Optimization: An Introduction to Mathematical Modelling and Numerical Simulation. Cary, NC, USA: Oxford University Press 2007. p 294.

Arora, J. S. (2007). Optimization of Structural and Mechanical Systems. Hackensack, NJ, World Scientific.

Arora, Jasbir S. (Editor). Optimization of Structural and Mechanical Systems. River Edge, NJ, USA: World Scientific 2007. p 2.

Chen, Der-San; Batson, Robert G.; Dang, Yu. Applied Integer Programming Modeling and Solution. Hoboken, NJ, USA: Wiley 2011. p 3-7.

Nocedal, Jorge; Wright, Stephen J.. Numerical Optimization. Secaucus, NJ, USA: Springer, 1999. p 1-2.

Pop, Petrica C.; Versita (Contribution by). De Gruyter Series in Discrete Mathematics and Applications, Volume 1 : Network Design Problems : Modeling and Optimization of Generalized Network Design Problems. Hawthorne, NY, USA: Walter de Gruyter 2012. p 1-3.

Rangaiah, G. P. (2010). Stochastic Global Optimization : Techniques and Applications in Chemical Engineering. Singapore, World Scientific.

Sarker, Ruhul (Editor); Mohammadian, Masoud (Editor); Yao, Xin (Editor). Evolutionary Optimization. Secaucus, NJ, USA: Kluwer Academic Publishers 2002. p 10.

Talbi, El-Ghazali. (2009). Metaheuristics. Wiley.

Urmila Diwekar. (2008). Introduction to Applied Optimization. Springer US.

Yalaoui, Alice; Chehade, Hicham; Yalaoui , Farouk; Amodeo, Lionel. (2013). Optimization of Logistics. Wiley-ISTE.

Yang, Xin-She. (2014). Nature-Inspired Optimization Algorithms. Elsevier.

Yang, Xin-She. Engineering Optimization: An Introduction with Metaheuristic Applications. Hoboken, NJ, USA: Wiley 2010. p xxviii.

Yang, Xin-She. Introduction to Mathematical Optimization: From Linear Programming to Metaheuristics. Cambridge, GBR: Cambridge International Science Publishing 2008. p 3-79.

Chapter 2. Typical Base Optimization Problems

This chapter briefly introduces typical pure base optimization problems of logistics systems. These problems are namely the assignment problem, knapsack problem, facility location problem, lot-sizing problem, vehicle routing problem, scheduling problems, shortest path problem, and the travelling salesman problem. Typical optimization problems are not limited to these pure base problems in this chapter. Indeed, the real world problems usually require much more complicated problem definitions and utilization of sophisticated solvers for finding optimal solutions.

In each section of this chapter, mathematical representation of the optimization problem is shown. Additionally, the literature on this subject is so vast. Therefore, in each section the titles of the most recent and highly cited studies are shown in tables. Readers might take a look at these studies and further familiarize themselves for the recent studies on the optimization topic.

2.1 Assignment Problem

Assignment problem is a special case of transportation problem and the transportation problem is a special case of linear programming problem, so it is a special case of linear programming problem (Mishra and Agarwal 2009). The assignment problem can be considered as a special case of transportation problem (Williams 2013). It can be considered a problem with n sources and n sinks (Williams 2013). The objective of the assignment problem is to assign a number of tasks or jobs to an equal number of persons or the machine to minimize cost or to maximize profit. Assignment problem is a minimization problem (Mishra and Agarwal 2009).

The assignment problem is to find a maximum profit assignment of n tasks to n machines such that each task (i = 1, 2, ..., n) is assigned to exactly one machine (j= 1, 2, ..., n) and each machine is assigned to exactly one task. The general assignment problem (GAP) is a generalization of the assignment problem that finds a maximum profitable assignment of m tasks to n (m > n) machines such that each task is assigned to exactly one machine and that each machine is allowed to be assigned to more than one task, subject to its capacity limitation (Chen *et al.* 2011).

The GAP is also known as the maximum weighted bipartite matching problem. Given a bipartite graph G(V,U,E) where V and U are two partitions and E edges between two partitions, the problem is the selection of a subset of the edges with maximum sum of weights such that each node $v \in V$ or $u \in U$ is connected to at most one edge.

The problem is finding a minimal cost (or maximum profit) assignment of n tasks over m capacity-constrained servers, whereby each task has to be processed by only one server.

Objective function

$$\text{Minimize } Z = \sum_{i=1}^{n}\sum_{j=1}^{m} C_{ij} x_{ij}$$

Subject to

$$\sum_{j=1}^{m} x_{ij} = 1, \qquad i = 1,\ldots,n$$

$$\sum_{j=1}^{m} a_{ij} x_{ij} \leq b_j, \qquad j = 1,\ldots,n$$

$$x_{ij} \in \{0,1\}, \qquad i = 1,\ldots,n, \quad j = 1,\ldots,m$$

where parameters are

n   = number of tasks

m   = number of servers

C_{ij} = cost of assigning task i to server j

b_j = units of resource available to server j

a_{ij} = units of resource required to perform task i by server j

and variables

x_{ij} = 1 if task i is assigned to server j, 0 otherwise

GAP is to find a minimal cost (or maximum profit) assignment of n tasks on servers with limited capabilities m, each task should be handled by a single server (Sarker 2008). The general assignment problem has a wide range of relevant fields. For example, one such area is the situation causing in container terminals (Cheung *et al.* 2002; Zhang *et al.* 2002). Given an n tasks on servers with limited capacity m can be considered assignment of n number of straddle carriers (SC) or trailers with special containers to minimize the total time (t) consumed or cost (c) when handling the containers.

An assignment problem is a discrete optimization problem. In discrete optimization problems, the size of the problems grows exponentially with the number of options along each dimension (Subbu and Sanderson 2004).

An extension of the complex assignment problem is the quadratic assignment problem. This problem occurs when the "cost of an assignment" is not independent from other assignments. The resulting problem can be seen as a problem of assigning a quadratic objective function (Williams 2013).

Various algorithms exist for the solution of general assignment problem. For solving problems of assignment in a container terminal, here two algorithms, i.e., Jonker-Volgenant and Hungarian algorithms are considered. Both algorithms provide the same solutions and the solutions are considered to be the best possible optimal solutions for the given probability matrix. For a performance comparison of the both algorithms, see Figure 2.1.

The most recent and highly cited studies about the assignment problem are shown in Table 2.1.

Table 2.1 Studies in the literature about the assignment problem

A computational study of exact knapsack separation for the generalized assignment problem (Avella *et al.* 2010).
Solving the class responsibility assignment problem in object-oriented analysis with multi-objective genetic algorithms (Bowman *et al.* 2010).
The multi-unit assignment problem: theory and evidence from course allocation at harvard (Budish and Cantillon 2012).
The wiener maximum quadratic assignment problem (cela *et al.* 2011).
An efficient solution algorithm for solving multi-class reliability-based traffic assignment problem (Chen *et al.* 2011).
The storage location assignment problem for outbound containers in a maritime terminal (Chen and Lu 2012).
An ant colony optimisation algorithm for solving the asymmetric traffic assignment problem (D'acierno *et al.* 2012).
Exploiting group symmetry in semidefinite programming relaxations of the quadratic assignment problem (de Klerk and Sotirov 2010).
State covariance assignment problem (Khaloozadeh and Baromand 2010).
A multi-objective evolutionary algorithm for the deployment and power assignment problem in wireless sensor networks (Konstantinidis *et al.* 2010).
Reliability optimization of component assignment problem for a multistate network in terms of minimal cuts (Lin and Yeh 2011).
Stability of user-equilibrium route flow solutions for the traffic assignment problem (Lu and Nie 2010).
Grasp with path-relinking for the generalized quadratic assignment problem (Mateus *et al.* 2011).
Multiobjective genetic algorithms for solving the impairment-aware routing and wavelength assignment problem (Monoyios and Vlachos 2011).
A class of bush-based algorithms for the traffic assignment problem (Nie 2010).
A cell-based merchant-nemhauser model for the system optimum dynamic traffic assignment problem (Nie 2011).
Solving the dynamic user optimal assignment problem considering queue spillback (Nie and Zhang 2010).
Bees algorithm for generalized assignment problem (Ozbakir *et al.* 2010).
A two phase method for multi-objective integer programming and its application to the assignment problem with three objectives (Przybylski *et al.* 2010).
Partial eigenvalue assignment problem of high order control systems using orthogonality relations (Ramadan and El-Sayed 2010).
A due-date assignment problem with learning effect and deteriorating jobs (Wang and Guo 2010).
Single-machine due-window assignment problem with learning effect and deteriorating jobs (Wang and Wang 2011).
Effective formulation reductions for the quadratic assignment problem (Zhang *et al.* 2010).
Routing and wavelength assignment problem in pce-based wavelength-switched optical networks (Zhao *et al.* 2010).
A novel global harmony search algorithm for task assignment problem (Zou *et al.* 2010).

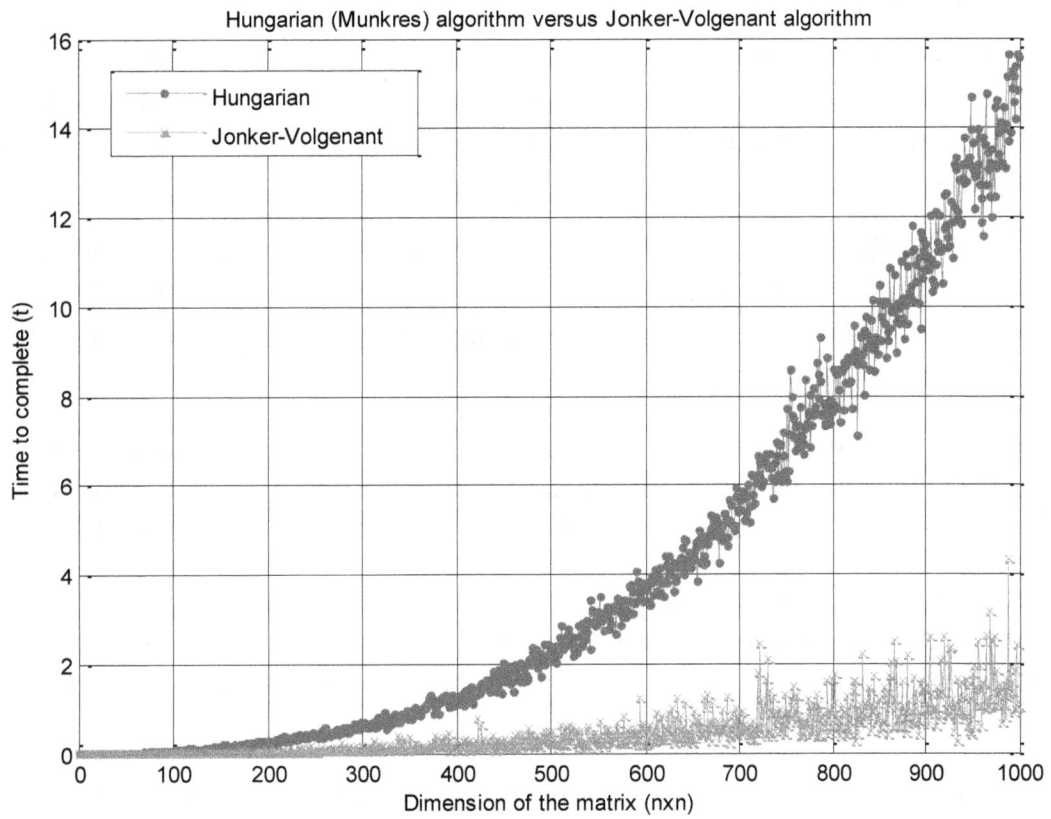

Figure 2.1 Hungarian (Munkres) algorithm versus Jonker-Volgenant algorithm

2.2 Knapsack Problem

The "knapsack" stems from the rather artificial application to try to fulfill a hiker's knapsack at the total maximum value application. Each item it considers taking with it has a certain value and a certain weight. Limiting overall weight is the single constraint (Williams 2013).

The knapsack problem has received considerable attention in the literature during the early development of operations research algorithms (1950-1970) (Chen *et al.* 2011). Dozens of specialized algorithms for knapsack problems have been developed (Chen *et al.* 2011).

The knapsack problem takes a set of items, each with a weight and profit, and tries to fit as many of these items in the knapsack to maximize profit while not exceeding the maximum weight the knapsack may contain (Rondeau and Bostian 2009).

The name is derived from a problem of decision facing a hiker who is to select a set of given elements to be included in his backpack (or knapsack) within a specified weight (Chen *et al.* 2011). The purpose of this decision problem is to maximize the total value of all the selected items (Chen *et al.* 2011).

The problem of the knapsack is sometimes called the problem of loading cargo when shipments of different weights are selected for loading on a vessel with a weight capacity limited (Chen *et al.* 2011).

Objective function

$$Maximize\ Z = \sum_{i=1}^{m}\sum_{j=1}^{n}v_j x_{ij}$$

Subject to

$$\sum_{i=1}^{m}x_{ij} \leq 1, \quad \forall j$$

$$\sum_{j=1}^{n}w_j x_{ij} \leq C_i, \quad \forall j$$

$$x_{ij} \in \{0,1\}, \quad \forall i, j$$

where parameters are

m = number of containers (index i)

n = number of items (index j)

w_j = weight of item j

v_j = value of item j

C_i = capacity of container i (weight)

The most recent and highly cited studies about the knapsack problem are shown in Table 2.2.

Table 2.2 Studies in the literature about the knapsack problem

Kernel search: a general heuristic for the multi-dimensional knapsack problem (Angelelli *et al.* 2010).
Identifying preferred solutions to multi-objective binary optimisation problems, with an application to the multi-objective knapsack problem (Argyris *et al.* 2011).
A multi-level search strategy for the 0-1 multidimensional knapsack problem (Boussier *et al.* 2010).
A knapsack problem as a tool to solve the production planning problem in small foundries (Camargo *et al.* 2012).
A column generation method for the multiple-choice multi-dimensional knapsack problem (Cherfi and Hifi 2010).
A heuristic approach for allocation of data to rfid tags: a data allocation knapsack problem (dakp) (Davis *et al.* 2012).
Revenue maximization in the dynamic knapsack problem (Dizdar *et al.* 2011).
Development of core to solve the multidimensional multiple-choice knapsack problem (Ghasemi and Razzazi 2011).
Sensor selection in distributed multiple-radar architectures for localization: a knapsack problem formulation (Godrich *et al.* 2012).
A ptas for the chance-constrained knapsack problem with random item sizes (Goyal and Ravi 2010).
Improved convergent heuristics for the 0-1 multidimensional knapsack problem (Hanafi and Wilbaut 2011).
Problem reduction heuristic for the 0-1 multidimensional knapsack problem (Hill *et al.* 2012).
An ant colony optimization approach for the multidimensional knapsack problem (Ke *et al.* 2010).
Fully polynomial approximation schemes for a symmetric quadratic knapsack problem and its scheduling applications (Kellerer and Strusevich 2010).
Upper bounds for the 0-1 stochastic knapsack problem and a b&b algorithm (Kosuch and Lisser 2010).
Assessing solution quality of biobjective 0-1 knapsack problem using evolutionary and heuristic algorithms (Kumar and Singh 2010).
Reoptimization in lagrangian methods for the 0-1 quadratic knapsack problem (Letocart *et al.* 2012).
Interdicting nuclear material on cargo containers using knapsack problem models (Mclay *et al.* 2011).
The multidimensional knapsack problem: structure and algorithms (Puchinger *et al.* 2010).
Fusing ant colony optimization with lagrangian relaxation for the multiple-choice multidimensional knapsack problem (Ren *et al.* 2012).
Dynamic programming based algorithms for the discounted {0-1} knapsack problem (Rong *et al.* 2012).
A two state reduction based dynamic programming algorithm for the bi-objective 0-1 knapsack problem (Rong *et al.* 2011).
A cooperative local search-based algorithm for the multiple-scenario max-min knapsack problem (Sbihi 2010).
An effective hybrid eda-based algorithm for solving multidimensional knapsack problem (Wang 2012).
Solving 0-1 knapsack problem by a novel global harmony search algorithm (Zou *et al.* 2011).

2.3 Facility location problem

Facility location problems deal with the number, location, equipment and size of new plants as well as the sale, removal or reduction of existing facilities (Ghiani 2013). In the logistics business, the process of site planning is to design the entire facility through which revenue from suppliers to demand points, while in the public sector, it is to determine all facilities from which users are served (Ghiani 2013).

One of the most important aspects of logistics is deciding where to locate new facilities such as retailers, warehouses or factories (Bramel and Simchi-Levi, 1997). These strategic decisions are a key determinant of whether the materials circulate efficiently in the distribution system (Bramel and Simchi-Levi, 1997). When location decisions are needed: the location of the installation decisions must obviously be taken when a logistics system is started from zero (Ghiani 2013). They are also required as a result of variations in the pattern of demand or the spatial distribution, or due to changes in materials, energy and labor costs.

Mathematical models of location are designed to address a number of issues, including (Daskin 2013)

- How many facilities should be located?
- Where each facility should be located?
- How large should each facility be?
- How should require the services of facilities allocated to installations?

In general, the goal is to find a set of facilities so that the total cost is minimized subject to a number of constraints that could include (Bramel and Simchi-Levi, 1997):

- each warehouse has a capacity that limits the area that can provide.
- each retailer receives deliveries of one and only one warehouse.
- each retailer must be at a fixed distance from the warehouse that supplies it, so that a reasonable delivery lead time is guaranteed.

Facility location is a wide range of mathematical models, methods and applications in operations research (Chvátal 2011). This is an interesting topic for theoretical studies, experimental research and real-world applications (Chvátal 2011).

The most recent and highly cited studies about the facility location problem are shown in Table 2.3.

Table 2.3 Studies in the literature about the facility location problem

A computational comparison of several formulations for the multi-period incremental service facility location problem (Albareda-Sambola *et al.* 2010).
The facility location problem with bernoulli demands (Albareda-Sambola *et al.* 2011).
P-hub approach for the optimal park-and-ride facility location problem (Aros-Vera *et al.* 2013).
Semi-lagrangian relaxation applied to the uncapacitated facility location problem (Beltran-Royo *et al.* 2012).
Solving conflicting bi-objective facility location problem by nsga ii evolutionary algorithm (Bhattacharya and Bandyopadhyay 2010).
An optimal bifactor approximation algorithm for the metric uncapacitated facility location problem (Byrka and Aardal 2010).
A computational study of a nonlinear minsum facility location problem (Carrizosa *et al.* 2012).
Strategic closed-loop facility location problem with carbon market trading (Diabat *et al.* 2013).
A primal-dual approximation algorithm for the facility location problem with submodular penalties (Du *et al.* 2012).
An approximation algorithm for the k-level capacitated facility location problem (Du *et al.* 2010).
A new approximation algorithm for the multilevel facility location problem (Gabor and van Ommeren 2010).
A generalized weiszfeld method for the multi-facility location problem (Iyigun and Ben-Israel 2010).
The ordered capacitated facility location problem (Kalcsics *et al.* 2010).
Competitive facility location problem with attractiveness adjustment of the follower a bilevel programming model and its solution (Kucukaydin *et al.* 2011).
Integrated use of fuzzy c-means and convex programming for capacitated multi-facility location problem (Kucukdeniz *et al.* 2012).
An efficient genetic algorithm for solving the multi-level uncapacitated facility location problem (Maric 2010).
The discrete facility location problem with balanced allocation of customers (Marin 2011).
An integer decomposition algorithm for solving a two-stage facility location problem with second-stage activation costs (Penuel *et al.* 2010).
On the structure of the solution set for the single facility location problem with average distances (Puerto and Rodriguez-Chia 2011).
The reliable facility location problem: formulations, heuristics, and approximation algorithms (Shen *et al.* 2011).
A tabu search heuristic procedure for the capacitated facility location problem (Sun 2012).
An improved benders decomposition algorithm for the logistics facility location problem with capacity expansions (Tang *et al.* 2013).
A genetic algorithm for the uncapacitated facility location problem (Tohyama *et al.* 2011).
An approximation algorithm for the k-level stochastic facility location problem (Wang *et al.* 2010).
A cut-and-solve based algorithm for the single-source capacitated facility location problem (Yang *et al.* 2012).

2.4 Lot sizing problem

Once the levels of inventory are determined, the next step is to calculate in what quantities the inventory will be replaced. This is called lot sizing. The lot size is the amount of material to be ordered from a supplier or internally produced to meet demand.

There are nine major types of lot-size methods, which fit into the following two categories:

Demand-based methods (static): Order quantities are kept constant.

- Fixed order quantity: min/max
- Economic order quantity (EOQ): is calculated periodically and used as fixed order quantity during interim

Discrete method (dynamic): Order quantities vary.

- Period order quantity
- Lot-for-lot
- Periods of supply
- Least unit cost
- Least total cost
- Part-period balancing
- Wagner-Whitin algorithm

Selecting the appropriate combination of methods will help reduce the ordering, setup and transportation costs, and reduce overall inventory levels of products being manufactured (Viale and Carrigan, 1996):

- **Fixed Order Quantity:** method of fixed order quantity orders will always suggest planned to be released for a predetermined fixed amount. The predetermined amount can be determined on the basis of experience and / or the use of the technique of the economic order quantity.
- **The economic order quantity (EOQ)** is the other type of formula based on demand or static. This calculation determines the amount to be purchased or made by determining the minimum cost of purchase or construction with the cost to carry inventory. The formula can be used to determine the minimum units to be built or purchased, or the minimum cost in dollars.
- **Period Order Quantity** Period order quantity is a lot-sizing technique where the lot size is equal to the requirements of a given number of periods in the future. The period order quantity is similar to the period of supply, with the exception of the order cycle is based on the calculation EOQ. The order frequency and the order quantities are scheduled using this method.

- **Lot-for-Lot** This is a technique commonly used in MRP lot-sizing technique in just-in-time (JIT) situations, in collaboration with the safety stock. In this method, the planned orders are generated at the height of the net requirements in each period.
- **Periods of supply** This method establishes -mainly by experience- an amount of the order to cover a predetermined period of time.
- **Least unit cost** The least unit cost method adds the cost of ordering and inventory-carrying cost for each trial lot size and divides the number of units in the lot size. Lot size with the lowest unit cost is chosen.
- **Least total cost** The least total cost lot-sizing technique calculates the order quantity by comparing the set (or order) costs and carrying costs for different lot sizes and selects the lot size where these costs are nearest equal.
- **Part-Period Balancing** This technique is similar to the method in the least cost method. However, this method uses a routine called look forward / looking back. When the function look forward / looking back is used, a lot quantity is calculated, and before it is firmed up, the next or previous period is examined to determine whether it would be economical to fit in the current lot.
- **Wagner-Whitin Algorithm** The last method is the algorithm of Wagner-Whitin. This is a very complex process that evaluates all possible means to meet the needs of each period of the planning horizon.

The most recent and highly cited studies about the lot-sizing problem are shown in Table 2.4.

Table 2.4 Studies in the literature about the lot sizing problem

Uncapacitated lot-sizing problem with production time windows, early productions, backlogs and lost sales (Absi *et al.* 2011).
Adaptive genetic algorithm for lot-sizing problem with self-adjustment operation rate: a discussion (Cardenas-Barron 2010).
A volume flexible economic production lot-sizing problem with imperfect quality and random machine failure in fuzzy-stochastic environment (Das *et al.* 2011).
A particle swarm optimization for solving joint pricing and lot-sizing problem with fluctuating demand and unit purchasing cost (Dye and Hsieh 2010).
A particle swarm optimization for solving joint pricing and lot-sizing problem with fluctuating demand and trade credit financing (Dye and Ouyang 2011).
Solving single-product economic lot-sizing problem with non-increasing setup cost, constant capacity and convex inventory cost in o(n log n) time (Feng *et al.* 2011).
A heuristic approach for a multi-product capacitated lot-sizing problem with pricing (Gonzalez-Ramirez *et al.* 2011).
Stochastic lot-sizing problem with inventory-bounds and constant order-capacities (Guan and Liu 2010).
A robust lot sizing problem with ill-known demands (Guillaume *et al.* 2012).
A simulation metamodelling based neural networks for lot-sizing problem in mto sector (Hachicha 2011).
A fix-and-optimize approach for the multi-level capacitated lot sizing problem (Helber and Sahling 2010).
A robust block-chain based tabu search algorithm for the dynamic lot sizing problem with product returns and remanufacturing (Li *et al.* 2014).
Integrating run-based preventive maintenance into the capacitated lot sizing problem with reliability constraint (Lu *et al.* 2013).
A new algorithmic approach for capacitated lot-sizing problem in flow shops with sequence-dependent setups (Mohammadi *et al.* 2010).
Grasp heuristic with path-relinking for the multi-plant capacitated lot sizing problem. (Nascimento *et al.* 2010).
A simple fptas for a single-item capacitated economic lot-sizing problem with a monotone cost structure (Ng *et al.* 2010).
An o(t-3) algorithm for the capacitated lot sizing problem with minimum order quantities (Okhrin and Richter 2011).
The economic lot-sizing problem with remanufacturing and one-way substitution (Pineyro and Viera 2010).
Solving the stochastic dynamic lot-sizing problem through nature-inspired heuristics (Piperagkas *et al.* 2012).
A new silver-meal based heuristic for the single-item dynamic lot sizing problem with returns and remanufacturing (Schulz 2011).
An efficient computational method for a stochastic dynamic lot-sizing problem under service-level constraints (Tarim *et al.* 2011).
Stochastic dynamic lot-sizing problem using bi-level programming base on artificial intelligence techniques (Wong *et al.* 2012).
An hnp-mp approach for the capacitated multi-item lot sizing problem with setup times (Wu *et al.* 2010).
An mip-based interval heuristic for the capacitated multi-level lot-sizing problem with setup times (Wu *et al.* 2012).
A lagrangian relaxation based approach for the capacitated lot sizing problem in closed-loop supply chain (Zhang *et al.* 2012).

2.5 Vehicle routing problems

Given the fluctuations and the upward trend in the price of oil, transportation costs represent a share of more and more of the final cost charged to the customer (about 10-20% of the overall cost of doing business) it becomes essential to control these costs within global supply chains (Jarboui *et al.* 2013). The class of problems that result from these studies is commonly called the vehicle routing problem (VRP) (Jarboui *et al.* 2013). The classic problem of the development of vehicle routing is to construct roads with minimal cost for a set of vehicles that can visit a set of customers geographically distributed exactly once (Jarboui *et al.* 2013).

The problem is to ascertain the operation plan satisfying the demand at various zones at minimum cost (Bish *et al.* 2001; Kim and Bae, 1998; Vis and De Koster 2003).

Objective function is

$$\text{Minimize } f_{obj} = \sum_{i=1}^{G} \sum_{j=1}^{Z} \sum_{k=1}^{F} C_{ijk} x_{ijk}$$

Subject to

$$\sum_{i=1}^{G} \sum_{k=1}^{F} L_k x_{ijk} \geq D_j \qquad \forall j$$

$$\sum_{j=1}^{Z} \sum_{k=1}^{F} L_k x_{ijk} \leq S_j \qquad \forall i$$

$$\sum_{j=1}^{Z} L_k x_{ijk} \leq U_{ki} \qquad \forall k, i$$

$$x_{ijk} \geq 0 \qquad \forall i, j, k$$

where parameters are

G = Number of source locations (index i)

Z = Number of receiving nodes for containers (index j)

F = Number of trailers available (index k)

L_k = Load capacity of trailer k

S_i = Quantity of available containers for transportation from location i

D_j = Quantity of containers required by zone j

C_{ijk} = Unit cost of transporting from location i to zone j by trailer k

U_{ik} = Maximum allowable containers that can be transported from location i by trailer k in a given period

and variables

x_{ijk} = the number of trips required by trailer k from location i to zone j

In the area of freight transport, both activities are generally distinguished (Jarboui *et al.* 2013):

- Full-load Transportation, the activity of a vehicle consists of a series of movements between the origins and destinations of the goods transported. For each movement, the goods to one customer is on board the vehicle.
- Less-than-truckload (LTL) transportation, each vehicle serves a range of customers from a warehouse where the goods are loaded. One of the central problems in the transportation of LTL is the vehicle routing problem (VRP)

Freight activities are essential in planning logistics systems. The reason is twofold (Ghiani 2013)

- First, they determine the most important part (often between one third and two thirds) of logistics costs; and,
- Second, they significantly affect the level of service provided to customers (Ghiani 2013).

Providing efficient a cost effective transport freight services results in an increase of the distance to the facilities of logistics system can be implemented economically (Ghiani 2013).

Problems related to the determination of the optimal routes for vehicles of one or more depots for a set of locations / customers are known as the vehicle routing problem (VRP) (Pop 2012). They have many practical applications in the field of distribution and logistics (Pop 2012).

The vehicle routing problem is, in practice, the most important extension of the traveling salesman problem where you have to schedule a number of vehicles, limited capacity, around a number of clients (Williams 2013). Thus, in addition to the sequence of customers to visit (the traveling salesman problem) one need to decide which vehicles visiting which customers (Williams 2013). Each customer has a known demand (assumed to be a commodity, but it can easily be extended to more than one product) (Williams 2013). Therefore, the limited capacities of vehicles need to be taken into account.

The distribution problem or vehicle routing (VRP) is often described as the problem in which vehicles based on a central depot need to visit geographically dispersed customers to meet the known demands of customers (Pardalos *et al.* 2002). The problem is to build a low-cost, feasible set of routes - one for each vehicle (Pardalos *et al.* 2002). A route is a sequence of locations that the vehicle has to go along with an indication of the service it provides (Pardalos *et al.* 2002). The vehicle must begin and end his tour at the station (Pardalos *et al.* 2002). We can say that the problem is a generalization of the traveling salesman problem (Pardalos *et al.* 2002). The traveling salesman problem (TSP) requires the determination of a minimum cost cycle that passes through each node of a given once (Pardalos *et al.* 2002) graph.

Due to the simplicity of the VRP, variations of the VRP, built on the basic VRP with additional features, proved more attractive to many researchers (Pop 2012):

- The Capacitated VRP, wherein each vehicle has a finite capacity and each location has a finite demand.
- The VRP with time windows, in which there is a specific opportunity in which to visit each location time window.
- VRP with multiple depots, which generalizes the idea of a depot, in such a way that there are several depots from which each customer can be served.
- The multi-commodity VRP wherein each location has associated demand for various commodities and each vehicle has a set of compartments in a single product which can be loaded. The problem then becomes one of deciding which products to place in which compartments to minimize the distance traveled.
- The general vehicles routing problem (GVRP) is the problem of the design of delivery or collection of optimal routes, one given to a number of nodes, predefined sets, mutually exclusive and exhaustive node-sets (clusters) subject to capacity restrictions. The GVRP can be seen as a particular kind of location-routing problem (see, e.g. Laporte, Nagy and Salhi) for which several heuristic algorithms, mostly exist.

The most recent and highly cited studies about the vehicle routing problem are shown in Table 2.5.

Table 2.5 Studies in the literature about the vehicle routing problem

An exact algorithm for a vehicle routing problem with time windows and multiple use of vehicles (Azi *et al.* 2010).
Some applications of the generalized vehicle routing problem (Baldacci *et al.* 2010).
New route relaxation and pricing strategies for the vehicle routing problem (Baldacci *et al.* 2011).
Metaheuristics for the waste collection vehicle routing problem with time windows, driver rest period and multiple disposal facilities (Benjamin and Beasley 2010).
A tabu search algorithm for the heterogeneous fixed fleet vehicle routing problem (Brandao 2011).
Iterated variable neighborhood descent algorithm for the capacitated vehicle routing problem (Chen *et al.* 2010).
Branch-and-price-and-cut for the split-delivery vehicle routing problem with time windows (Desaulniers 2010).
A multi-start evolutionary local search for the two-dimensional loading capacitated vehicle routing problem (Duhamel *et al.* 2011).
An iterative route construction and improvement algorithm for the vehicle routing problem with soft time windows (Figliozzi 2010).
An improved multi-objective evolutionary algorithm for the vehicle routing problem with time windows (Garcia-najera and Bullinaria 2011).
Using simulated annealing to minimize fuel consumption for the time-dependent vehicle routing problem (Kuo 2010).
Application of genetic algorithms to solve the multidepot vehicle routing problem (Lau *et al.* 2010).
An enhanced ant colony optimization (eaco) applied to capacitated vehicle routing problem (lee *et al.* 2010).
A hybrid genetic - particle swarm optimization algorithm for the vehicle routing problem (Marinakis and Marinaki 2010).
A hybrid particle swarm optimization algorithm for the vehicle routing problem (Marinakis *et al.* 2010).
A memetic algorithm for the multi-compartment vehicle routing problem with stochastic demands (Mendoza *et al.* 2010).
A penalty-based edge assembly memetic algorithm for the vehicle routing problem with time windows (Nagata *et al.* 2010).
An effective memetic algorithm for the cumulative capacitated vehicle routing problem (Ngueveu *et al.* 2010).
The two-echelon capacitated vehicle routing problem: models and math-based heuristics (Perboli *et al.* 2011).
A hybrid evolution strategy for the open vehicle routing problem (Repoussis *et al.* 2010).
An adaptive large neighborhood search heuristic for the cumulative capacitated vehicle routing problem (Ribeiro and Laporte 2012).
A parallel heuristic for the vehicle routing problem with simultaneous pickup and delivery (Subramanian *et al.* 2010).
An artificial bee colony algorithm for the capacitated vehicle routing problem (Szeto *et al.* 2011).
An ant colony optimization model: the period vehicle routing problem with time windows (Yu and Yang 2011).
A parallel improved ant colony optimization for multi-depot vehicle routing problem (Yu *et al.* 2011).

2.6 Scheduling problem

The main activity of an industrial company in the short term is to organize production on a time frame (Yalaoui *et al.* 2013). Scheduling of production orders is to decide which machine(s) to use and in what order to achieve the production (Yalaoui *et al.* 2013). Because of the diversity in production processes and management styles that vary across industries, many methods for optimizing the allocation of tasks have been developed (Yalaoui *et al.* 2013).

In a scheduling problem, four basic concepts are involved: tasks (jobs), resources, constraints and objectives. A job is defined by a set of operations that must be performed. It is characterized in time by a start date and end date. A resource is a machine or a human involved in performing the work (Yalaoui *et al.* 2013). Constraints represent limitations in time, technology and resources. The objectives are the criteria to be optimized in terms of time, resources, costs or output (Yalaoui *et al.* 2013).

The purpose of the supply chain scheduling is to optimize short-medium term decisions in supply chains, taking into account the balance between the tangible economic objectives such as cost reduction or profit maximization and less tangible goals such as customer satisfaction or level of customer service (Sawik 2011). In addition to production scheduling, scheduling for the supply chain considering the manufacture and supply of materials and distribution of finished products, and includes other decisions related to the functional, spatial, and intertemporal integration and coordination of schedules for these activities (Sawik 2011).

The scheduling problem is known to be intractable computation in many cases (El-Abd and Rewini-El-Barr 2005). Fast optimal algorithms can only be obtained when restrictions are imposed on models representing the program and the distributed system (El-Abd and Rewini-El-Barr 2005). Solving the general problem in a reasonable time requires the use of heuristic algorithms (El-Abd and Rewini-El-Barr 2005). These heuristics do not guarantee optimal solutions to the problem, but they are trying to find solutions near optimal (El-Abd and Rewini-El-Barr 2005).

Quay Crane/Yard Crane Scheduling

The scheduling problem with the assumption is that there are n jobs and m machines. Each job must be processed on all machines (i.e. cranes) in a given order. A machine (i.e. crane) can only process one job at a time, and once a job is started on any machine (i.e. crane), it must be processed to completion. The objective is to minimize the sum of the completion times of all the jobs.

Objective function

$$\text{Minimize } Z = \sum_{j=1}^{n} t_{j(m),j}$$

Subject to

$$t_{j(r+1),j} \geq t_{j(r),j} + P_{j(r),j} \quad \text{for } r = 1, 2, ..., m-1 \text{ and } \forall j$$

$$t_{ij} - t_{ik} \leq -P_{ij} + U(1 - x_{ijk}) \quad \forall i, j, k$$

$$t_{ik} - t_{ij} \leq -P_{ik} + Ux_{ijk} \quad \forall i, j, k$$

$$t_{ij} \geq 0 \quad \forall i, j$$

$$x_{ijk} \in \{0, 1\} \quad \forall i, j, k$$

where parameters are

n = the number of jobs

m = the number of machines

P_{ij} = the processing time of job j on machine i

j(r) = the order of machines/operations for job j (for example, job j must be processed on machine 2 first (r=1, i=2), and then machine 4 (r=2, i=4), and so on). For any job j, r = m means the last operation of the job.

and variables:

t_{ij} = the start time of job j on machine i

x_{ijk} = 1 if job j precedes job k on machine i, 0 otherwise (i.e., if job k precedes job j on machine i)

Scheduling (Employees, Stevedore, etc.)

The problem is to determine the number of employees required to meet the different daily work force necessities of seaport terminal while minimizing the general scheduling cost.

Objective function

$$\text{Minimize } Z = \sum_{i=1}^{N} C_i x_i$$

Subject to

29

$$\sum_{i \in M_j} x_i \geq R_j \quad \forall j$$

$$x_i \geq 0 \quad \forall i$$

where parameters are

N   = the total number of roster type

M_j = the set of roster types that will allow working on a day j

R_j = the number of employees required on each day j

C_i = weekly cost per employee assigned to roster type i

and variables

x_i = the number of employees assigned to roster type i

The most recent and highly cited studies about the scheduling problem are shown in Table 2.6.

Table 2.6 Studies in the literature about the scheduling problem

A neurogenetic approach for the resource-constrained project scheduling problem (Agarwal *et al.* 2011).
An artificial immune algorithm for the flexible job-shop scheduling problem (Bagheri *et al.* 2010).
A two-agent single-machine scheduling problem with truncated sum-of-processing-times-based learning considerations (Cheng *et al.* 2011).
An improved genetic algorithm for the distributed and flexible job-shop scheduling problem (de Giovanni and Pezzella 2010).
New multi-objective method to solve reentrant hybrid flow shop scheduling problem (Dugardin *et al.* 2010).
A novel competitive co-evolutionary quantum genetic algorithm for stochastic job shop scheduling problem (Gu *et al.* 2010).
A proactive approach for simultaneous berth and quay crane scheduling problem with stochastic arrival and handling time (Han *et al.* 2010).
A survey of variants and extensions of the resource-constrained project scheduling problem (Hartmann and Briskorn 2010).
A two-machine flowshop scheduling problem with deteriorating jobs and blocking (Lee *et al.* 2010).
A single-machine scheduling problem with two-agent and deteriorating jobs (Lee *et al.* 2010).
A single-machine learning effect scheduling problem with release times (Lee *et al.* 2010).
A hybrid tabu search algorithm with an efficient neighborhood structure for the flexible job shop scheduling problem (Li *et al.* 2011).
A pareto approach to multi-objective flexible job-shop scheduling problem using particle swarm optimization and local search (Moslehi and Mahnam 2011).
A local-best harmony search algorithm with dynamic sub-harmony memories for lot-streaming flow shop scheduling problem (Pan *et al.* 2011).
A discrete artificial bee colony algorithm for the lot-streaming flow shop scheduling problem (Pan *et al.* 2011).
An iterated greedy algorithm for the flowshop scheduling problem with blocking (Ribas *et al.* 2011).
The hybrid flow shop scheduling problem (Ruiz and Vazquez-Rodriguez 2010).
Total flow time minimization in a flowshop sequence-dependent group scheduling problem (Salmasi *et al.* 2010).
A genetic algorithm for the unrelated parallel machine scheduling problem with sequence dependent setup times (Vallada and Ruiz 2011).
A genetic algorithm for the preemptive and non-preemptive multi-mode resource-constrained project scheduling problem (van Peteghem and Vanhoucke 2010).
A multi-objective genetic algorithm based on immune and entropy principle for flexible job-shop scheduling problem (Wang *et al.* 2010).
A multi-objective ant colony system algorithm for flow shop scheduling problem (Yagmahan and Yenisey 2010).
An efficient ant colony system based on receding horizon control for the aircraft arrival sequencing and scheduling problem (Zhan *et al.* 2010).
An effective genetic algorithm for the flexible job-shop scheduling problem (Zhang *et al.* 2011).
A hybrid immune simulated annealing algorithm for the job shop scheduling problem (Zhang and Wu 2010).

2.7 Shortest Path Problems

A shortest path in a graph defines the shortest possible edges that can be traversed to reach a target node and minimize cost (Mastorakis 2011). Shortest path problems can be classified into Single Source Shortest Path, where the problem of finding a path between two vertices such that the sum of the weights of its constituent edges is minimized (Mastorakis 2011).

The solution to a shortest path problem must contain both information (Daskin 2013): the cost of the shortest path between origin s and destination t and the actual sequence of links (or nodes) that are traversed in going from s to t (Daskin 2013).

Shortest path problem at one short time fraction with an intense terminal traffic conditions and thus, dynamically assigned path nodes for dynamic yard operations (nodes network) can be modeled by a graph $G = (V, E)$ where it comprises a set of vertices or nodes V and a set of E of edges or lines. A tour at the yard area within the dynamically assigned path nodes can be represented via a permutation $\tau = (\tau_1, \tau_2, ..., \tau_n)$. The shortest distance at yard area is formulated as,

Objective function is

$$\text{Minimize Z} = \sum_{(i,j) \in A} C_{ij} x_{ij}$$

Subject to

$$\sum_{\{j:(j,i) \in A\}} x_{ji} - \sum_{\{i:(i,j) \in A\}} x_{ij} = -1 \ if \ i = s, \ 0 \ if \ i \neq s \ or \ d \quad \forall i \in N, 1 \ if \ i = d$$

$$x_{ij} \geq 0 \quad \forall (i, j) \in A$$

where

N   set of number of nodes at seaport terminal (seaside nodes and yard area/stacking area nodes)

A   set of existing arcs (i,j)

C_{ij} arc length (or arc cost) united with each arc (i,j)

i   = s for source node, or i = d for destination node

x_{ij} is the flow from node i to node j

The objective function is to minimize the total distance that is dynamically defined on the seaside and yard area. Constraints ensure that the every point (nodes) visited only once and all these points are included in a tour.

32

The most recent and highly cited studies about the shortest path problem are shown in Table 2.7.

Table 2.7 Studies in the literature about the shortest path problem

An evolutionary solution for multimodal shortest path problem in metropolises (Abbaspour and Samadzadegan 2010).
An extended shortest path problem: a data envelopment analysis approach (Amirteimoori 2012).
An enhanced exact procedure for the absolute robust shortest path problem (Bruni and Guerriero 2010).
Bicriterion shortest path problem with a general nonadditive cost (Chen and Nie 2013).
Fuzzy dijkstra algorithm for shortest path problem under uncertain environment (Deng *et al.* 2012).
Tight analysis of the (1+1)-ea for the single source shortest path problem (Doerr *et al.* 2011).
Solving the fuzzy shortest path problem using multi-criteria decision method based on vague similarity measure (Dou *et al.* 2012).
New models for the robust shortest path problem: complexity, resolution and generalization (Gabrel *et al.* 2013).
Shortest path problem with uncertain arc lengths (Gao 2011).
An ant colony optimization algorithm for the bi-objective shortest path problem (Ghoseiri and Nadjari 2010).
Exploring the runtime of an evolutionary algorithm for the multi-objective shortest path problem (Horoba 2010).
The shortest path problem on a fuzzy time-dependent network (Huang and Ding 2012).
An aggregate label setting policy for the multi-objective shortest path problem (Iori *et al.* 2010).
Solving the constrained shortest path problem using random search strategy (Li *et al.* 2010).
On an exact method for the constrained shortest path problem (Lozano and Medaglia 2013).
An efficient dynamic model for solving the shortest path problem (Nazemi and Omidi 2013).
On algorithms for the tricriteria shortest path problem with two bottleneck objective functions (Pinto and Pascoal 2010).
A computational study of solution approaches for the resource constrained elementary shortest path problem (Pugliese and Guerriero 2012).
Dynamic programming approaches to solve the shortest path problem with forbidden paths (Pugliese and Guerriero 2013).
Shortest path problem with forbidden paths: the elementary version (Pugliese and Guerriero 2013).
A shortest path problem in an urban transportation network based on driver perceived travel time (Ramazani *et al.* 2010).
K constrained shortest path problem (Shi 2010).
A simplified algorithm for the all pairs shortest path problem with o(n(2) log n) expected time (Takaoka 2013).
A physarum polycephalum optimization algorithm for the bi-objective shortest path problem (Zhang *et al.* 2014).
A novel algorithm for all pairs shortest path problem based on matrix multiplication and pulse coupled neural network (Zhang *et al.* 2011).

2.7.1 Snapshot of an example solved by using Dijkstra's algorithm for the shortest path problem

The solution is depicted in Figure 2.1.

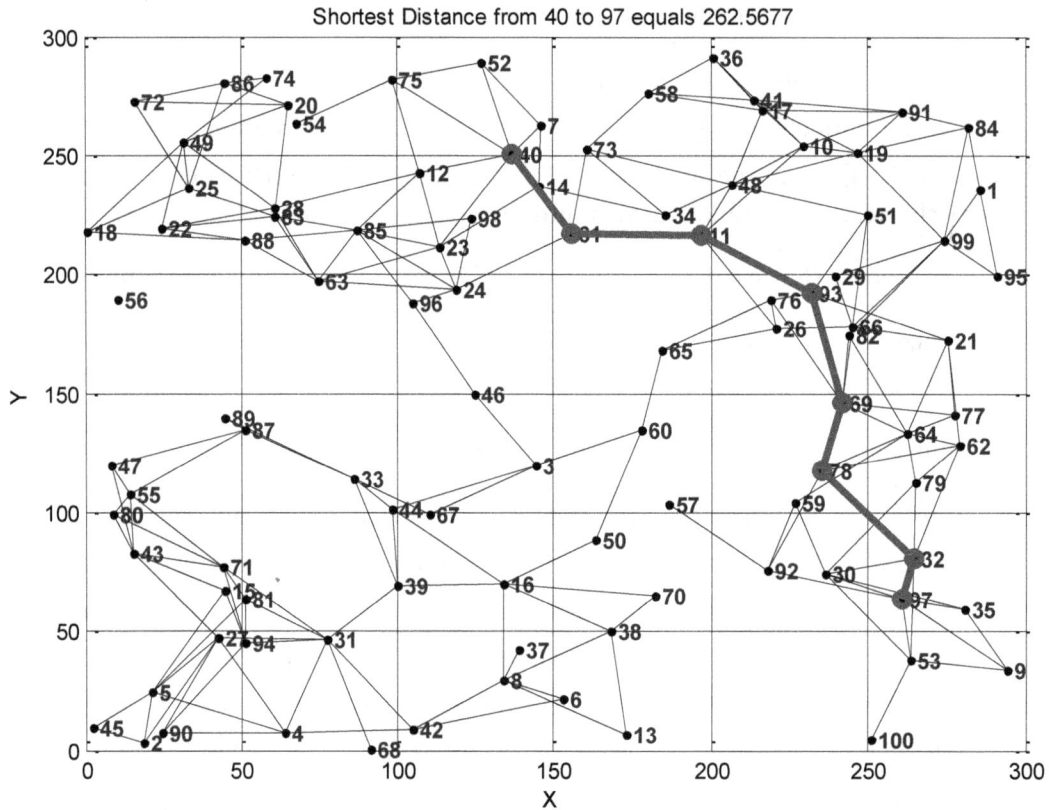

Figure 2.1 The Dijkstra's algortihm's solution for the shortest distance from node 40 to node 97.

Solution parameters are below.

start id   = 40
finish id   = 97
distance   = 262.5677
path   = [40 61 11 93 69 78 32 97]

2.8 Travelling Salesman Problem

The traveling salesman problem is a problem in the field of operations research that is reminiscent of Hamiltonian cycles; again, we know of no general solution method (Ore and Wilson, 1990). Suppose a traveling salesman must visit a number of cities before returning home (Ore and Wilson, 1990). Naturally, it is interested in doing this in the shortest possible time, or perhaps he may be concerned to do as cheap as possible (Ore and Wilson, 1990).

The traveling salesman problem (TSP) is to find a route of a salesman starting from a location of the house, went to a prescribed set of cities and returns to the original location so that the total distance is minimum and each city is visited exactly once (Gutin and Punnen 2002).

Traveling salesman problem (TSP) is an optimization problem for a given number of cities (e.g., n) and their locations (Yang 2008). Nodes of a graph represent the cities and the distance between the two cities are represented as the weight of an edge or route made between the two cities (Yang 2008). The goal is to find a path that visits each city once and return to the departure city, minimizing the total distance (Yang 2008).

Routing problem I: The problem is to ascertain the operation plan satisfying the demand at various zones at minimum cost.

Objective function is

$$\text{Minimize } f_{obj} = \sum_{i=1}^{G} \sum_{j=1}^{Z} \sum_{k=1}^{F} C_{ijk} x_{ijk}$$

Subject to

$$\sum_{i=1}^{G} \sum_{k=1}^{F} L_k x_{ijk} \geq D_j \qquad \forall j$$

$$\sum_{j=1}^{Z} \sum_{k=1}^{F} L_k x_{ijk} \leq S_j \qquad \forall i$$

$$\sum_{j=1}^{Z} L_k x_{ijk} \leq U_{ki} \qquad \forall k, i$$

$$x_{ijk} \geq 0 \qquad \forall i, j, k$$

where parameters are

G = Number of source locations (index i)

Z = Number of receiving nodes for containers (index j)

F = Number of trailers available (index k)

L_k = Load capacity of trailer k

S_i = Quantity of available containers for transportation from location i

D_j = Quantity of containers required by zone j

C_{ijk} = Unit cost of transporting from location i to zone j by trailer k

U_{ik} = Maximum allowable containers that can be transported from location i by trailer k in a given period

and variables

x_{ijk} = the number of trips required by trailer k from location i to zone j

Routing Problem II: A generic model that practitioners encounter in many planning and decision processes. For instance, the delivery and collection of containers/cargos, etc.

Objective function is

$$\text{Minimize } Z = \sum_{k=1}^{K} \sum_{(i,j) \in A} C_{ij} x_{kij}$$

Subject to

$$\sum_{i=1}^{n} y_{ij} = 1, \qquad j = 2, 3, \ldots, n$$

$$\sum_{j=1}^{n} y_{ij} = 1, \qquad i = 2, 3, \ldots, n$$

$$\sum_{j=1}^{n} y_{1j} = K$$

$$\sum_{j=1}^{n} y_{i1} = K$$

$$\sum_{i=1}^{n} \sum_{j=2}^{n} D_j x_{kij} \leq U, \qquad k = 1, 2, \ldots, K$$

$$\sum_{k=1}^{K} x_{kij} = y_{ij} \qquad \forall i, j$$

$$\sum_{(i,j) \in S \times S} y_{ij} \leq |S| - 1, \qquad \text{for all subsets } S \text{ of } \{2, 3, \ldots, n\}$$

$$x_{kij} = 0 \text{ or } 1 \quad \forall (i, j) \in A \text{ and } \forall k$$

$$y_{ij} = 0 \text{ or } 1 \quad \forall (i, j) \in A$$

- A fleet of M capacitated vehicles located in a depot (i=1)
- A set of target zones (of size N-1), each having a demand D_j (j=2,...,N)
- A cost C_{ij} of traveling from location i to location j
- The problem is to find a set of routes for delivering / picking up goods to/from the target zones at minimum possible cost.

The vehicle fleet is homogeneous and that each vehicle has a capacity of U units.

and variables:

x_{kij} = 1 if the vehicle k travels on the arc i to j, 0 otherwise

y_{ij} = 1 if any vehicle travels on the arc (i,j), 0 otherwise

The TSP is known to be NP-hard combinatorial optimization problem, which implies that there is no algorithm to solve all cases of problems in polynomial time (Mester *et al.* 2010). Heuristics are often the only alternative for providing high quality, but not necessarily optimal solutions (Mester *et al.* 2010).

The most recent and highly cited studies about the travelling salesman problem are shown in Table 2.8. Various example TSP problems with their optimal solutions are depicted in Figures 2.2 through 2.8.

Table 2.8 Studies in the literature about the traveling salesman problem

Development a new mutation operator to solve the traveling salesman problem by aid of genetic algorithms (Albayrak and Allahverdi 2011).
Amoeba-based neurocomputing for 8-city traveling salesman problem (Aono *et al.* 2011).
Estimation-based metaheuristics for the probabilistic traveling salesman problem (Balaprakash *et al.* 2010).
The traveling salesman problem with pickups, deliveries, and handling costs (Battarra *et al.* 2010).
A memetic algorithm with a large neighborhood crossover operator for the generalized traveling salesman problem (Bontoux *et al.* 2010).
Experimental demonstration of a quantum annealing algorithm for the traveling salesman problem in a nuclear-magnetic-resonance quantum simulator (Chen *et al.* 2011).
Parallelized genetic ant colony systems for solving the traveling salesman problem (Chen *et al.* 2011).
Solving the traveling salesman problem based on the genetic simulated annealing ant colony system with particle swarm optimization techniques (Chen and Chien 2011).
Branch-and-cut for the pickup and delivery traveling salesman problem with fifo loading (Cordeau *et al.* 2010).
A branch-and-cut algorithm for the pickup and delivery traveling salesman problem with lifo loading (Cordeau *et al.* 2010).
The traveling salesman problem with pickup and delivery: polyhedral results and a branch-and-cut algorithm (Dumitrescu *et al.* 2010).
An application of the self-organizing map in the non-euclidean traveling salesman problem (Faigl *et al.* 2011).
Eugenic bacterial memetic algorithm for fuzzy road transport traveling salesman problem (Foldesi *et al.* 2011).
Solving the traveling salesman problem based on an adaptive simulated annealing algorithm with greedy search (Geng *et al.* 2011).
Verification and rectification of the physical analogy of simulated annealing for the solution of the traveling salesman problem (Hasegawa 2011).
A concise guide to the traveling salesman problem (Laporte 2010).
Different initial solution generators in genetic algorithms for solving the probabilistic traveling salesman problem (Liu 2010).
Two-phase pareto local search for the biobjective traveling salesman problem (Lust and Teghem 2010).
A hybrid multi-swarm particle swarm optimization algorithm for the probabilistic traveling salesman problem (Marinakis and Marinaki 2010).
Honey bees mating optimization algorithm for the euclidean traveling salesman problem (Marinakis *et al.* 2011).
Traveling salesman problem heuristics: leading methods, implementations and latest advances (Rego *et al.* 2011).
An ensemble of discrete differential evolution algorithms for solving the generalized traveling salesman problem (Tasgetiren *et al.* 2010).
Solving traveling salesman problem in the adleman-lipton model (Wang *et al.* 2012).
Chaotic ant swarm for the traveling salesman problem (Wei *et al.* 2011).
A parallel immune algorithm for traveling salesman problem and its application on cold rolling scheduling (Zhao *et al.* 2011).

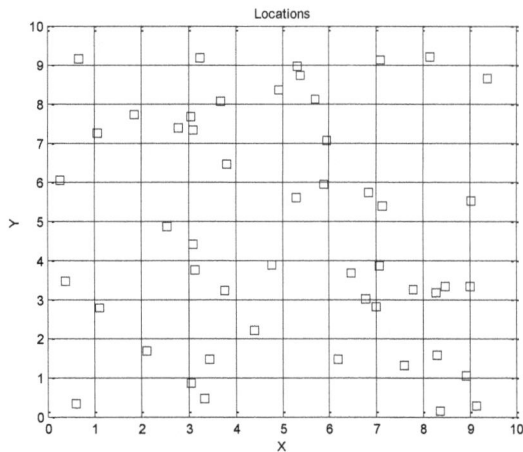

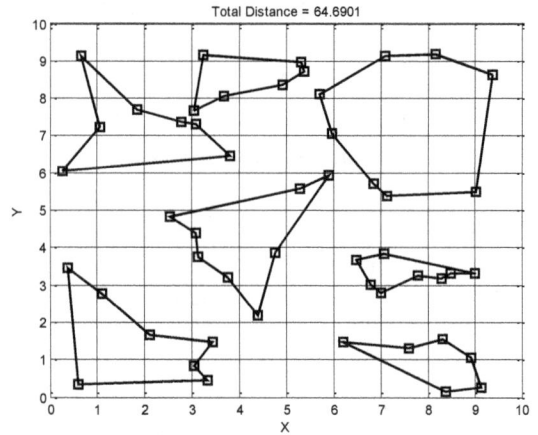

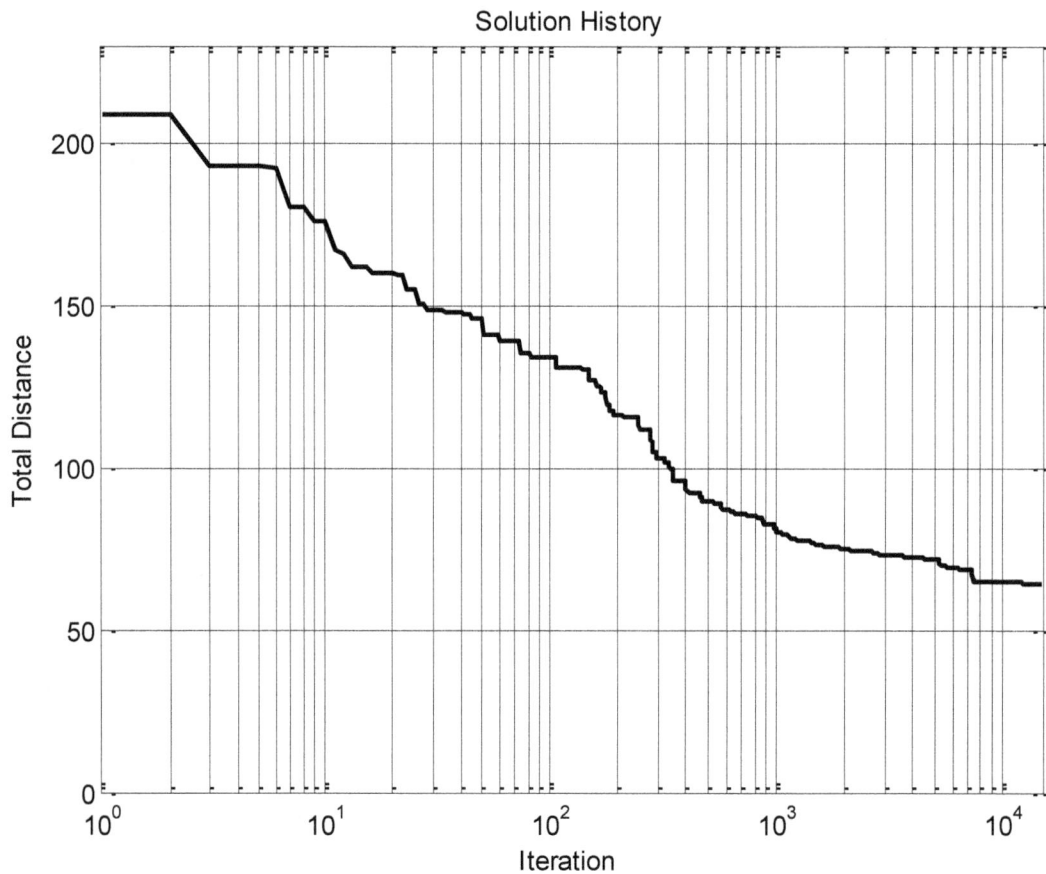

Figure 2.2 Multiple Traveling Salesmen Problem (MTSP) – Each salesman travels to a unique set of cities and completes the route by returning to the city he started from. Each city is visited by exactly one salesman – 7 salesmen, 50 locations

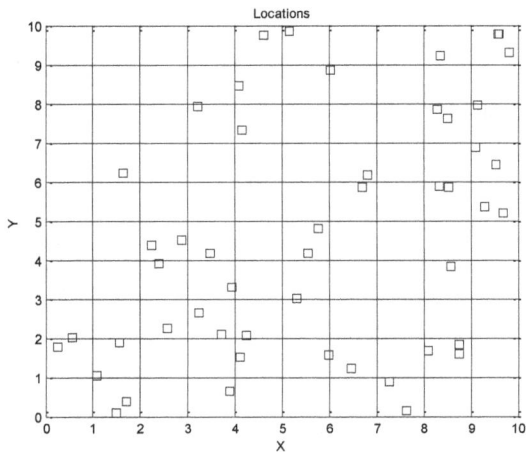

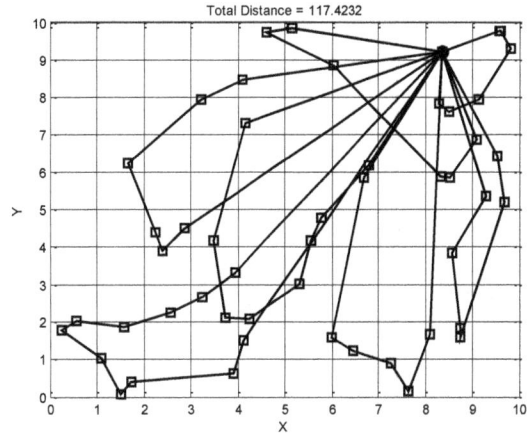

Figure 2.3 Fixed Multiple Traveling Salesmen Problem (MTSPF) – Each salesman starts at the first point, and ends at the first point, but travels to a unique set of cities in between. Except for the first, each city is visited by exactly one salesman – 7 salesmen, 50 locations

41

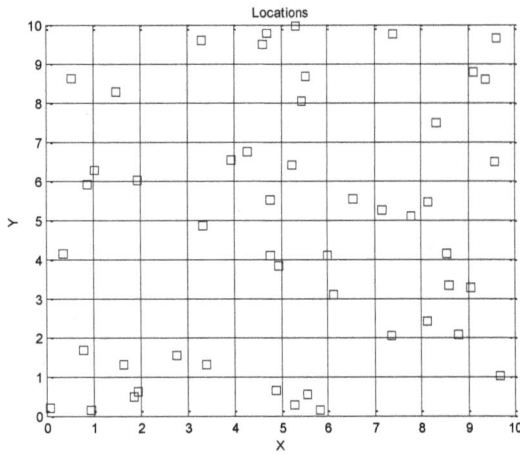

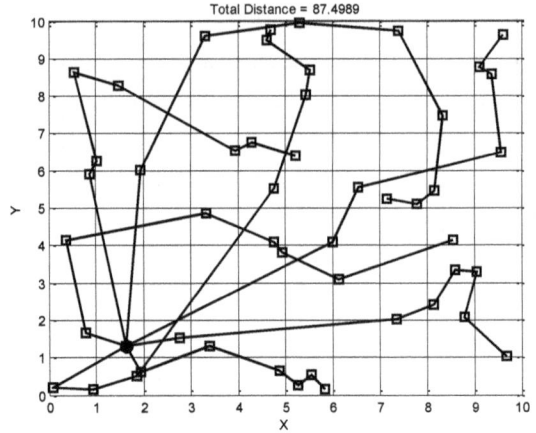

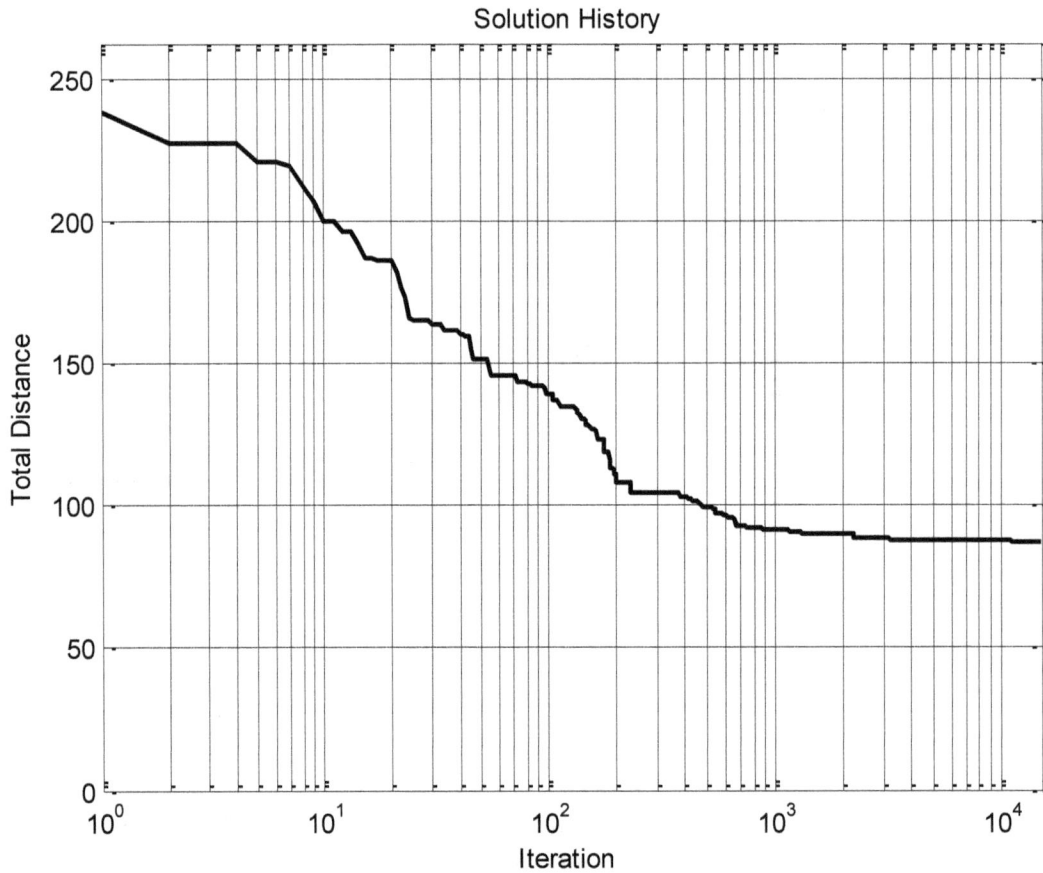

Figure 2.4 Fixed Start Open Multiple Traveling Salesmen Problem (MTSPOFS) – Each salesman starts at the first point and travels to a unique set of cities after that. Except for the first, each city is visited by exactly one salesman – 7 salesmen, 50 locations

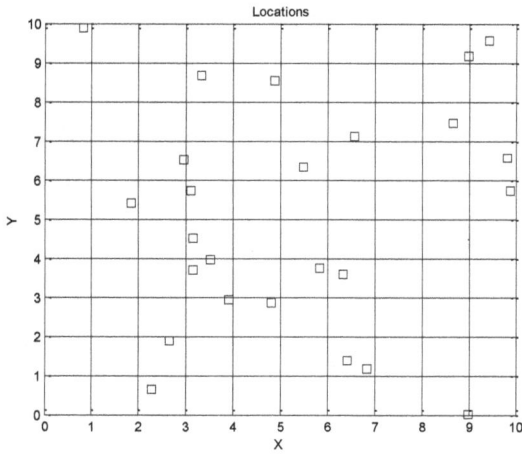

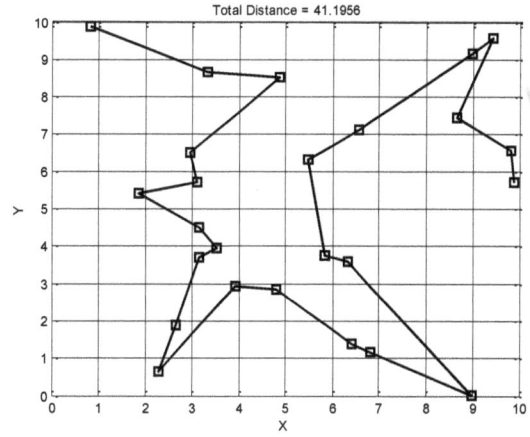

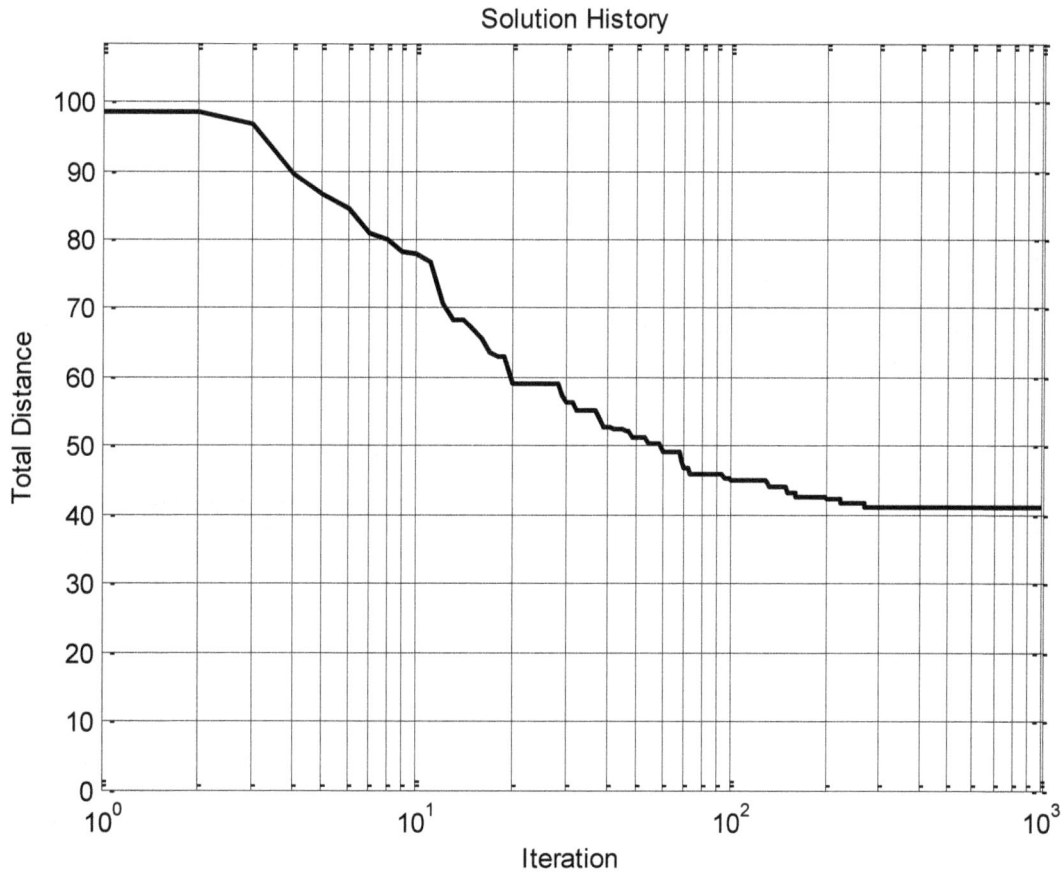

Figure 2.5 Open Traveling Salesman Problem (TSPO) – A single salesman travels to each of the cities but does not close the loop by returning to the city he started from. Each city is visited by the salesman exactly once.

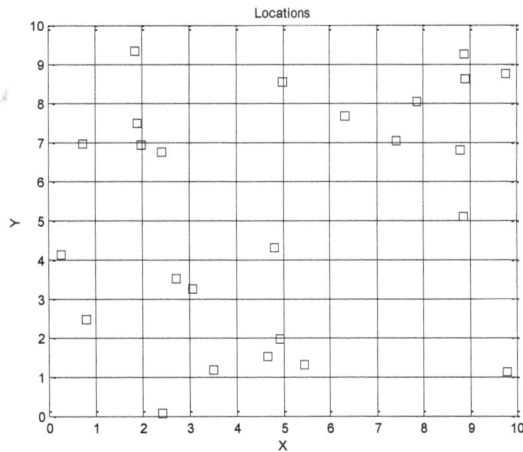

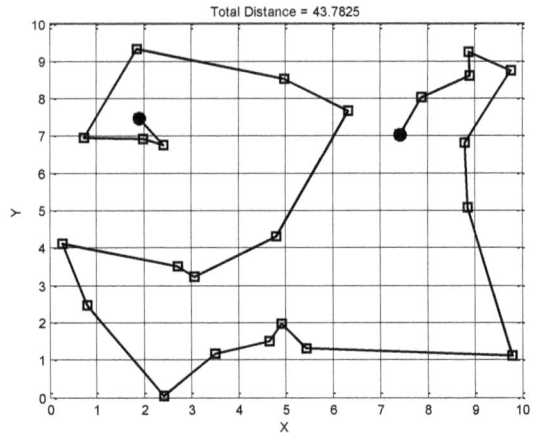

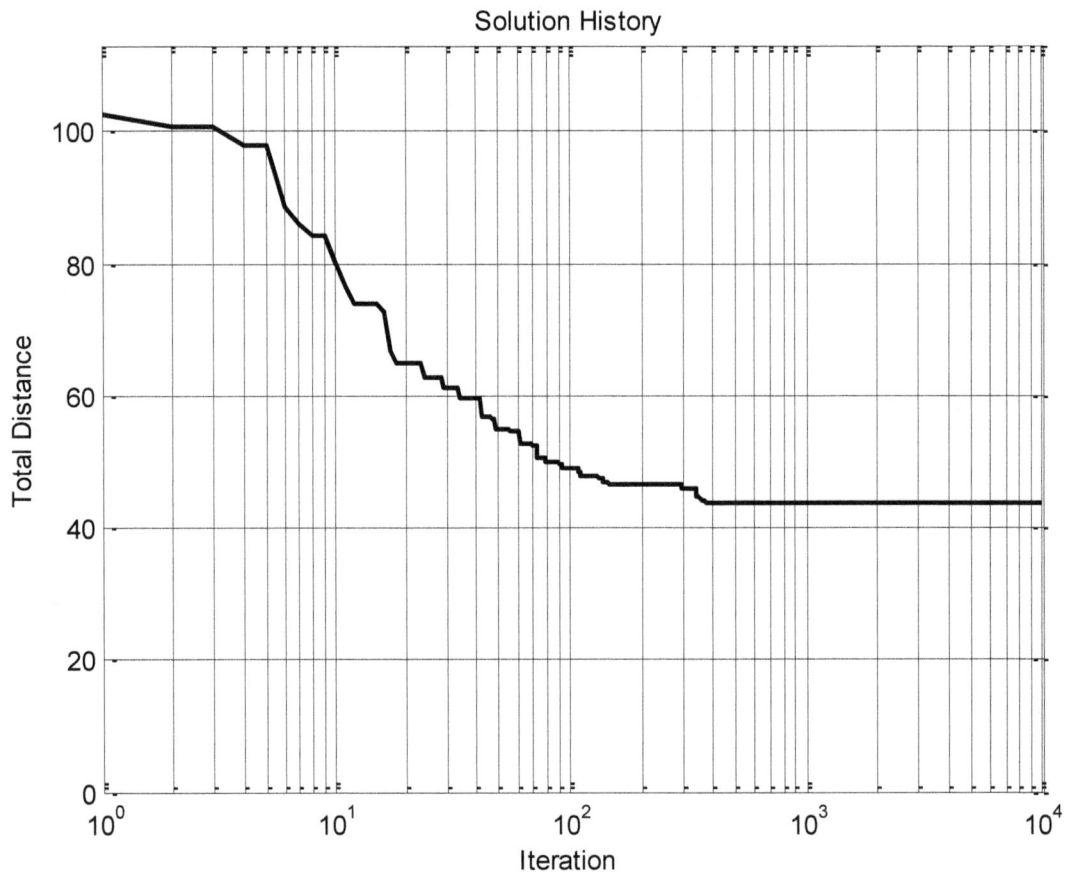

Figure 2.6 Fixed Open Traveling Salesman Problem (TSPOF) – A single salesman starts at the first point, ends at the last point, and travels to each of the remaining cities in between. Each city is visited by the salesman exactly once.

44

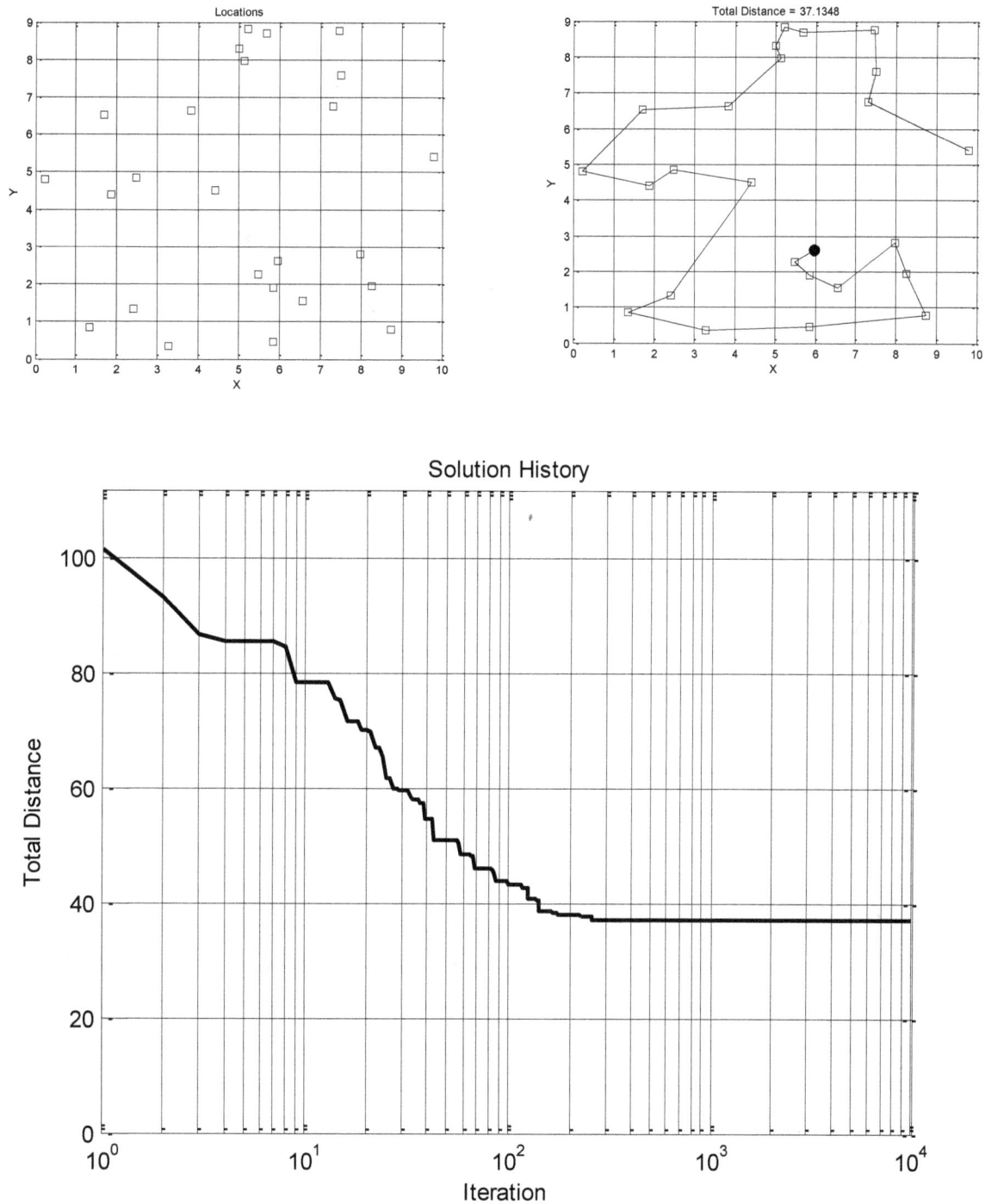

Figure 2.7 Fixed Start Open Traveling Salesman Problem (TSPOFS) – A single salesman starts at the first point and travels to each of the remaining cities but does not close the loop by returning to the city he started from. Each city is visited by the salesman exactly once.

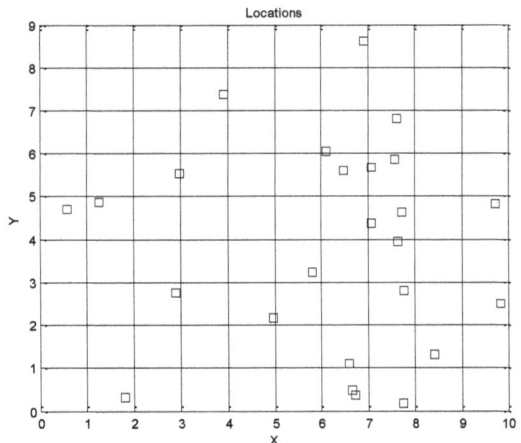

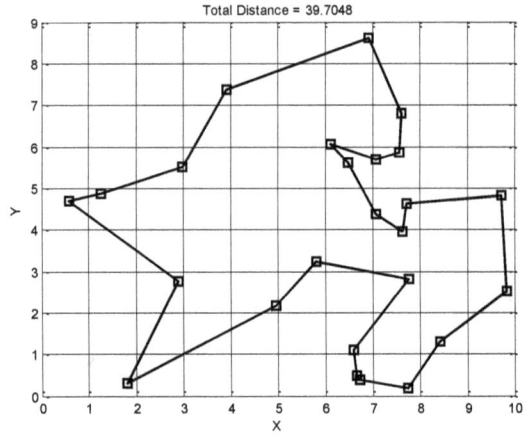

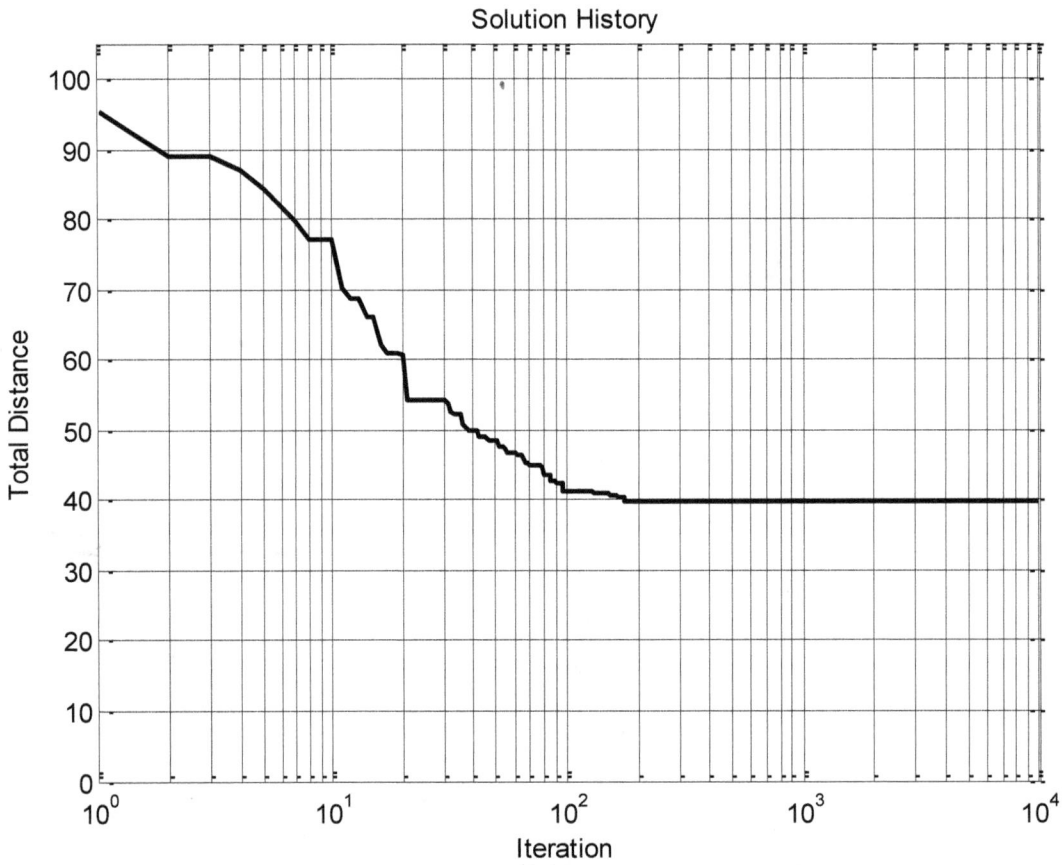

Figure 2.8 Traveling Salesman Problem (TSP) – A single salesman travels to each of the cities and completes the route by returning to the city he started from. Each city is visited by the salesman exactly once.

References

Abbaspour, R. A., & Samadzadegan, F. (2010). An Evolutionary Solution for Multimodal Shortest Path Problem in Metropolises. *Computer Science and Information Systems, 7*(4), 789-811. doi: 10.2298/csis090710024a

Absi, N., Kedad-Sidhoum, S., & Dauzere-Peres, S. (2011). Uncapacitated lot-sizing problem with production time windows, early productions, backlogs and lost sales. *International Journal of Production Research, 49*(9), 2551-2566. doi: 10.1080/00207543.2010.532920

Agarwal, A., Colak, S., & Erenguc, S. (2011). A Neurogenetic approach for the resource-constrained project scheduling problem. *Computers & Operations Research, 38*(1), 44-50. doi: 10.1016/j.cor.2010.01.007

Albareda-Sambola, M., Alonso-Ayuso, A., Escudero, L. F., Fernandez, E., Hinojosa, Y., & Pizarro-Romero, C. (2010). A computational comparison of several formulations for the multi-period incremental service facility location problem. *Top, 18*(1), 62-80. doi: 10.1007/s11750-009-0106-3

Albareda-Sambola, M., Fernandez, E., & Saldanha-da-Gama, F. (2011). The facility location problem with Bernoulli demands. *Omega-International Journal of Management Science, 39*(3), 335-345. doi: 10.1016/j.omega.2010.08.002

Albayrak, M., & Allahverdi, N. (2011). Development a new mutation operator to solve the Traveling Salesman Problem by aid of Genetic Algorithms. *Expert Systems with Applications, 38*(3), 1313-1320. doi: 10.1016/j.eswa.2010.07.006

Amirteimoori, A. (2012). An extended shortest path problem: A data envelopment analysis approach. *Applied Mathematics Letters, 25*(11), 1839-1843. doi: 10.1016/j.aml.2012.02.042

Angelelli, E., Mansini, R., & Speranza, M. G. (2010). Kernel search: A general heuristic for the multi-dimensional knapsack problem. *Computers & Operations Research, 37*(11) 2017-2026. doi: 10.1016/j.cor.2010.02.002

Aono, M., Zhu, L. P., & Hara, M. (2011). Amoeba-based Neurocomputing for 8-City Traveling Salesman Problem. *International Journal of Unconventional Computing, 7*(6), 463-480.

Argyris, N., Figueira, J. R., & Morton, A. (2011). Identifying preferred solutions to Multi-Objective Binary Optimisation problems, with an application to the Multi-Objective Knapsack Problem. *Journal of Global Optimization, 49*(2), 213-235. doi: 10.1007/s10898-010-9541-9

Aros-Vera, F., Marianov, V., & Mitchell, J. E. (2013). p-Hub approach for the optimal park-and-ride facility location problem. European Journal of Operational Research, 226(2), 277-285. doi: 10.1016/j.ejor.2012.11.006

Avella, P., Boccia, M., & Vasilyev, I. (2010). A computational study of exact knapsack separation for the generalized assignment problem. Computational Optimization and Applications, 45(3), 543-555. doi: 10.1007/s10589-008-9183-8

Azi, N., Gendreau, M., & Potvin, J. Y. (2010). An exact algorithm for a vehicle routing problem with time windows and multiple use of vehicles. European Journal of Operational Research 202(3), 756-763. doi: 10.1016/j.ejor.2009.06.034

Bagheri, A., Zandieh, M., Mahdavi, I., & Yazdani, M. (2010). An artificial immune algorithm for the flexible job-shop scheduling problem. Future Generation Computer Systems-the International Journal of Grid Computing and Escience, 26(4), 533-541. doi: 10.1016/j.future.2009.10.004

Balaprakash, P., Birattari, M., Stutzle, T., & Dorigo, M. (2010). Estimation-based metaheuristics for the probabilistic traveling salesman problem. Computers & Operations Research, 37(11), 1939-1951. doi: 10.1016/j.cor.2009.12.005

Baldacci, R., Bartolini, E., & Laporte, G. (2010). Some applications of the generalized vehicle routing problem. Journal of the Operational Research Society, 61(7), 1072-1077. doi: 10.1057/jors.2009.51

Baldacci, R., Mingozzi, A., & Roberti, R. (2011). New Route Relaxation and Pricing Strategies for the Vehicle Routing Problem. Operations Research, 59(5), 1269-1283. doi: 10.1287/opre.1110.0975

Battarra, M., Erdogan, G., Laporte, G., & Vigo, D. (2010). The Traveling Salesman Problem with Pickups, Deliveries, and Handling Costs. Transportation Science, 44(3), 383-399. doi: 10.1287/trsc.1100.0316

Beltran-Royo, C., Vial, J. P., & Alonso-Ayuso, A. (2012). Semi-Lagrangian relaxation applied to the uncapacitated facility location problem. Computational Optimization and Applications, 51(1), 387-409. doi: 10.1007/s10589-010-9338-2

Benjamin, A. M., & Beasley, J. E. (2010). Metaheuristics for the waste collection vehicle routing problem with time windows, driver rest period and multiple disposal facilities. Computers & Operations Research, 37(12), 2270-2280. doi: 10.1016/j.cor.2010.03.019

Bhattacharya, R., & Bandyopadhyay, S. (2010). Solving conflicting bi-objective facility location problem by NSGA II evolutionary algorithm. International Journal of Advanced Manufacturing Technology, 51(1-4), 397-414. doi: 10.1007/s00170-

010-2622-6

Bontoux, B., Artigues, C., & Feillet, D. (2010). A Memetic Algorithm with a large neighborhood crossover operator for the Generalized Traveling Salesman Problem. Computers & Operations Research, 37(11), 1844-1852. doi: 10.1016/j.cor.2009.05.004

Boussier, S., Vasquez, M., Vimont, Y., Hanafi, S., & Michelon, P. (2010). A multi-level search strategy for the 0-1 Multidimensional Knapsack Problem. Discrete Applied Mathematics, 158(2), 97-109. doi: 10.1016/j.dam.2009.08.007

Bowman, M., Briand, L. C., & Labiche, Y. (2010). Solving the Class Responsibility Assignment Problem in Object-Oriented Analysis with Multi-Objective Genetic Algorithms. Ieee Transactions on Software Engineering, 36(6), 817-837. doi: 10.1109/tse.2010.70

Bramel, Julien; Simchi-Levi, David. Logic of Logistics : Theory, Algorithms, and Applications for Logistics Management. Secaucus, NJ, USA: Springer, 1997. p 203.

Brandao, J. (2011). A tabu search algorithm for the heterogeneous fixed fleet vehicle routing problem. Computers & Operations Research, 38(1), 140-151. doi: 10.1016/j.cor.2010.04.008

Bruni, M. E., & Guerriero, F. (2010). An enhanced exact procedure for the absolute robust shortest path problem. International Transactions in Operational Research, 17(2) 207-220. doi: 10.1111/j.1475-3995.2009.00702.x

Budish, E., & Cantillon, E. (2012). The Multi-unit Assignment Problem: Theory and Evidence from Course Allocation at Harvard. American Economic Review, 102(5), 2237-2271. doi: 10.1257/aer.102.5.2237

Byrka, J., & Aardal, K. (2010). An Optimal Bifactor Approximation Algorithm For The Metric Uncapacitated Facility Location Problem. Siam Journal on Computing, 39(6), 2212-2231. doi: 10.1137/070708901

Camargo, V. C. B., Mattiolli, L., & Toledo, F. M. B. (2012). A knapsack problem as a tool to solve the production planning problem in small foundries. Computers & Operations Research, 39(1), 86-92. doi: 10.1016/j.cor.2010.10.023

Cardenas-Barron, L. E. (2010). Adaptive genetic algorithm for lot-sizing problem with self-adjustment operation rate: A discussion. International Journal of Production Economics, 123(1), 243-245. doi: 10.1016/j.ijpe.2009.07.007

Carrizosa, E., Ushakov, A., & Vasilyev, I. (2012). A computational study of a nonlinear minsum facility location problem. Computers & Operations Research, 39(11),

2625-2633. doi: 10.1016/j.cor.2012.01.009

Cela, E., Schmuck, N. S., Wimer, S., & Woeginger, G. J. (2011). The Wiener maximum quadratic assignment problem. Discrete Optimization, 8(3), 411-416. doi: 10.1016/j.disopt.2011.02.002

Chen, B. Y., Lam, W. H. K., Sumalee, A., & Shao, H. (2011). An efficient solution algorithm for solving multi-class reliability-based traffic assignment problem. Mathematical and Computer Modelling, 54(5-6), 1428-1439. doi: 10.1016/j.mcm.2011.04.015

Chen, Der-San; Batson, Robert G.; Dang, Yu. Applied Integer Programming Modeling and Solution. Hoboken, NJ, USA: Wiley 2011. p 30-345.

Chen, H. W., Kong, X., Chong, B., Qin, G., Zhou, X. Y., Peng, X. H., & Du, J. F. (2011). Experimental demonstration of a quantum annealing algorithm for the traveling salesman problem in a nuclear-magnetic-resonance quantum simulator. Physical Review A, 83(3). doi: 10.1103/PhysRevA.83.032314

Chen, L., & Lu, Z. Q. (2012). The storage location assignment problem for outbound containers in a maritime terminal. International Journal of Production Economics, 135(1), 73-80. doi: 10.1016/j.ijpe.2010.09.019

Chen, P., & Nie, Y. (2013). Bicriterion shortest path problem with a general nonadditive cost. Transportation Research Part B-Methodological, 57, 419-435. doi: 10.1016/j.trb.2013.05.008

Chen, P., Huang, H. K., & Dong, X. Y. (2010). Iterated variable neighborhood descent algorithm for the capacitated vehicle routing problem. Expert Systems with Applications, 37(2), 1620-1627. doi: 10.1016/j.eswa.2009.06.047

Chen, S. M., & Chien, C. Y. (2011). Parallelized genetic ant colony systems for solving the traveling salesman problem. Expert Systems with Applications, 38(4), 3873-3883. doi: 10.1016/j.eswa.2010.09.048

Chen, S. M., & Chien, C. Y. (2011). Solving the traveling salesman problem based on the genetic simulated annealing ant colony system with particle swarm optimization techniques. Expert Systems with Applications, 38(12), 14439-14450. doi: 10.1016/j.eswa.2011.04.163

Cheng, T. C. E., Cheng, S. R., Wu, W. H., Hsu, P. H., & Wu, C. C. (2011). A two-agent single-machine scheduling problem with truncated sum-of-processing-times-based learning considerations. Computers & Industrial Engineering, 60(4), 534-541. doi: 10.1016/j.cie.2010.12.008

Cherfi, N., & Hifi, M. (2010). A column generation method for the multiple-choice multi-dimensional knapsack problem. Computational Optimization and Applications, 46(1), 51-73. doi: 10.1007/s10589-008-9184-7

Chvátal, V. (Editor). Combinatorial Optimization : Methods and Applications. Amsterdam, NLD: IOS Press 2011. p 97.

Cordeau, J. F., Dell'Amico, M., & Iori, M. (2010). Branch-and-cut for the pickup and delivery traveling salesman problem with FIFO loading. Computers & Operations Research, 37(5), 970-980. doi: 10.1016/j.cor.2009.08.003

Cordeau, J. F., Iori, M., Laporte, G., & Gonzalez, J. J. S. (2010). A Branch-and-Cut Algorithm for the Pickup and Delivery Traveling Salesman Problem with LIFO Loading. Networks, 55(1), 46-59. doi: 10.1002/net.20312

D'Acierno, L., Gallo, M., & Montella, B. (2012). An Ant Colony Optimisation algorithm for solving the asymmetric traffic assignment problem. European Journal of Operational Research, 217(2), 459-469. doi: 10.1016/j.ejor.2011.09.035

Das, D., Roy, A., & Kar, S. (2011). A volume flexible economic production lot-sizing problem with imperfect quality and random machine failure in fuzzy-stochastic environment. Computers & Mathematics with Applications, 61(9), 2388-2400. doi: 10.1016/j.camwa.2011.02.015

Daskin, Mark S.. Network and Discrete Location : Models, Algorithms, and Applications (2nd Edition). Somerset, NJ, USA: Wiley 2013. p 3-66.

Davis, L., Samanlioglu, F., Jiang, X. C., Mota, D., & Stanfield, P. (2012). A heuristic approach for allocation of data to RFID tags: A data allocation knapsack problem (DAKP). Computers & Operations Research, 39(1), 93-104. doi: 10.1016/j.cor.2011.01.019

De Giovanni, L., & Pezzella, F. (2010). An Improved Genetic Algorithm for the Distributed and Flexible Job-shop Scheduling problem. European Journal of Operational Research 200(2), 395-408. doi: 10.1016/j.ejor.2009.01.008

de Klerk, E., & Sotirov, R. (2010). Exploiting group symmetry in semidefinite programming relaxations of the quadratic assignment problem. Mathematical Programming, 122(2), 225-246. doi: 10.1007/s10107-008-0246-5

Deng, Y., Chen, Y. X., Zhang, Y. J., & Mahadevan, S. (2012). Fuzzy Dijkstra algorithm for shortest path problem under uncertain environment. Applied Soft Computing, 12(3), 1231-1237. doi: 10.1016/j.asoc.2011.11.011

Desaulniers, G. (2010). Branch-and-Price-and-Cut for the Split-Delivery Vehicle

Routing Problem with Time Windows. Operations Research, 58(1), 179-192. doi: 10.1287/opre.1090.0713

Diabat, A., Abdallah, T., Al-Refaie, A., Svetinovic, D., & Govindan, K. (2013). Strategic Closed-Loop Facility Location Problem With Carbon Market Trading. Ieee Transactions on Engineering Management, 60(2), 398-408. doi: 10.1109/tem.2012.2211105

Dizdar, D., Gershkov, A., & Moldovanu, B. (2011). Revenue maximization in the dynamic knapsack problem. Theoretical Economics, 6(2), 157-184. doi: 10.3982/te700

Doerr, B., Happ, E., & Klein, C. (2011). Tight Analysis of the (1+1)-EA for the Single Source Shortest Path Problem. Evolutionary Computation, 19(4), 673-691.

Dou, Y. L., Zhu, L. C., & Wang, H. S. (2012). Solving the fuzzy shortest path problem using multi-criteria decision method based on vague similarity measure. Applied Soft Computing, 12(6), 1621-1631. doi: 10.1016/j.asoc.2012.03.013

Du, D. L., Lu, R. X., & Xu, D. C. (2012). A Primal-Dual Approximation Algorithm for the Facility Location Problem with Submodular Penalties. Algorithmica, 63(1-2), 191-200. doi: 10.1007/s00453-011-9526-1

Du, D. L., Wang, X., & Xu, D. C. (2010). An approximation algorithm for the k-level capacitated facility location problem. Journal of Combinatorial Optimization 20(4), 361-368. doi: 10.1007/s10878-009-9213-1

Dugardin, F., Yalaoui, F., & Amodeo, L. (2010). New multi-objective method to solve reentrant hybrid flow shop scheduling problem. European Journal of Operational Research 203(1), 22-31. doi: 10.1016/j.ejor.2009.06.031

Duhamel, C., Lacomme, P., Quilliot, A., & Toussaint, H. (2011). A multi-start evolutionary local search for the two-dimensional loading capacitated vehicle routing problem. Computers & Operations Research, 38(3), 617-640. doi: 10.1016/j.cor.2010.08.017

Dumitrescu, I., Ropke, S., Cordeau, J. F., & Laporte, G. (2010). The traveling salesman problem with pickup and delivery: polyhedral results and a branch-and-cut algorithm. Mathematical Programming, 121(2), 269-305. doi: 10.1007/s10107-008-0234-9

Dye, C. Y., & Hsieh, T. P. (2010). A particle swarm optimization for solving joint pricing and lot-sizing problem with fluctuating demand and unit purchasing cost. Computers & Mathematics with Applications, 60(7), 1895-1907. doi: 10.1016/j.camwa.2010.07.023

Dye, C. Y., & Ouyang, L. Y. (2011). A particle swarm optimization for solving joint pricing and lot-sizing problem with fluctuating demand and trade credit financing. Computers & Industrial Engineering, 60(1), 127-137. doi: 10.1016/j.cie.2010.10.010

El-Rewini, Hesham; Abd-El-Barr, Mostafa. Advanced Computer Architecture and Parallel Processing. Hoboken, NJ, USA: Wiley 2005. p 251-252.

Faigl, J., Kulich, M., Vonasek, V., & Preucil, L. (2011). An application of the self-organizing map in the non-Euclidean Traveling Salesman Problem. Neurocomputing, 74(5), 671-679. doi: 10.1016/j.neucom.2010.08.026

Feng, Y., Chen, S. X., Kumar, A., & Lin, B. (2011). Solving single-product economic lot-sizing problem with non-increasing setup cost, constant capacity and convex inventory cost in O(N log N) time. Computers & Operations Research, 38(4), 717-722. doi: 10.1016/j.cor.2010.08.009

Figliozzi, M. A. (2010). An iterative route construction and improvement algorithm for the vehicle routing problem with soft time windows. Transportation Research Part C-Emerging Technologies, 18(5), 668-679. doi: 10.1016/j.trc.2009.08.005

Foldesi, P., Botzheim, J., & Koczy, L. T. (2011). Eugenic Bacterial Memetic Algorithm For Fuzzy Road Transport Traveling Salesman Problem. International Journal of Innovative Computing Information and Control, 7(5B), 2775-2798.

Gabor, A. F., & van Ommeren, J. (2010). A new approximation algorithm for the multilevel facility location problem. Discrete Applied Mathematics, 158(5), 453-460. doi: 10.1016/j.dam.2009.11.007

Gabrel, V., Murat, C., & Wu, L. (2013). New models for the robust shortest path problem: complexity, resolution and generalization. Annals of Operations Research 207(1), 97-120. doi: 10.1007/s10479-011-1004-2

Gao, Y. (2011). Shortest path problem with uncertain arc lengths. Computers & Mathematics with Applications, 62(6), 2591-2600. doi: 10.1016/j.camwa.2011.07.058

Garcia-Najera, A., & Bullinaria, J. A. (2011). An improved multi-objective evolutionary algorithm for the vehicle routing problem with time windows. Computers & Operations Research, 38(1), 287-300. doi: 10.1016/j.cor.2010.05.004

Geng, X. T., Chen, Z. H., Yang, W., Shi, D. Q., & Zhao, K. (2011). Solving the traveling salesman problem based on an adaptive simulated annealing algorithm with greedy search. Applied Soft Computing, 11(4), 3680-3689. doi:

10.1016/j.asoc.2011.01.039

Ghasemi, T., & Razzazi, M. (2011). Development of core to solve the multidimensional multiple-choice knapsack problem. Computers & Industrial Engineering, 60(2), 349-360. doi: 10.1016/j.cie.2010.12.001

Ghiani, Gianpaolo. Wiley Essentials in Operations Research and Management Science : Introduction to Logistics Systems Management (2nd Edition). Somerset, NJ, USA: Wiley 2013. p 146-342.

Ghoseiri, K., & Nadjari, B. (2010). An ant colony optimization algorithm for the bi-objective shortest path problem. Applied Soft Computing, 10(4), 1237-1246. doi: 10.1016/j.asoc.2009.09.014

Godrich, H., Petropulu, A. P., & Poor, H. V. (2012). Sensor Selection in Distributed Multiple-Radar Architectures for Localization: A Knapsack Problem Formulation. Ieee Transactions on Signal Processing, 60(1), 247-260. doi: 10.1109/tsp.2011.2170170

Gonzalez-Ramirez, R. G., Smith, N. R., & Askin, R. G. (2011). A heuristic approach for a multi-product capacitated lot-sizing problem with pricing. International Journal of Production Research, 49(4), 1173-1196. doi: 10.1080/00207540903524482

Goyal, V., & Ravi, R. (2010). A PTAS for the chance-constrained knapsack problem with random item sizes. Operations Research Letters, 38(3), 161-164. doi: 10.1016/j.orl.2010.01.003

Gu, J. W., Gu, M. Z., Cao, C. W., & Gu, X. S. (2010). A novel competitive co-evolutionary quantum genetic algorithm for stochastic job shop scheduling problem. Computers & Operations Research, 37(5), 927-937. doi: 10.1016/j.cor.2009.07.002

Guan, Y. P., & Liu, T. M. (2010). Stochastic lot-sizing problem with inventory-bounds and constant order-capacities. European Journal of Operational Research 207(3), 1398-1409. doi: 10.1016/j.ejor.2010.07.003

Guillaume, R., Kobylanski, P., & Zielinski, P. (2012). A robust lot sizing problem with ill-known demands. Fuzzy Sets and Systems 206, 39-57. doi: 10.1016/j.fss.2012.01.015

Gutin, Gregory; Punnen, Abraham P.. Travelling Salesman Problem and Its Variations. Secaucus, NJ, USA: Kluwer Academic Publishers 2002. p 20.

Hachicha, W. (2011). A Simulation Metamodelling Based Neural Networks For Lot-Sizing Problem In Mto Sector. International Journal of Simulation Modelling,

10(4), 191-203. doi: 10.2507/ijsimm10(4)3.188

Han, X. L., Lu, Z. Q., & Xi, L. F. (2010). A proactive approach for simultaneous berth and quay crane scheduling problem with stochastic arrival and handling time. European Journal of Operational Research 207(3), 1327-1340. doi: 10.1016/j.ejor.2010.07.018

Hanafi, S., & Wilbaut, C. (2011). Improved convergent heuristics for the 0-1 multidimensional knapsack problem. Annals of Operations Research, 183(1), 125-142. doi: 10.1007/s10479-009-0546-z

Hartmann, S., & Briskorn, D. (2010). A survey of variants and extensions of the resource-constrained project scheduling problem. European Journal of Operational Research 207(1), 1-14. doi: 10.1016/j.ejor.2009.11.005

Hasegawa, M. (2011). Verification and rectification of the physical analogy of simulated annealing for the solution of the traveling salesman problem. Physical Review E, 83(3). doi: 10.1103/PhysRevE.83.036708

Helber, S., & Sahling, F. (2010). A fix-and-optimize approach for the multi-level capacitated lot sizing problem. International Journal of Production Economics, 123(2), 247-256. doi: 10.1016/j.ijpe.2009.08.022

Hill, R. R., Cho, Y. K., & Moore, J. T. (2012). Problem reduction heuristic for the 0-1 multidimensional knapsack problem. Computers & Operations Research, 39(1), 19-26. doi: 10.1016/j.cor.2010.06.009

Horoba, C. (2010). Exploring the Runtime of an Evolutionary Algorithm for the Multi-Objective Shortest Path Problem. Evolutionary Computation, 18(3), 357-381. doi: 10.1162/EVCO_a_00014

Huang, W., & Ding, L. X. (2012). The Shortest Path Problem on a Fuzzy Time-Dependent Network. Ieee Transactions on Communications, 60(11), 3376-3385. doi: 10.1109/tcomm.2012.090512.100570

Iori, M., Martello, S., & Pretolani, D. (2010). An aggregate label setting policy for the multi-objective shortest path problem. European Journal of Operational Research 207(3), 1489-1496. doi: 10.1016/j.ejor.2010.06.035

Iyigun, C., & Ben-Israel, A. (2010). A generalized Weiszfeld method for the multi-facility location problem. Operations Research Letters, 38(3) 207-214. doi: 10.1016/j.orl.2009.11.005

Jarboui, Bassem (Editor); Siarry, Patrick (Editor); Teghem, Jacques (Editor). Metaheuristics for Production Scheduling. Somerset, NJ, USA: Wiley-ISTE 2013. p

373-433.

Kalcsics, J., Nickel, S., Puerto, J., & Rodriguez-Chia, A. (2010). The ordered capacitated facility location problem. Top, 18(1) 203-222. doi: 10.1007/s11750-009-0089-0

Ke, L. J., Feng, Z. R., Ren, Z. G., & Wei, X. L. (2010). An ant colony optimization approach for the multidimensional knapsack problem. Journal of Heuristics, 16(1), 65-83. doi: 10.1007/s10732-008-9087-x

Kellerer, H., & Strusevich, V. A. (2010). Fully Polynomial Approximation Schemes for a Symmetric Quadratic Knapsack Problem and its Scheduling Applications. Algorithmica, 57(4), 769-795. doi: 10.1007/s00453-008-9248-1

Khaloozadeh, H., & Baromand, S. (2010). State covariance assignment problem. Iet Control Theory and Applications, 4(3), 391-402. doi: 10.1049/iet-cta.2008.0359

Konstantinidis, A., Yang, K., Zhang, Q. F., & Zeinalipour-Yazti, D. (2010). A multi-objective evolutionary algorithm for the deployment and power assignment problem in wireless sensor networks. Computer Networks, 54(6), 960-976. doi: 10.1016/j.comnet.2009.08.010

Kosuch, S., & Lisser, A. (2010). Upper bounds for the 0-1 stochastic knapsack problem and a B&B algorithm. Annals of Operations Research, 176(1), 77-93. doi: 10.1007/s10479-009-0577-5

Kucukaydin, H., Aras, N., & Altinel, I. K. (2011). Competitive facility location problem with attractiveness adjustment of the follower A bilevel programming model and its solution. European Journal of Operational Research 208(3) 206-220. doi: 10.1016/j.ejor.2010.08.009

Kucukdeniz, T., Baray, A., Ecerkale, K., & Esnaf, S. (2012). Integrated use of fuzzy c-means and convex programming for capacitated multi-facility location problem. Expert Systems with Applications, 39(4), 4306-4314. doi: 10.1016/j.eswa.2011.09.102

Kumar, R., & Singh, P. K. (2010). Assessing solution quality of biobjective 0-1 knapsack problem using evolutionary and heuristic algorithms. Applied Soft Computing, 10(3), 711-718. doi: 10.1016/j.asoc.2009.08.037

Kuo, Y. Y. (2010). Using simulated annealing to minimize fuel consumption for the time-dependent vehicle routing problem. Computers & Industrial Engineering, 59(1), 157-165. doi: 10.1016/j.cie.2010.03.012

Laporte, G. (2010). A concise guide to the Traveling Salesman Problem. Journal of the Operational Research Society, 61(1), 35-40. doi: 10.1057/jors.2009.76

Lau, H. C. W., Chan, T. M., Tsui, W. T., & Pang, W. K. (2010). Application of Genetic Algorithms to Solve the Multidepot Vehicle Routing Problem. Ieee Transactions on Automation Science and Engineering, 7(2), 383-392. doi: 10.1109/tase.2009.2019265

Lee, C. Y., Lee, Z. J., Lin, S. W., & Ying, K. C. (2010). An enhanced ant colony optimization (EACO) applied to capacitated vehicle routing problem. Applied Intelligence, 32(1), 88-95. doi: 10.1007/s10489-008-0136-9

Lee, W. C., Shiau, Y. R., Chen, S. K., & Wu, C. C. (2010). A two-machine flowshop scheduling problem with deteriorating jobs and blocking. International Journal of Production Economics, 124(1), 188-197. doi: 10.1016/j.ijpe.2009.11.001

Lee, W. C., Wang, W. J., Shiau, Y. R., & Wu, C. C. (2010). A single-machine scheduling problem with two-agent and deteriorating jobs. Applied Mathematical Modelling, 34(10), 3098-3107. doi: 10.1016/j.apm.2010.01.015

Lee, W. C., Wu, C. C., & Hsu, P. H. (2010). A single-machine learning effect scheduling problem with release times. Omega-International Journal of Management Science, 38(1-2), 3-11. doi: 10.1016/j.omega.2009.01.001

Letocart, L., Nagih, A., & Plateau, G. (2012). Reoptimization in Lagrangian methods for the 0-1 quadratic knapsack problem. Computers & Operations Research, 39(1), 12-18. doi: 10.1016/j.cor.2010.10.027

Li, J. Q., Pan, Q. K., Suganthan, P. N., & Chua, T. J. (2011). A hybrid tabu search algorithm with an efficient neighborhood structure for the flexible job shop scheduling problem. International Journal of Advanced Manufacturing Technology, 52(5-8), 683-697. doi: 10.1007/s00170-010-2743-y

Li, K. P., Gao, Z. Y., Tang, T., & Yang, L. X. (2010). Solving the constrained shortest path problem using random search strategy. Science China-Technological Sciences, 53(12), 3258-3263. doi: 10.1007/s11431-010-4105-2

Li, X. Y., Baki, F., Tian, P., & Chaouch, B. A. (2014). A robust block-chain based tabu search algorithm for the dynamic lot sizing problem with product returns and remanufacturing. Omega-International Journal of Management Science, 42(1), 75-87. doi: 10.1016/j.omega.2013.03.003

Lin, Y. K., & Yeh, C. T. (2011). Reliability Optimization Of Component Assignment Problem For A Multistate Network In Terms Of Minimal Cuts. Journal of Industrial and Management Optimization, 7(1), 211-227. doi: 10.3934/jimo.2011.7.211

Liu, Y. H. (2010). Different initial solution generators in genetic algorithms for solving the probabilistic traveling salesman problem. Applied Mathematics and Computation, 216(1), 125-137. doi: 10.1016/j.amc.2010.01.021

Lozano, L., & Medaglia, A. L. (2013). On an exact method for the constrained shortest path problem. Computers & Operations Research, 40(1), 378-384. doi: 10.1016/j.cor.2012.07.008

Lu, S., & Nie, Y. (2010). Stability of user-equilibrium route flow solutions for the traffic assignment problem. Transportation Research Part B-Methodological, 44(4), 609-617. doi: 10.1016/j.trb.2009.09.003

Lu, Z. Q., Zhang, Y. J., & Han, X. L. (2013). Integrating run-based preventive maintenance into the capacitated lot sizing problem with reliability constraint. International Journal of Production Research, 51(5), 1379-1391. doi: 10.1080/00207543.2012.693637

Lust, T., & Teghem, J. (2010). Two-phase Pareto local search for the biobjective traveling salesman problem. Journal of Heuristics, 16(3), 475-510. doi: 10.1007/s10732-009-9103-9

Maric, M. (2010). An Efficient Genetic Algorithm For Solving The Multi-Level Uncapacitated Facility Location Problem. Computing and Informatics, 29(2), 183-201.

Marin, A. (2011). The discrete facility location problem with balanced allocation of customers. European Journal of Operational Research, 210(1), 27-38. doi: 10.1016/j.ejor.2010.10.012

Marinakis, Y., & Marinaki, M. (2010). A hybrid genetic - Particle Swarm Optimization Algorithm for the vehicle routing problem. Expert Systems with Applications, 37(2), 1446-1455. doi: 10.1016/j.eswa.2009.06.085

Marinakis, Y., & Marinaki, M. (2010). A Hybrid Multi-Swarm Particle Swarm Optimization algorithm for the Probabilistic Traveling Salesman Problem. Computers & Operations Research, 37(3), 432-442. doi: 10.1016/j.cor.2009.03.004

Marinakis, Y., Marinaki, M., & Dounias, G. (2010). A hybrid particle swarm optimization algorithm for the vehicle routing problem. Engineering Applications of Artificial Intelligence, 23(4), 463-472. doi: 10.1016/j.engappai.2010.02.002

Marinakis, Y., Marinaki, M., & Dounias, G. (2011). Honey bees mating optimization algorithm for the Euclidean traveling salesman problem. Information Sciences, 181(20), 4684-4698. doi: 10.1016/j.ins.2010.06.032

Mastorakis, Nikos E.. Computer Science, Technology and Applications : Pathway Modeling and Algorithm Research. New York, NY, USA: Nova Science Publishers, Inc. 2011. p 46.

Mateus, G. R., Resende, M. G. C., & Silva, R. M. A. (2011). GRASP with path-relinking for the generalized quadratic assignment problem. Journal of Heuristics, 17(5), 527-565. doi: 10.1007/s10732-010-9144-0

McLay, L. A., Lloyd, J. D., & Niman, E. (2011). Interdicting nuclear material on cargo containers using knapsack problem models. Annals of Operations Research, 187(1), 185-205. doi: 10.1007/s10479-009-0667-4

Mendoza, J. E., Castanier, B., Gueret, C., Medaglia, A. L., & Velasco, N. (2010). A memetic algorithm for the multi-compartment vehicle routing problem with stochastic demands. Computers & Operations Research, 37(11), 1886-1898. doi: 10.1016/j.cor.2009.06.015

Menlo Park, CA, USA: Course Technology / Cengage Learning, 1996. p 51-56.

Mester, D. (Editor); Ronin, D. (Editor); Frenkel, M. (Editor). Genetics - Research and Issues : Discrete Optimization for Some TSP-like Genome Mapping Problems. New York, NY, USA: Nova Science Publishers, Inc. 2010. p 6.

Mishra, D.N.; Agarwal, S.K.. Operation Research. Lucknow, IND: Global Media 2009. p 107.

Mohammadi, M., Torabi, S. A., Ghomi, S., & Karimi, B. (2010). A new algorithmic approach for capacitated lot-sizing problem in flow shops with sequence-dependent setups. International Journal of Advanced Manufacturing Technology, 49(1-4) 201-211. doi: 10.1007/s00170-009-2366-3

Monoyios, D., & Vlachos, K. (2011). Multiobjective Genetic Algorithms for Solving the Impairment-Aware Routing and Wavelength Assignment Problem. Journal of Optical Communications and Networking, 3(1), 40-47. doi: 10.1364/jocn.3.000040

Moslehi, G., & Mahnam, M. (2011). A Pareto approach to multi-objective flexible job-shop scheduling problem using particle swarm optimization and local search. International Journal of Production Economics, 129(1), 14-22. doi: 10.1016/j.ijpe.2010.08.004

Nagata, Y., Braysy, O., & Dullaert, W. (2010). A penalty-based edge assembly memetic algorithm for the vehicle routing problem with time windows. Computers & Operations Research, 37(4), 724-737. doi: 10.1016/j.cor.2009.06.022

Nascimento, M. C. V., Resende, M. G. C., & Toledo, F. M. B. (2010). GRASP heuristic with path-relinking for the multi-plant capacitated lot sizing problem. European Journal of Operational Research 200(3), 747-754. doi: 10.1016/j.ejor.2009.01.047

Nazemi, A., & Omidi, F. (2013). An efficient dynamic model for solving the shortest path problem. Transportation Research Part C-Emerging Technologies, 26, 1-19. doi: 10.1016/j.trc.2012.07.005

Ng, C. T., Kovalyov, M. Y., & Cheng, T. C. E. (2010). A simple FPTAS for a single-item capacitated economic lot-sizing problem with a monotone cost structure. European Journal of Operational Research 200(2), 621-624. doi: 10.1016/j.ejor.2009.01.040

Ngueveu, S. U., Prins, C., & Calvo, R. W. (2010). An effective memetic algorithm for the cumulative capacitated vehicle routing problem. Computers & Operations Research, 37(11), 1877-1885. doi: 10.1016/j.cor.2009.06.014

Nie, Y. (2010). A class of bush-based algorithms for the traffic assignment problem. Transportation Research Part B-Methodological, 44(1), 73-89. doi: 10.1016/j.trb.2009.06.005

Nie, Y. (2011). A cell-based Merchant-Nemhauser model for the system optimum dynamic traffic assignment problem. Transportation Research Part B-Methodological, 45(2), 329-342. doi: 10.1016/j.trb.2010.07.001

Nie, Y., & Zhang, H. M. (2010). Solving the Dynamic User Optimal Assignment Problem Considering Queue Spillback. Networks & Spatial Economics, 10(1), 49-71. doi: 10.1007/s11067-007-9022-y

Okhrin, I., & Richter, K. (2011). An O(T-3) algorithm for the capacitated lot sizing problem with minimum order quantities. European Journal of Operational Research, 211(3), 507-514. doi: 10.1016/j.ejor.2011.01.007

Ore, Oystein; Wilson, Robin J. (Revised by). Anneli Lax New Mathematical Library, Volume 34 : Graphs and Their Uses. Washington, DC, USA: Mathematical Association of America, 1990. p 33.

Ozbakir, L., Baykasoglu, A., & Tapkan, P. (2010). Bees algorithm for generalized assignment problem. Applied Mathematics and Computation, 215(11), 3782-3795. doi: 10.1016/j.amc.2009.11.018

Pan, Q. K., Suganthan, P. N., Liang, J. J., & Tasgetiren, M. F. (2011). A local-best harmony search algorithm with dynamic sub-harmony memories for lot-streaming flow shop scheduling problem. Expert Systems with Applications, 38(4), 3252-3259. doi: 10.1016/j.eswa.2010.08.111

Pan, Q. K., Tasgetiren, M. F., Suganthan, P. N., & Chua, T. J. (2011). A discrete artificial bee colony algorithm for the lot-streaming flow shop scheduling problem. Information Sciences, 181(12), 2455-2468. doi: 10.1016/j.ins.2009.12.025

Pardalos, Panos M. (Editor); Migdalas, Athanasios (Editor); Burkard, Rainer E. (Editor). Combinatorial and Global Optimization. River Edge, NJ, USA: World Scientific 2002. p 10p.

Penuel, J., Smith, J. C., & Yuan, Y. (2010). An Integer Decomposition Algorithm for Solving a Two-Stage Facility Location Problem with Second-Stage Activation Costs. Naval Research Logistics, 57(5), 391-402. doi: 10.1002/nav.20401

Perboli, G., Tadei, R., & Vigo, D. (2011). The Two-Echelon Capacitated Vehicle Routing Problem: Models and Math-Based Heuristics. Transportation Science, 45(3), 364-380. doi: 10.1287/trsc.1110.0368

Pineyro, P., & Viera, O. (2010). The economic lot-sizing problem with remanufacturing and one-way substitution. International Journal of Production Economics, 124(2), 482-488. doi: 10.1016/j.ijpe.2010.01.007

Pinto, L. L., & Pascoal, M. M. B. (2010). On algorithms for the tricriteria shortest path problem with two bottleneck objective functions. Computers & Operations Research, 37(10), 1774-1779. doi: 10.1016/j.cor.2010.01.005

Piperagkas, G. S., Konstantaras, I., Skouri, K., & Parsopoulos, K. E. (2012). Solving the stochastic dynamic lot-sizing problem through nature-inspired heuristics. Computers & Operations Research, 39(7), 1555-1565. doi: 10.1016/j.cor.2011.09.004

Pop, P. C. (2012). Generalized Network Design Problems: Modeling and Optimization. Berlin, De Gruyter.

Pop, Petrica C.; Versita (Contribution by). De Gruyter Series in Discrete Mathematics and Applications, Volume 1 : Network Design Problems : Modeling and Optimization of Generalized Network Design Problems. Hawthorne, NY, USA: Walter de Gruyter 2012. p 128.

Przybylski, A., Gandibleux, X., & Ehrgott, M. (2010). A two phase method for multi-objective integer programming and its application to the assignment problem with three objectives. Discrete Optimization, 7(3), 149-165. doi: 10.1016/j.disopt.2010.03.005

Puchinger, J., Raidl, G. R., & Pferschy, U. (2010). The Multidimensional Knapsack Problem: Structure and Algorithms. Informs Journal on Computing, 22(2), 250-265. doi: 10.1287/ijoc.1090.0344

Puerto, J., & Rodriguez-Chia, A. M. (2011). On the structure of the solution set for the single facility location problem with average distances. Mathematical Programming, 128(1-2), 373-401. doi: 10.1007/s10107-009-0308-3

Pugliese, L. D., & Guerriero, F. (2012). A computational study of solution approaches for the resource constrained elementary shortest path problem. Annals of Operations Research 201(1), 131-157. doi: 10.1007/s10479-012-1162-x

Pugliese, L. D., & Guerriero, F. (2013). Dynamic programming approaches to solve the shortest path problem with forbidden paths. Optimization Methods & Software, 28(2), 221-255. doi: 10.1080/10556788.2011.630077

Pugliese, L. D., & Guerriero, F. (2013). Shortest path problem with forbidden paths: The elementary version. European Journal of Operational Research, 227(2), 254-267. doi: 10.1016/j.ejor.2012.11.010

Ramadan, M. A., & El-Sayed, E. A. (2010). Partial eigenvalue assignment problem of high order control systems using orthogonality relations. Computers & Mathematics with Applications, 59(6), 1918-1928. doi: 10.1016/j.camwa.2009.07.063

Ramamurthy, P.. Operations Research. Daryaganj, Delhi, IND: New Age International 2007. p 214.

Ramazani, H., Shafahi, Y., & Seyedabrishami, S. E. (2010). A Shortest Path Problem in an Urban Transportation Network Based on Driver Perceived Travel Time. Scientia Iranica Transaction a-Civil Engineering, 17(4), 285-296.

Rego, C., Gamboa, D., Glover, F., & Osterman, C. (2011). Traveling salesman problem heuristics: Leading methods, implementations and latest advances. European Journal of Operational Research, 211(3), 427-441. doi: 10.1016/j.ejor.2010.09.010

Ren, Z. G., Feng, Z. R., & Zhang, A. M. (2012). Fusing ant colony optimization with Lagrangian relaxation for the multiple-choice multidimensional knapsack problem. Information Sciences, 182(1), 15-29. doi: 10.1016/j.ins.2011.07.033

Repoussis, P. P., Tarantilis, C. D., Braysy, O., & Ioannou, G. (2010). A hybrid evolution strategy for the open vehicle routing problem. Computers & Operations Research, 37(3), 443-455. doi: 10.1016/j.cor.2008.11.003

Ribas, I., Companys, R., & Tort-Martorell, X. (2011). An iterated greedy algorithm for the flowshop scheduling problem with blocking. Omega-International Journal of Management Science, 39(3), 293-301. doi: 10.1016/j.omega.2010.07.007

Ribeiro, G. M., & Laporte, G. (2012). An adaptive large neighborhood search heuristic for the cumulative capacitated vehicle routing problem. Computers & Operations Research, 39(3), 728-735. doi: 10.1016/j.cor.2011.05.005

Rondeau, Thomas W.; Bostian, Charles W.. Artificial Intelligence in Wireless Communications. Norwood, MA, USA: Artech House 2009. p 78.

Rong, A. Y., Figueira, J. R., & Klamroth, K. (2012). Dynamic programming based algorithms for the discounted {0-1} knapsack problem. Applied Mathematics and Computation, 218(12), 6921-6933. doi: 10.1016/j.amc.2011.12.068

Rong, A. Y., Figueira, J. R., & Pato, M. V. (2011). A two state reduction based dynamic programming algorithm for the bi-objective 0-1 knapsack problem. Computers & Mathematics with Applications, 62(8), 2913-2930. doi: 10.1016/j.camwa.2011.07.067

Ruiz, R., & Vazquez-Rodriguez, J. A. (2010). The hybrid flow shop scheduling problem. European Journal of Operational Research 205(1), 1-18. doi: 10.1016/j.ejor.2009.09.024

Salmasi, N., Logendran, R., & Skandari, M. R. (2010). Total flow time minimization in a flowshop sequence-dependent group scheduling problem. Computers & Operations Research, 37(1), 199-212. doi: 10.1016/j.cor.2009.04.013

Sawik, Tadeusz. Scheduling in Supply Chains Using Mixed Integer Programming. Hoboken, NJ, USA: Wiley 2011. p 29.

Sbihi, A. (2010). A cooperative local search-based algorithm for the Multiple-Scenario Max-Min Knapsack Problem. European Journal of Operational Research 202(2), 339-346. doi: 10.1016/j.ejor.2009.05.033

Schulz, T. (2011). A new Silver-Meal based heuristic for the single-item dynamic lot sizing problem with returns and remanufacturing. International Journal of Production Research, 49(9), 2519-2533. doi: 10.1080/00207543.2010.532916

Shen, Z. J. M., Zhan, R. L., & Zhang, J. W. (2011). The Reliable Facility Location Problem: Formulations, Heuristics, and Approximation Algorithms. Informs Journal on Computing, 23(3), 470-482. doi: 10.1287/ijoc.1100.0414

Shi, N. (2010). K Constrained Shortest Path Problem. Ieee Transactions on Automation Science and Engineering, 7(1), 15-23. doi: 10.1109/tase.2009.2012434

Subbu, Raj; Sanderson, Arthur C.. Network-Based Distributed Planning Using Coevolutionary Algorithms. River Edge, NJ, USA: World Scientific 2004. p 3.

Subramanian, A., Drummond, L. M. A., Bentes, C., Ochi, L. S., & Farias, R. (2010). A parallel heuristic for the Vehicle Routing Problem with Simultaneous Pickup and Delivery. Computers & Operations Research, 37(11), 1899-1911. doi: 10.1016/j.cor.2009.10.011

Sun, M. H. (2012). A tabu search heuristic procedure for the capacitated facility location problem. Journal of Heuristics, 18(1), 91-118. doi: 10.1007/s10732-011-9157-3

Szeto, W. Y., Wu, Y. Z., & Ho, S. C. (2011). An artificial bee colony algorithm for the capacitated vehicle routing problem. European Journal of Operational Research, 215(1), 126-135. doi: 10.1016/j.ejor.2011.06.006

Takaoka, T. (2013). A simplified algorithm for the all pairs shortest path problem with O(n(2) log n) expected time. Journal of Combinatorial Optimization, 25(2), 326-337. doi: 10.1007/s10878-012-9550-3

Tang, L. X., Jiang, W., & Saharidis, G. K. D. (2013). An improved Benders decomposition algorithm for the logistics facility location problem with capacity expansions. Annals of Operations Research, 210(1), 165-190. doi: 10.1007/s10479-011-1050-9

Tarim, S. A., Dogru, M. K., Ozen, U., & Rossi, R. (2011). An efficient computational method for a stochastic dynamic lot-sizing problem under service-level constraints. European Journal of Operational Research, 215(3), 563-571. doi: 10.1016/j.ejor.2011.06.034

Tasgetiren, M. F., Suganthan, P. N., & Pan, Q. K. (2010). An ensemble of discrete differential evolution algorithms for solving the generalized traveling salesman problem. Applied Mathematics and Computation, 215(9), 3356-3368. doi: 10.1016/j.amc.2009.10.027

Tohyama, H., Ida, K., & Matsueda, J. (2011). A Genetic Algorithm for the Uncapacitated Facility Location Problem. Electronics and Communications in Japan, 94(5), 47-54. doi: 10.1002/ecj.10180

Vallada, E., & Ruiz, R. (2011). A genetic algorithm for the unrelated parallel machine scheduling problem with sequence dependent setup times. European Journal of Operational Research, 211(3), 612-622. doi: 10.1016/j.ejor.2011.01.011

Van Peteghem, V., & Vanhoucke, M. (2010). A genetic algorithm for the preemptive and non-preemptive multi-mode resource-constrained project scheduling problem. European Journal of Operational Research 201(2), 409-418. doi: 10.1016/j.ejor.2009.03.034

Viale, J. David; Carrigan, Christopher (Editor). Inventory Management : From

Warehouse to Distribution Center.

Viale, J. David; Carrigan, Christopher (Editor). Inventory Management : From Warehouse to Distribution Center. Menlo Park, CA, USA: Course Technology / Cengage Learning, 1996. p 50-57.

Wang, J. B., & Guo, Q. (2010). A due-date assignment problem with learning effect and deteriorating jobs. Applied Mathematical Modelling, 34(2), 309-313. doi: 10.1016/j.apm.2009.04.020

Wang, J. B., & Wang, C. (2011). Single-machine due-window assignment problem with learning effect and deteriorating jobs. Applied Mathematical Modelling, 35(8), 4017-4022. doi: 10.1016/j.apm.2011.02.023

Wang, L., Wang, S. Y., & Xu, Y. (2012). An effective hybrid EDA-based algorithm for solving multidimensional knapsack problem. Expert Systems with Applications, 39(5), 5593-5599. doi: 10.1016/j.eswa.2011.11.058

Wang, X. J., Gao, L., Zhang, C. Y., & Shao, X. Y. (2010). A multi-objective genetic algorithm based on immune and entropy principle for flexible job-shop scheduling problem. International Journal of Advanced Manufacturing Technology, 51(5-8), 757-767. doi: 10.1007/s00170-010-2642-2

Wang, Z. C., Zhang, Y. M., Zhou, W. H., & Liu, H. F. (2012). Solving traveling salesman problem in the Adleman-Lipton model. Applied Mathematics and Computation, 219(4), 2267-2270. doi: 10.1016/j.amc.2012.08.073

Wang, Z., Du, D. L., Gabor, A. F., & Xu, D. C. (2010). An approximation algorithm for the k-level stochastic facility location problem. Operations Research Letters, 38(5), 386-389. doi: 10.1016/j.orl.2010.04.010

Wei, Z., Ge, F. Z., Lu, Y., Li, L. X., & Yang, Y. X. (2011). Chaotic ant swarm for the traveling salesman problem. Nonlinear Dynamics, 65(3), 271-281. doi: 10.1007/s11071-010-9889-x

Williams, H. Paul. Model Building in Mathematical Programming (5th Edition). Somerset, NJ, USA: Wiley 2013. p 109-220.

Wong, J. T., Su, C. T., & Wang, C. H. (2012). Stochastic dynamic lot-sizing problem using bi-level programming base on artificial intelligence techniques. Applied Mathematical Modelling, 36(5) 2003-2016. doi: 10.1016/j.apm.2011.08.017

Wu, T., Shi, L. Y., & Duffie, N. A. (2010). An HNP-MP Approach for the Capacitated Multi-Item Lot Sizing Problem With Setup Times. Ieee Transactions on Automation Science and Engineering, 7(3), 500-511. doi:

10.1109/tase.2009.2039134

Wu, T., Shi, L. Y., & Song, J. (2012). An MIP-based interval heuristic for the capacitated multi-level lot-sizing problem with setup times. Annals of Operations Research, 196(1), 635-650. doi: 10.1007/s10479-011-1026-9

Yagmahan, B., & Yenisey, M. M. (2010). A multi-objective ant colony system algorithm for flow shop scheduling problem. Expert Systems with Applications, 37(2), 1361-1368. doi: 10.1016/j.eswa.2009.06.105

Yalaoui, Alice;Chehade, Hicham;Yalaoui , Farouk;Amodeo, Lionel. (2013). Optimization of Logistics. Wiley-ISTE.

Yang, Xin-She. Introduction to Mathematical Optimization : From Linear Programming to Metaheuristics. Cambridge, GBR: Cambridge International Science Publishing 2008. p 93.

Yang, Z., Chu, F., & Chen, H. X. (2012). A cut-and-solve based algorithm for the single-source capacitated facility location problem. European Journal of Operational Research, 221(3), 521-532. doi: 10.1016/j.ejor.2012.03.047

Yu, B., & Yang, Z. Z. (2011). An ant colony optimization model: The period vehicle routing problem with time windows. Transportation Research Part E-Logistics and Transportation Review, 47(2), 166-181. doi: 10.1016/j.tre.2010.09.010

Yu, B., Yang, Z. Z., & Xie, J. X. (2011). A parallel improved ant colony optimization for multi-depot vehicle routing problem. Journal of the Operational Research Society, 62(1), 183-188. doi: 10.1057/jors.2009.161

Zhan, Z. H., Zhang, J., Li, Y., Liu, O., Kwok, S. K., Ip, W. H., & Kaynak, O. (2010). An Efficient Ant Colony System Based on Receding Horizon Control for the Aircraft Arrival Sequencing and Scheduling Problem. Ieee Transactions on Intelligent Transportation Systems, 11(2), 399-412. doi: 10.1109/tits.2010.2044793

Zhang, G. H., Gao, L., & Shi, Y. (2011). An effective genetic algorithm for the flexible job-shop scheduling problem. Expert Systems with Applications, 38(4), 3563-3573. doi: 10.1016/j.eswa.2010.08.145

Zhang, H. Z., Beltran-Royo, C., & Constantino, M. (2010). Effective formulation reductions for the quadratic assignment problem. Computers & Operations Research, 37(11) 2007-2016. doi: 10.1016/j.cor.2010.02.001

Zhang, R., & Wu, C. (2010). A hybrid immune simulated annealing algorithm for the job shop scheduling problem. Applied Soft Computing, 10(1), 79-89. doi: 10.1016/j.asoc.2009.06.008

Zhang, X. G., Wang, Q., Chan, F. T. S., Mahadevan, S., & Deng, Y. (2014). A Physarum Polycephalum Optimization Algorithm for the Bi-objective Shortest Path Problem. International Journal of Unconventional Computing, 10(1-2), 143-162.

Zhang, Y. D., Wu, L. N., Wei, G., & Wang, S. H. (2011). A novel algorithm for all pairs shortest path problem based on matrix multiplication and pulse coupled neural network. Digital Signal Processing, 21(4), 517-521. doi: 10.1016/j.dsp.2011.02.004

Zhang, Z. H., Jiang, H., & Pan, X. Z. (2012). A Lagrangian relaxation based approach for the capacitated lot sizing problem in closed-loop supply chain. International Journal of Production Economics, 140(1), 249-255. doi: 10.1016/j.ijpe.2012.01.018

Zhao, J., Liu, Q. L., Wang, W., Wei, Z. Q., & Shi, P. (2011). A parallel immune algorithm for traveling salesman problem and its application on cold rolling scheduling. Information Sciences, 181(7), 1212-1223. doi: 10.1016/j.ins.2010.12.003

Zhao, Y. L., Zhang, J. E., Ji, Y. F., & Gu, W. Y. (2010). Routing and Wavelength Assignment Problem in PCE-Based Wavelength-Switched Optical Networks. Journal of Optical Communications and Networking, 2(4), 196-205. doi: 10.1364/jocn.2.000196

Zou, D. X., Gao, L. Q., Li, S., & Wu, J. H. (2011). Solving 0-1 knapsack problem by a novel global harmony search algorithm. Applied Soft Computing, 11(2), 1556-1564. doi: 10.1016/j.asoc.2010.07.019

Zou, D. X., Gao, L. Q., Li, S., Wu, J. H., & Wang, X. (2010). A novel global harmony search algorithm for task assignment problem. Journal of Systems and Software, 83(10), 1678-1688. doi: 10.1016/j.jss.2010.04.070

Chapter 3. Algorithms

This chapter is about algorithms used for solving optimization problems. In this chapter, the fundamentals of ant colony optimization, the cross-entropy method, the Dijkstra, the Bellman-Ford algorithm, the genetic algorithm, the Hungarian, the Jonker-Volgenant algorithm, the particle swarm, the simulated annealing method, and the Wagner-Whitin algorithm are introduced.

There exist many other numerous algorithms for solving optimization problems. This chapter only introduces few of the most popular algorithms.

3.1 Ant colony optimization

The ant colony optimization (ACO) was introduced by Marco Dorigo in the early 1990s (Sandou 2013). This algorithm is based on the social behavior of ants and allows us to solve complex optimization problems, especially integers programming problems (Sandou 2013). The ant colony optimization is a metaheuristic in which a colony of artificial ants cooperates in finding good solutions to difficult discrete optimization problems (Dorigo and Stützle 2004). Cooperation is a key element in the design of ACO algorithms: The choice is to allocate computational resources to a relatively simple set of agents (artificial ants) that communicate indirectly by stigmergy is mediated by indirect communication by the environment (Dorigo and Stützle 2004).

This algorithm is based on the social behavior of ants when they are foraging (Sandou 2013). Ant colonies, and societies of social insects in general, are systems that, despite the simplicity of their individuals, have a highly structured social organization (Dorigo and Stützle 2004) distributed. ACO metaheuristic embodies a large class of algorithms whose design is based mainly on the foraging behavior of real ants (Dehuri 2011). The route search initial random food, the concentration ranges of pheromone and follow the path of ants of the higher concentration of pheromone and pheromone is enhanced by increasing number of ants. As more and more ants follow the same path, it becomes the preferred route (Yang 2008). The amount of pheromone deposited, which may depend on the quantity and quality of the food, will guide other ants to the food source (Cho *et al.* 2011). Indirect communication between the ants via pheromone trails allows them to find the shortest path between their nest and food sources (Cho *et al.* 2011). This characteristic of real ant colonies is exploited inartificial ant colonies to solve NP-hard problems (Cho *et al.* 2011). For example, some preferred (often the shortest or most efficient) route emerges (Yang 2008). This is in fact a positive feedback mechanism (Yang 2008). ACO algorithms were originally designed and have a long tradition in solving a specific type of combinatorial optimization problems (ie, problems for which the construction process of the solution can be implemented by simulating a walk through a construction graph) (Dehuri 2011).

The key works using the ACO meta-heuristic has been devoted to the Travelling Salesperson Problem (TSP), a classic NP-complete problem whose main features can easily be manipulated to show the applicability of this metaheuristic (Dehuri 2011).

Ants that perform well in a given iteration influence exploration ants in future iterations. Because ants explore alternatives, the resulting pheromone trail is the result of different views on the solution space (Bonabeau *et al.*, 1999). Even when only the best ant execution is allowed to strengthen its solution, there is a cooperative effect in time because the ants in the next iteration using the pheromone trail to guide their exploration (Bonabeau *et al.*, 1999).

Algorithm – Outline of the ACO metaheuristic (Dehuri 2011):
1: Initialize();
2: while termination-condition is NOT TRUE do
3: BuildSolutions();
4: PheromoneUpdate();
5: DaemonActions(); // Optional
6: end while

It should be emphasized that the algorithms of ant colonies are the right tool for discrete and combinatorial optimization (Yang 2008). They have the advantages over other stochastic algorithms such as genetic algorithms and simulated annealing in the treatment of dynamic network routing problems (Yang 2008). For continuous decision variables, its performance is still being investigated (Yang 2008).

The most recent and highly cited studies about the ant colony optimization are shown in Table 3.1.

Table 3.1 Studies about the ant colony optimization

A parameter free continuous ant colony optimization algorithm for the optimal design of storm sewer networks: constrained and unconstrained approach (Afshar 2010).
A two-stage ant colony optimization algorithm to minimize the makespan on unrelated parallel machines with sequence-dependent setup times (Arnaout *et al.* 2010).
In vivo diagnosis of gastric cancer using raman endoscopy and ant colony optimization techniques (Bergholt *et al.* 2011).
Bi-objective ant colony optimization approach to optimize production and maintenance scheduling (Berrichi *et al.* 2010).
Optimizing discounted cash flows in project scheduling - An ant colony optimization approach (Chen *et al.* 2010).
A rough set approach to feature selection based on ant colony optimization (Chen *et al.* 2010).
Scheduling resource-constrained projects with ant colony optimization artificial agents (Christodoulou 2010).
Designing fuzzy-rule-based systems using continuous ant-colony optimization (Juang and Chang 2010).
An improved ant colony optimization for constrained engineering design problems (Kaveh and Talatahari 2010).
An improved ant colony optimization for the design of planar steel frames (Kaveh and Talatahari 2010).
An enhanced ant colony optimization (eaco) applied to capacitated vehicle routing problem (lee *et al.* 2010).
Integrated process planning and scheduling by an agent-based ant colony optimization (Leung *et al.* 2010).
A survey: ant colony optimization based recent research and implementation on several engineering domain (Mohan and Baskaran 2012).
Ant colony optimization algorithm to solve for the transportation problem of cross-docking network (Musa *et al.* 2010).
Ant colony optimization and the minimum spanning tree problem (Neumann and Witt 2010).
Power load forecasting using support vector machine and ant colony optimization (Niu *et al.* 2010).
A survey on parallel ant colony optimization (Pedemonte *et al.* 2011).
Comparing ant colony optimization and genetic algorithm approaches for solving traffic signal coordination under oversaturation conditions (Putha *et al.* 2012).
An improved ant colony optimization based algorithm for the capacitated arc routing problem (Santos *et al.* 2010).
Ant colony optimization for wavelet-based image interpolation using a three-component exponential mixture model (tian *et al.* 2011).
Knowledge-based ant colony optimization for flexible job shop scheduling problems (Xing *et al.* 2010).
An improved ant colony optimization algorithm for solving a complex combinatorial optimization problem (Yang and Zhuang 2010).
An ant colony optimization model: the period vehicle routing problem with time windows (Yu and Yang 2011).
A parallel improved ant colony optimization for multi-depot vehicle routing problem (Yu *et al.* 2011).
End member extraction of hyperspectral remote sensing images based on the ant colony optimization (ACO) algorithm (Zhang *et al.* 2011).

3.2 The cross-entropy (CE) method

The cross entropy method (CE) is a powerful technique to solve estimation and optimization of difficult problems, based on Kullback-Leibler (or cross-entropy) minimization (Rubinstein *et al.* 2013). It was launched in 1999 by Rubinstein, as an adaptive importance sampling to estimate the probability of rare-events (Rubinstein *et al.* 2013). Subsequent work has shown that in many optimization problems can be converted into a problem of estimation of rare events (Rubinstein *et al.* 2013). Accordingly, the method of CE can be used as randomized algorithm for optimization (Rubinstein *et al.* 2013).

3.3 The Dijkstra and the Bellman–Ford algorithm

An efficient method was developed in 1959 by the Dutch mathematician Edsger W. Dijkstra, a pioneer in the art of computer programming (Koshy 2003). Dijkstra's algorithm can find the shortest path from any vertex to any vertex in the digraph, if it exists (Koshy 2003).

An overview of Dijkstra's algorithm is described as follows (Tatipamula *et al.* 2012):

> **Step 1:** Set the distance for the source node to zero, and to all other nodes to infinity.
>
> **Step 2:** Mark all nodes as unvisited. Set the source node as a current.
>
> **Step 3:** For current node, consider all its unvisited neighbors and calculate their distance from the source node. If the distance is less than the distance previously recorded, the previous distance is replaced by the new one in the record. The last hop is also updated with the new leap in the previous case.
>
> **Step 4:** Select the unvisited node whose distance is shortest, and set it to a current node. Mark previous current node visited. Visited a node will not be checked over. The recorded distance of the current node is the smallest among the unvisited nodes, and it is final.
>
> **Step 5:** If all nodes have been visited, the process is complete. If not, repeat step 3.

The Bellman-Ford: Before describing the algorithm in detail, a few preliminaries are necessary (Benvenuto and Zorzi 2011). The Bellman-Ford algorithm derives its solution iteratively by building sequences of nodes which are temporary best walks (Benvenuto Zorzi and 2011). We say "walks" because we do not necessarily advocate visit the same node more than once. Thus, the algorithm seems at first to find walks, not paths (Benvenuto and Zorzi 2011). Recall that a path is always one walk (and thus a shortest path is an optimal walk), but not vice versa; it depends if walking contains cycles or not (only in the latter case is the walk path also a path) (Benvenuto and Zorzi 2011).

Dijkstra's algorithm: Dijkstra's algorithm is similar to the Bellman-Ford algorithm, although it has a computational complexity of lower the worst-case (Benvenuto and Zorzi 2011).

Dijkstra's algorithm uses an iterative approach, the evaluation of path costs and update (Benvenuto and Zorzi 2011). Bellman-Ford algorithm blindly reiterated all the temporary costs, and recognized that there was no need for further updates only after an iteration that left unchanged all costs (Benvenuto and Zorzi 2011). Dijkstra's algorithm takes a slightly different approach: it determines the shortest to an order of increasing cost paths (Benvenuto and Zorzi 2011).

3.4 The genetic algorithm (GA)

The genetic algorithm (GA), developed by John Holland and his colleagues in the 1960s and 1970s, is a model or abstraction of biological evolution based on Charles Darwin's theory of natural selection (Yang 2010). Holland was the first to use the crossover and recombination, mutation and selection in the study of adaptive and artificial systems (Yang 2010).

GA encodes the decision variables or input parameters of the problem solution strings of finite length (Rao and Savsani 2012). Although the traditional optimization techniques work directly with decision variables or input parameters, genetic algorithms usually work with coding. Genetic algorithms start looking in a population of encoded solutions instead of a single point in the solution space (Rao and Savsani 2012). The initial population of individuals is created randomly. Genetic algorithms use genetic operators to create global optimal solutions based on the solutions to the current population (Rao and Savsani 2012). The most popular genetic operators are (1) selection, (2) crossover and (3) mutation (Rao and Savsani 2012). The newly generated individuals replace the old population, and the product of evolutionary processes until some termination criteria are satisfied (Rao and Savsani 2012). These genetic operators form the core of the genetic algorithm as a strategy for problem solving (Yang 2010). Since then, many variations of genetic algorithms have been developed and applied to a wide range of optimization problems, graph coloring pattern recognition, discrete systems such as the traveling salesman problem for continuous systems such the effective wing design in aerospace engineering, and financial market multiobjective engineering optimization (Yang 2010).

There are many advantages of genetic algorithms optimization algorithms, and two most notable benefits include: the ability to cope with complex problems of optimization and parallelism (Yang 2010). Genetic algorithms can cope with different types of optimization if the objective function (fitness) is stationary or non-stationary (changing over time), linear or nonlinear, continuous or discontinuous, or random noise (Yang 2010). As multiple offsprings in an act of the population as independent agents, population (or subgroup) can explore the search space in several directions simultaneously (Yang 2010). This feature makes it ideal for parallel algorithms implementation (Yang 2010). Different parameters and even different groups of encoded strings can be handled simultaneously (Yang 2010).

However, genetic algorithms also have some disadvantages (Yang 2010). The formulation of the fitness function, using the population size, the choice of important parameters such as the rate of mutation and crossover, and the criteria for selection of the new population should be carefully conducted (Yang 2010). Any wrong choice, it will be difficult for the algorithm to converge, or it simply produces impossible results (Yang 2010). Despite these disadvantages, genetic algorithms are one of the optimization algorithms widely used in modern nonlinear optimization (Yang 2010).

This is often done by the following procedure (Yang 2010):

- Encoding objectives or optimization
- The initialization of a population of individuals; functions; Defining a fitness function or selection criterion;
- The evaluation of the fitness of all individuals in the population;
- Creation of a new population by crossing performance, and mutation, fitness proportional reproduction etc;
- Changes in population until some stopping criteria are met;
- Decoding results for the solution to this problem.

The most recent and highly cited studies about the genetic algorithm optimization are shown in Table 3.2.

Table 3.2 Studies about the optimization by using genetic algorithm

Exergoenvironmental analysis and optimization of a cogeneration plant system using multimodal genetic algorithm (mga) (Ahmadi and Dincer 2010).
Application of pso (particle swarm optimization) and ga (genetic algorithm) techniques on demand estimation of oil in Iran (Assareh *et al.* 2010).
A comparison of feature selection models utilizing binary particle swarm optimization and genetic algorithm in determining coronary artery disease using support vector machine (Babaoglu *et al.* 2010).
Exergoeconomic analysis and optimization of an integrated solar combined cycle system (isccs) using genetic algorithm (Baghernejad and Yaghoubi 2011).
Rsm and ann modeling for electrocoagulation of copper from simulated wastewater: multi objective optimization using genetic algorithm approach (Bhatti *et al.* 2011).
Tracing sediment loss from eroding farm tracks using a geochemical fingerprinting procedure combining local and genetic algorithm optimisation (Collins *et al.* 2010).
Damping of power system oscillations using genetic algorithm and particle swarm optimization (Eslami *et al.* 2010).
Genetic algorithm optimization in drug design qsar: bayesian-regularized genetic neural networks (brgnn) and genetic algorithm-optimized support vectors machines (ga-svm) (Fernandez *et al.* 2011).
Mathematical modeling and genetic algorithm optimization of clove oil extraction with supercritical carbon dioxide (Hatami *et al.* 2010).
Modeling and multi-objective exergy based optimization of a combined cycle power plant using a genetic algorithm (Kaviri *et al.* 2012).
A hybrid of genetic algorithm and particle swarm optimization for solving bi-level linear programming problem - a case study on supply chain model (Kuo and Han 2011).
Multiobjective optimization of building design using trnsys simulations, genetic algorithm, and artificial neural network (Magnier and Haghighat 2010).
A combination of genetic algorithm and particle swarm optimization for optimal dg location and sizing in distribution systems (Moradi and Abedini 2012).
Optimization of an artificial neural network topology using coupled response surface methodology and genetic algorithm for fluidized bed drying (Nazghelichi *et al.* 2011).
Comparing ant colony optimization and genetic algorithm approaches for solving traffic signal coordination under oversaturation conditions (Putha *et al.* 2012).
A genetic-algorithm-aided stochastic optimization model for regional air quality management under uncertainty (Qin *et al.* 2010).
Thermal-economic multi-objective optimization of plate fin heat exchanger using genetic algorithm (Sanaye and Hajabdollahi 2010).
A case study on optimization of biomass flow during single-screw extrusion cooking using genetic algorithm (ga) and response surface method (rsm) (Shankar *et al.* 2010).
Simultaneous optimization of luminance and color chromaticity of phosphors using a nondominated sorting genetic algorithm (Sharma *et al.* 2010).
Artificial neural network modeling and genetic algorithm based medium optimization for the improved production of marine biosurfactant (Sivapathasekaran *et al.* 2010).
Parametric optimization design for supercritical co2 power cycle using genetic algorithm and artificial neural network (Wang *et al.* 2010).
Optimization of capacity and operation for cchp system by genetic algorithm (Wang *et al.* 2010).
Accounting for greenhouse gas emissions in multiobjective genetic algorithm optimization of water distribution systems (Wu *et al.* 2010).
An adaptive reanalysis method for genetic algorithm with application to fast truss optimization (Xu *et al.* 2010).
Artificial neural network-genetic algorithm based optimization for the immobilization of cellulase on the smart polymer eudragit l-100 (Zhang *et al.* 2010).

3.5 *The Hungarian algorithm & the Jonker-Volgenant algorithm*

The algorithm is defined here for minimizing the total cost of assignment. The agents form the rows of the matrix and the tasks form the columns with the entry c_{ij} being the cost of using agent i to perform task j. If the matrix is not square, we make it square by adding row(s) or column(s) of zeroes when necessary. A maximum assignment can be converted into a minimum assignment by replacing each entry c_{ij} with $C - c_{ij}$, where C is the maximum value in the assignment matrix. The Hungarian algorithm proceeds in the following steps:

- Subtract the minimum number in each row from each entry in the entire row.
- Subtract the minimum number in each column from each entry in the entire column.
- Cover all zeroes in the matrix with as few lines (horizontal and/or vertical only) as possible. Let k be the number of lines and n the size of the matrix.
- If k < n, let m be the minimum uncovered number. Subtract m from every uncovered number and add m to every number covered by two lines. Go back to step 3.
- If k = n, go to step 4.
- Starting with the top row, work your way downwards as you make assignments. An assignment can be (uniquely) made when there is exactly one zero in a row. Once an assignment is made, delete that row and column from the matrix.

The linear assignment problem (LAP) is useful as a relaxation for difficult combinatorial optimization problems like quadratic assignment, and traveling salesman. Furthermore, theoretical developments for the LAP can often be extended to other problems, such as minimum cost flow and transportation.

The computational results show that the average computation times of the algorithm LAPJV are uniformly lower than the best of other algorithms. The code is of moderate size, and the memory requirements are small. The algorithm is suited for both dense and sparse assignment problems, and its sensitivity to cost range is relatively low.

3.6 The particle swarm optimization

Particle swarm optimization (PSO) was developed by Kennedy and Eberhart in 1995 based on the behavior of swarms such as fish schooling and birds in nature (Yang 2008). Many algorithms (such as ant colony algorithms and virtual ants algorithms) use the behavior of the so-called swarm intelligence (Yang 2008). Although particle swarm optimization has many similarities with genetic algorithms and virtual ants algorithms, but it is much simpler because it does not use mutation / crossover operators or pheromone (Yang 2008). Swarm intelligence with the collective behavior of systems with many interacting locally with one another and with their environment offers, and the ways of using decentralized control and self-organization to achieve their goals (Cho *et al.* 2011). In computing, particle swarm optimization (PSO) is a calculation method that optimizes a problem by iteratively trying to improve a candidate solution with regard to a given measure of quality (Garrett 2012). These methods are commonly known as metaheuristics as they make few or no assumptions about the problem optimized and can search very large spaces of candidate solutions (Garrett 2012). However, meta-heuristics such as PSO do not guarantee an optimal solution is ever found (Garrett 2012).

PSO is a search algorithm based on population-based simulation of the social behavior of birds within a flock (Cho *et al.* 2011). Originally, it was adopted for the formation of neural networks and optimization of nonlinear function, and has quickly become a popular global optimizer, especially in problems where the decision variables are real numbers (Cho *et al.* 2011). This algorithm search space of an objective function by adjusting the trajectories of individual agents, called particles, such as the path of pieces formed by position vectors in a quasi-random manner (Yang 2008). Particle motion has two main components: a stochastic component and a deterministic component (Yang 2008). The particle is attracted towards the position of the current global best while at the same time; they tend to move randomly (Yang 2008). When a particle is a place that is better than any of the locations previously found, it updates as the best new current of the particle i. It is better aware of all the particles n (Yang 2008). The goal is to find the best in all the world's best current until the goal is not to improve or after a certain number of iterations (Yang 2008).

PSO can also be used on optimization problems that are partially irregular, noisy, change over time, etc. (Garrett 2012). PSO optimizes a problem by having a population of candidate solutions, here dubbed particles, and moving the particles around the search space by simple mathematical formulas (Garrett 2012). The movements of the particles are guided by the most found in the search space positions are updated as better positions are found by particles (Garrett 2012).

PSO has become so popular because its main algorithm is relatively simple and easy to implement (Cho *et al.* 2011). It is also simple and has proven to be very effective in a wide variety of applications with very good results at a very low computational cost (Cho *et al.* 2011).

Particle Swarm Optimization Algorithm (Cho *et al.* 2011)
    (1) BEGIN
    (2) Parameter settings and initialization of swarm.
    (3) Evaluate fitness and locate the leader (i.e., initialize p best and g best).
    (4) I = 0 /* I = Iteration count */
    (5) WHILE (the stopping criterion is not met, say, I<Imax)
    (6) DO
    (7) FOR each particle
    (8) Update position & velocity (flight) as per equations
    (9) Evaluate fitness
    (10) Update p best
    (11) END FOR
    (12) Update leader (i.e.,g best)
    (13) I++
    (14) END WHILE
    (15) END

It has the attributes common evolutionary computation, including the initialization with a population of random solutions and the search for an optimum by updating generations (Rao and Savsani 2012). Possible solutions, called particles, are then transported through the problem space by following the current optimum particles (Rao and Savsani 2012). The concept of particle swarm was originally a simulation of a simplified social system (Rao and Savsani 2012). The original intent was to graphically simulate the graceful but unpredictable choreography of a flock of birds. Each particle keeps track of its coordinates in the problem space, which are associated with the best solution (fitness) it has achieved so far (Rao and Savsani 2012).

The most recent and highly cited studies about the particle swarm optimization are shown in Table 3.3.

Table 3.3 Studies about the particle swarm optimization

A novel set-based particle swarm optimization method for discrete optimization problems (Chen *et al.* 2010).
Gaussian quantum-behaved particle swarm optimization approaches for constrained engineering design problems (Coelho 2010).
An improved particle swarm optimization (pso)-based mppt for pv with reduced steady-state oscillation (Ishaque *et al.* 2012).
Niching without niching parameters: particle swarm optimization using a ring topology (Li 2010).
An automatic group composition system for composing collaborative learning groups using enhanced particle swarm optimization (Lin *et al.* 2010).
Hybridizing particle swarm optimization with differential evolution for constrained numerical and engineering optimization (Liu *et al.* 2010).
Quantum-inspired particle swarm optimization for valve-point economic load dispatch (Meng *et al.* 2010).
A combination of genetic algorithm and particle swarm optimization for optimal dg location and sizing in distribution systems (Moradi and Abedini 2012).
A pareto approach to multi-objective flexible job-shop scheduling problem using particle swarm optimization and local search (Moslehi and Mahnam 2011).
A novel particle swarm optimization algorithm with adaptive inertia weight (Nickabadi *et al.* 2011).
A new fuzzy adaptive hybrid particle swarm optimization algorithm for non-linear, non-smooth and non-convex economic dispatch problem (Niknam 2010).
A new fuzzy adaptive particle swarm optimization for non-smooth economic dispatch (Niknam *et al.* 2010).
An improved particle swarm optimization for nonconvex economic dispatch problems (Park *et al.* 2010).
Simplifying particle swarm optimization (Pedersen and Chipperfield 2010).
Handling sideband radiations in time-modulated arrays through particle swarm optimization (Poli *et al.* 2010).
Cellular particle swarm optimization (Shi *et al.* 2011).
A discrete particle swarm optimization method for feature selection in binary classification problems (Unler and Murat 2010).
An improved evolutionary method with fuzzy logic for combining particle swarm optimization and genetic algorithms (Valdez *et al.* 2011).
Real-time pid control strategy for maglev transportation system via particle swarm optimization (Wai *et al.* 2011).
Particle swarm optimization for redundant building cooling heating and power system (Wang *et al.* 2010).
Self-adaptive learning based particle swarm optimization (Wang *et al.* 2011).
Crystal structure prediction via particle-swarm optimization (Wang *et al.* 2010).
Prediction of parkinson's disease tremor onset using a radial basis function neural network based on particle swarm optimization (Wu *et al.* 2010).
A hybrid-forecasting model based on gaussian support vector machine and chaotic particle swarm optimization (Wu 2010).
Orthogonal learning particle swarm optimization (Zhan *et al.* 2011).

3.6.1 Snapshots of examples solved by using PSO

The Himmelblau test function

The Himmelblau test function is defined as

$$F(x1, x2) = (x1^2 + x2 - 11)^2 + (x1 + x2^2 - 7)^2$$

This function is depicted in Figure 3.1 below.

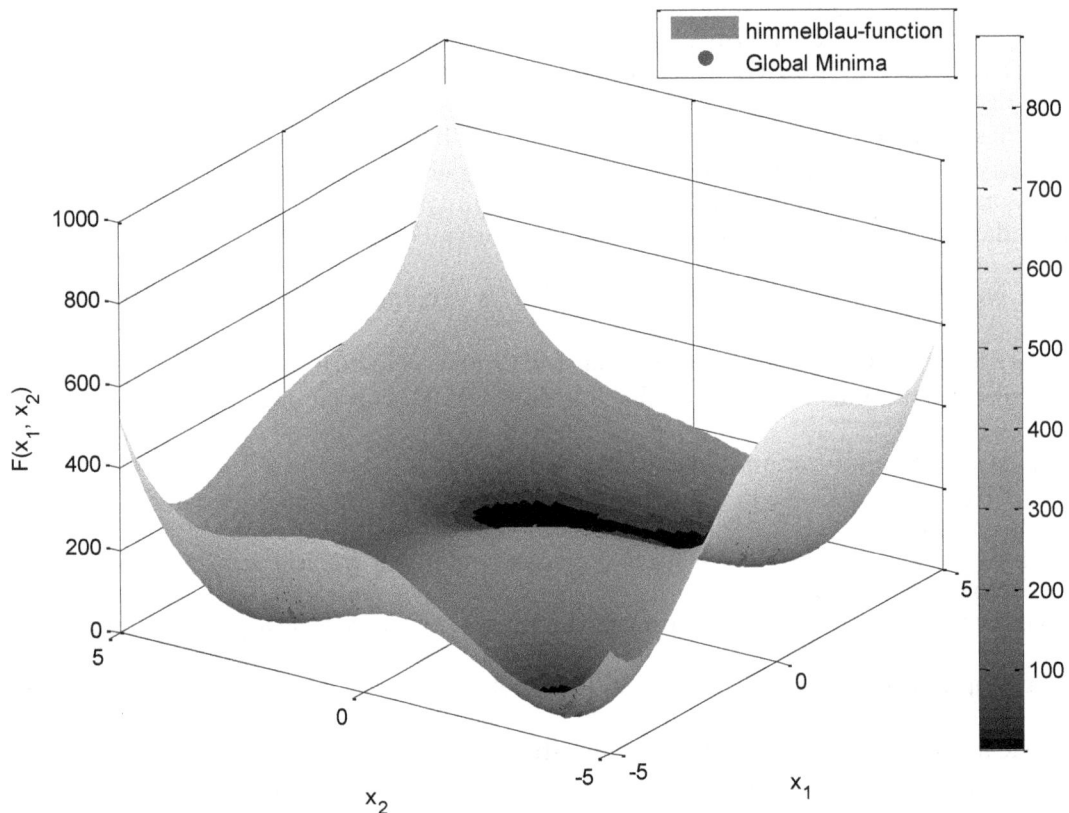

Figure 3.1 The Himmelblau test function

Optimal solution for the Himmelblau test function is obtained by PSO. See Figure 3.2.

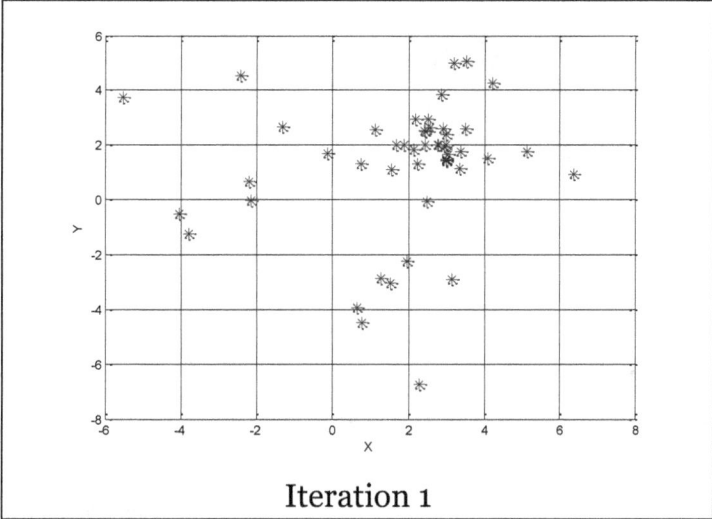

Iteration 1

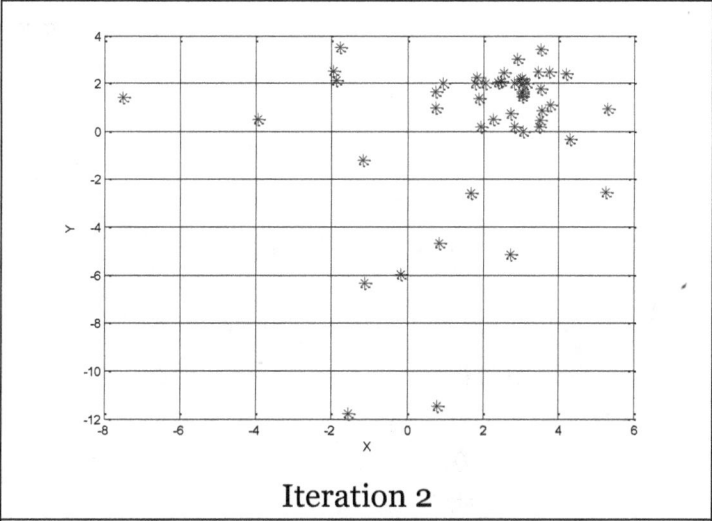

Iteration 2

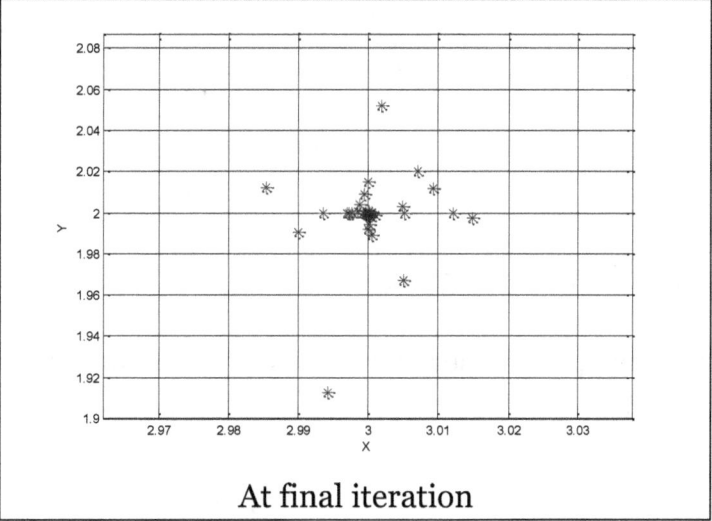

At final iteration

Figure 3.2 PSO Iterations

83

The "pen holder" test function

The "pen holder" test function is defined as

$$F(x1, x2) = -e^{\left(-\left|\cos(x1)\cos(x2)\, e^{\left|1 - \frac{\sqrt{x1^2 + x2^2}}{\pi}\right|}\right|^{-1}\right)}$$

The function is depicted in Figure 3.3 below.

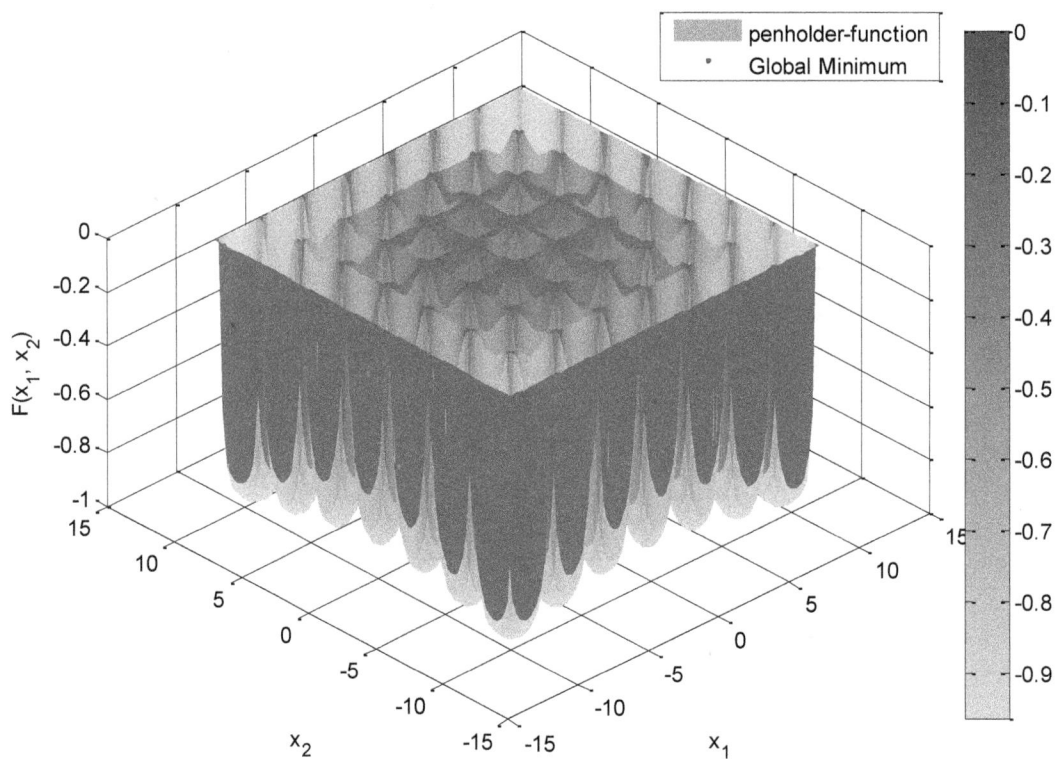

Figure 3.3 The "pen holder" test function

Optimal solution for the "pen holder" test function is obtained by PSO. See Figure 3.4.

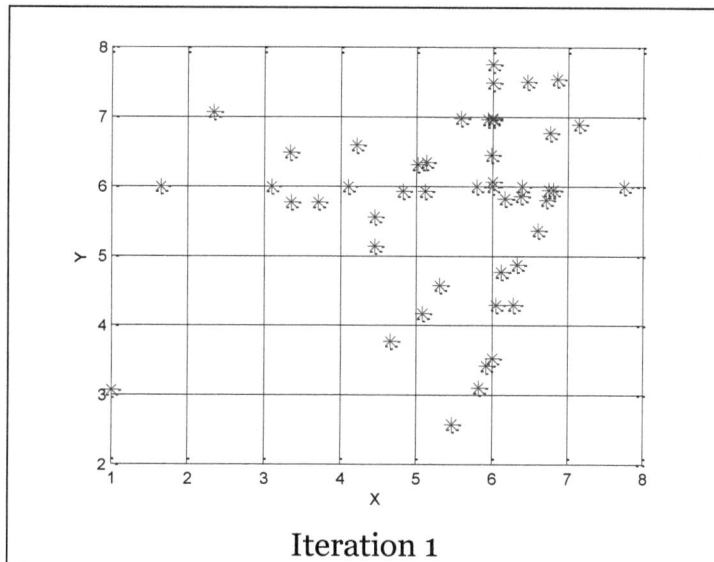

Iteration 1

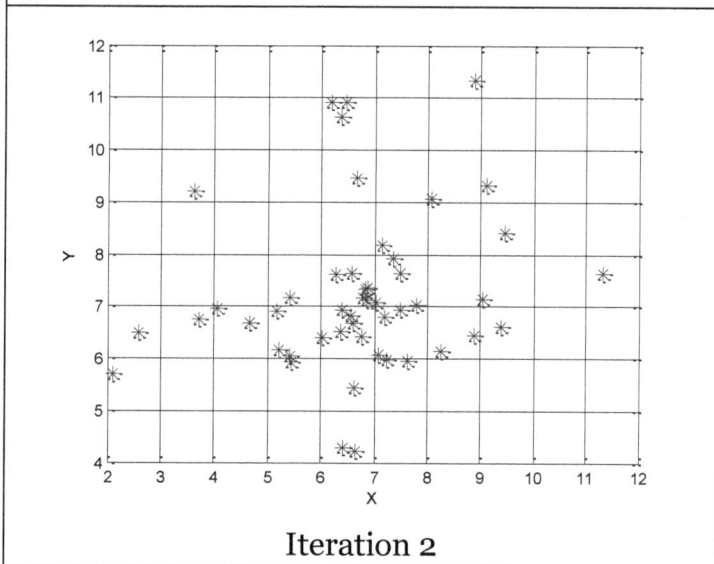

Iteration 2

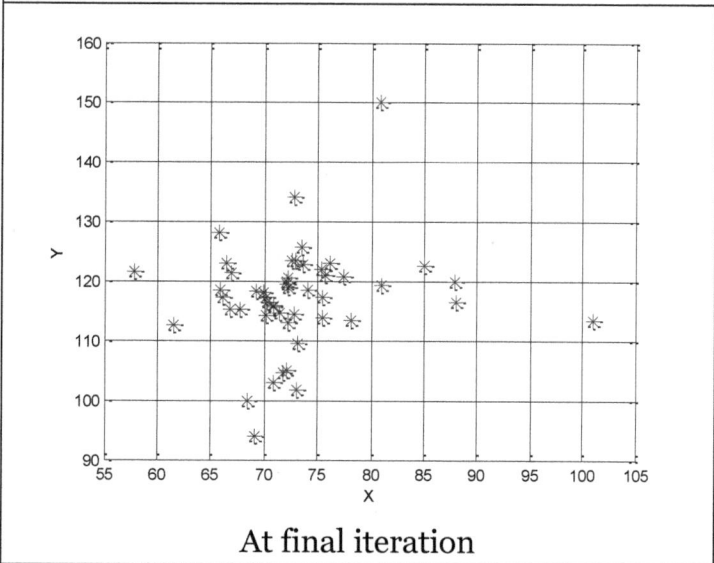

At final iteration

Figure 3.4 PSO Iterations

85

3.7 The simulated annealing method

Simulated annealing (SA) is a random search technique for global optimization problems, and it mimics the annealing process in the treatment of materials where the metal cools and solidifies in a crystalline state than the minimum energy and a larger to reduce crystal defects in metallic structures (Yang 2008). The method comprises annealing the precise control of the temperature and cooling rate (often called annealing schedule) (Yang 2008).

Simulated annealing (SA) is a generic probabilistic metaheuristic for the global optimization problem of applied mathematics, namely locating a good approximation to the global optimum of a given in a large search space (Garrett 2012) function. It is often used when the search space is discrete (eg, all tours that visit a set of cities) (Garrett 2012).

For some problems, simulated annealing may be more effective than exhaustive enumeration provided that the goal is simply to find a good solution in an acceptable period of time, rather than the best possible solution (Garrett 2012).

The most recent and highly cited studies about the simulated annealing optimization are shown in Table 3.4.

Table 3.4 Studies about the optimization by using simulated annealing

Prediction of principal ground-motion parameters using a hybrid method coupling artificial neural networks and simulated annealing (Alavi and Gandomi 2011).
Robust optimization with simulated annealing (Bertsimas and Nohadani 2010).
Multi-objective optimization of a stochastic assembly line balancing: a hybrid simulated annealing algorithm (Cakir *et al.* 2011).
Optimization of wire electrical discharge machining for pure tungsten using a neural network integrated simulated annealing approach (Chen *et al.* 2010).
Solving the traveling salesman problem based on the genetic simulated annealing ant colony system with particle swarm optimization techniques (Chen and Chien 2011).
Size optimization of a pv/wind hybrid energy conversion system with battery storage using simulated annealing (Ekren and Ekren 2010).
Solving the traveling salesman problem based on an adaptive simulated annealing algorithm with greedy search (Geng *et al.* 2011).
Solving a single-machine scheduling problem with maintenance, job deterioration and learning effect by simulated annealing (Ghodratnama *et al.* 2010).
Traffic flow forecasting by seasonal svr with chaotic simulated annealing algorithm (Hong 2011).
Simulated annealing for optimal ship routing (Kosmas and Vlachos 2012).
Using simulated annealing to minimize fuel consumption for the time-dependent vehicle routing problem (Kuo 2010).
Branch-and-bound and simulated annealing algorithms for a two-agent scheduling problem (Lee *et al.* 2010).
Electromagnetism-like mechanism and simulated annealing algorithms for flowshop scheduling problems minimizing the total weighted tardiness and makespan (Naderi *et al.* 2010).
Balancing stochastic two-sided assembly lines: a chance-constrained, piecewise-linear, mixed integer program and a simulated annealing algorithm (Ozcan 2010).
Reverse logistics network design using simulated annealing (Pishvaee *et al.* 2010).
Fuzzy control systems with reduced parametric sensitivity based on simulated annealing (Precup *et al.* 2012).
3d face recognition using simulated annealing and the surface interpenetration measure (Queirolo *et al.* 2010).
C-psa: constrained pareto simulated annealing for constrained multi-objective optimization (Singh *et al.* 2010).
Intelligent energy resource management considering vehicle-to-grid: a simulated annealing approach (Sousa *et al.* 2012).
Fast and accurate protein substructure searching with simulated annealing and gpus (Stivala *et al.* 2010).
A monte carlo/simulated annealing algorithm for sequential resonance assignment in solid state nmr of uniformly labeled proteins with magic-angle spinning (Tycko and Hu 2010).
Coupled simulated annealing (Xavier-de-Souza *et al.* 2010).
A simulated annealing heuristic for the capacitated location routing problem (Yu *et al.* 2010).
A hybrid immune simulated annealing algorithm for the job shop scheduling problem (Zhang and Wu 2010).
A simulated annealing algorithm based on block properties for the job shop scheduling problem with total weighted tardiness objective (Zhang and Wu 2011).

3.7.1 Snapshots of some examples solved by using simulated annealing method

The "test tube holder" sample test function

The "test tube holder" sample test function is defined as

$$F(x1, x2) = -4 \left| \sin(x1) \cos(x2) \; \mathbf{e}^{\left| \cos\left(\frac{x1^2}{200} + \frac{x2^2}{200} \right) \right|} \right|$$

This function is depicted in Figure 3.5 below.

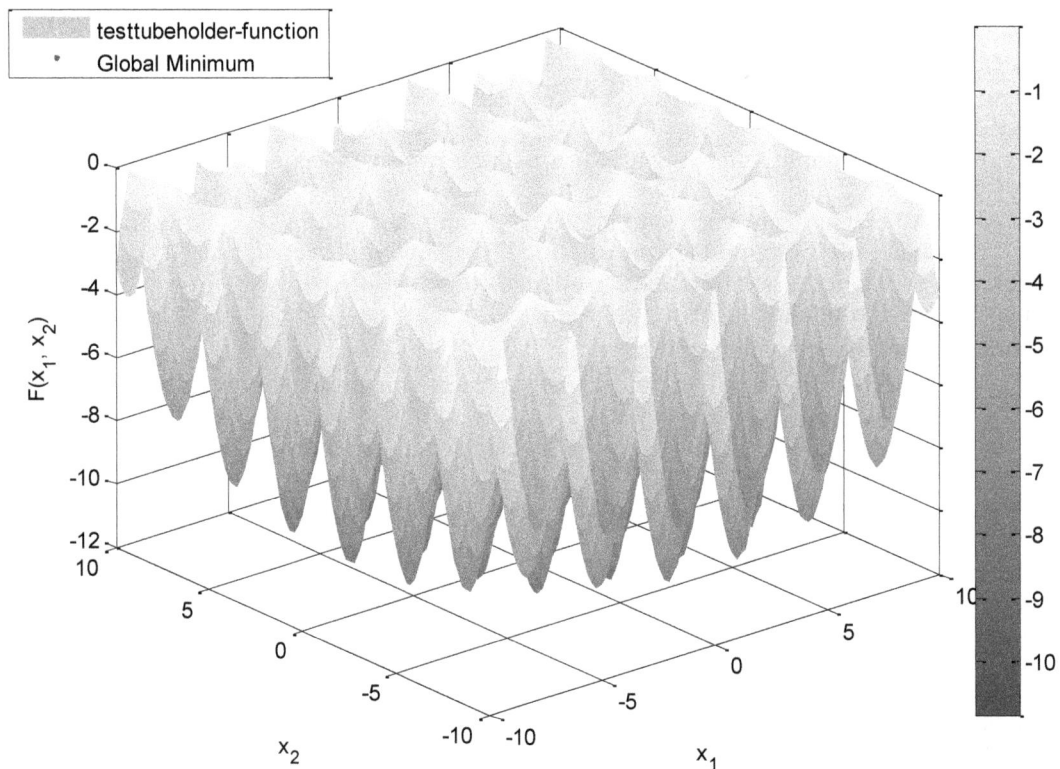

Figure 3.5 The "test tube holder" test function

The solution parameters are shown below.

Initial temperature	: 1
Final temperature	: 2.5711e-007
Consecutive rejections	: 1239
Number of function calls	: 6836
Total final loss	: -10.8723

x = 1.5707 -0.0000

f = -10.8723

The Levi13 sample test function

The Levi13 test function is defined as

$$F(x1, x2) = \sin(3\pi x1)^2 + (x1 - 1)^2 (1 + \sin(3\pi x2)^2)$$
$$+ (x2 - 1)^{2.} (1 + \sin(2\pi x2)^2)$$

The function is depicted in Figure 3.6 below.

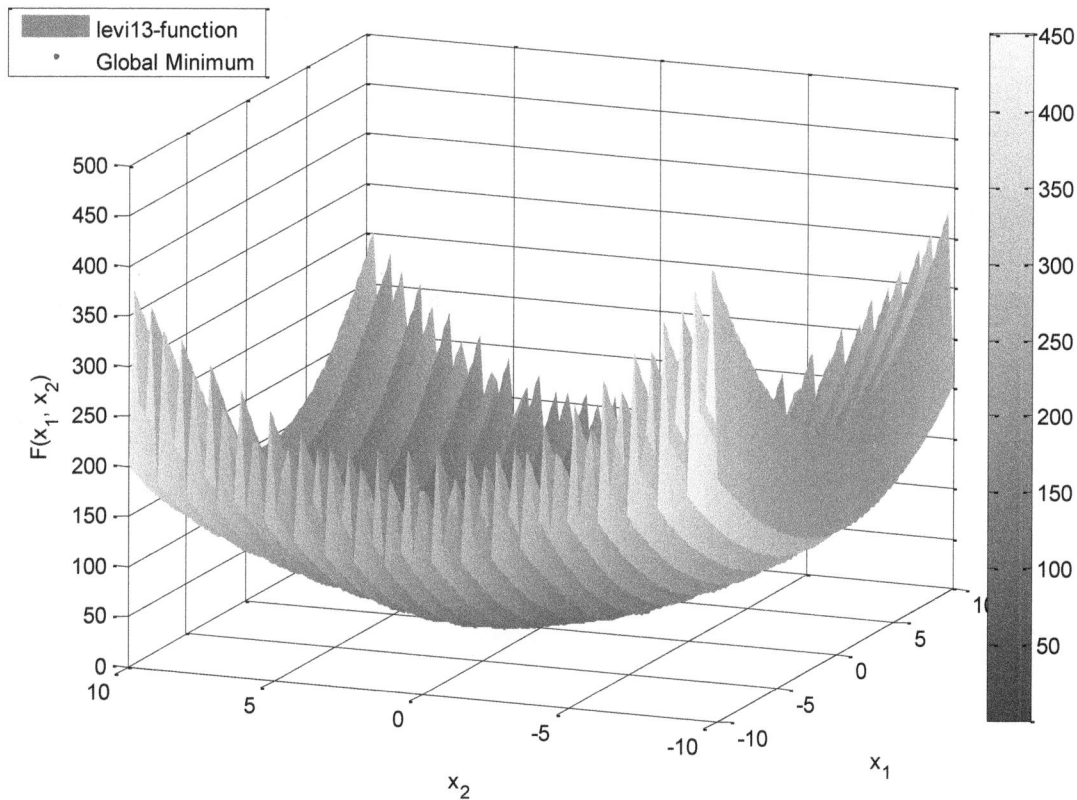

Figure 3.6 The Levi13 test function

The solution parameters by using SA are shown below.

Initial temperature   : 1
Final temperature   : 5.02168e-007
Consecutive rejections   : 1085
Number of function calls   : 9342
Total final loss   : 1.98092
x =   0.0113   0.0079
f =   1.9809

3.8 The Wagner-Whitin algorithm

Wagner and Whitin studied the uncapacitated model with set-up of fixed costs and linear inventory and production costs (Woodruff 2002). Their main contribution was to demonstrate that an optimal replenishment policy is one in which production is achieved when inventory is zero (Woodruff 2002). In addition, they proposed an algorithm for dynamic programming before effective to solve the problem (Woodruff 2002).

Wagner and Whitin (1958) have shown that the algorithm optimal solution in any period is either (Wee 2011):

1) the initial stock is zero and lot size is positive,
2) or the initial stock is positive and the lot size is zero,
3) or initial inventory both lot-size are zero

This property is known as the Wagner and Whitin property (Wee 2011). All heuristic procedures listed above are lot size for each period that has property W-W (Wee 2011). This implies that the lot size in any period must be exactly the requirements of the proceeding several periods (Wee 2011).

References

Afshar, M. H. (2010). A parameter free Continuous Ant Colony Optimization Algorithm for the optimal design of storm sewer networks: Constrained and unconstrained approach. Advances in Engineering Software, 41(2), 188-195. doi: 10.1016/j.advengsoft.2009.09.009

Ahmadi, P., & Dincer, I. (2010). Exergoenvironmental analysis and optimization of a cogeneration plant system using Multimodal Genetic Algorithm (MGA). Energy, 35(12), 5161-5172. doi: 10.1016/j.energy.2010.07.050

Alavi, A. H., & Gandomi, A. H. (2011). Prediction of principal ground-motion parameters using a hybrid method coupling artificial neural networks and simulated annealing. Computers & Structures, 89(23-24), 2176-2194. doi: 10.1016/j.compstruc.2011.08.019

Arnaout, J. P., Rabadi, G., & Musa, R. (2010). A two-stage Ant Colony Optimization algorithm to minimize the makespan on unrelated parallel machines with sequence-dependent setup times. Journal of Intelligent Manufacturing, 21(6), 693-701. doi: 10.1007/s10845-009-0246-1

Assareh, E., Behrang, M. A., Assari, M. R., & Ghanbarzadeh, A. (2010). Application of PSO (particle swarm optimization) and GA (genetic algorithm) techniques on demand estimation of oil in Iran. Energy, 35(12), 5223-5229. doi: 10.1016/j.energy.2010.07.043

Babaoglu, I., Findik, O., & Ulker, E. (2010). A comparison of feature selection models utilizing binary particle swarm optimization and genetic algorithm in determining coronary artery disease using support vector machine. Expert Systems with Applications, 37(4), 3177-3183. doi: 10.1016/j.eswa.2009.09.064

Baghernejad, A., & Yaghoubi, M. (2011). Exergoeconomic analysis and optimization of an Integrated Solar Combined Cycle System (ISCCS) using genetic algorithm. Energy Conversion and Management, 52(5), 2193-2203. doi: 10.1016/j.enconman.2010.12.019

Benvenuto, Nevio (Editor); Zorzi, Michele (Editor). Principles of Communications Networks and Systems. Hoboken, NJ, USA: Wiley 2011. p 723.

Benvenuto, Nevio (Editor); Zorzi, Michele (Editor). Principles of Communications Networks and Systems. Hoboken, NJ, USA: Wiley 2011. p 726.

Benvenuto, Nevio (Editor); Zorzi, Michele (Editor). Principles of Communications Networks and Systems. Hoboken, NJ, USA: Wiley 2011. p 727.

91

Bergholt, M. S., Zheng, W., Lin, K., Ho, K. Y., Teh, M., Yeoh, K. G., . . . Huang, Z. W. (2011). In vivo diagnosis of gastric cancer using Raman endoscopy and ant colony optimization techniques. International Journal of Cancer, 128(11), 2673-2680. doi: 10.1002/ijc.25618

Berrichi, A., Yalaoui, F., Amodeo, L., & Mezghiche, M. (2010). Bi-Objective Ant Colony Optimization approach to optimize production and maintenance scheduling. Computers & Operations Research, 37(9), 1584-1596. doi: 10.1016/j.cor.2009.11.017

Bertsimas, D., & Nohadani, O. (2010). Robust optimization with simulated annealing. Journal of Global Optimization, 48(2), 323-334. doi: 10.1007/s10898-009-9496-x

Bhatti, M. S., Kapoor, D., Kalia, R. K., Reddy, A. S., & Thukral, A. K. (2011). RSM and ANN modeling for electrocoagulation of copper from simulated wastewater: Multi objective optimization using genetic algorithm approach. Desalination, 274(1-3), 74-80. doi: 10.1016/j.desal.2011.01.083

Bonabeau, Eric; Theraulaz, Guy; Dorigo, Marco. Swarm Intelligence: From Natural to Artificial Systems. Cary, NC, USA: Oxford University Press, 1999. p 54.

Bonabeau, Eric; Theraulaz, Guy; Dorigo, Marco. Swarm Intelligence: From Natural to Artificial Systems. Cary, NC, USA: Oxford University Press, 1999. p 54.

Cakir, B., Altiparmak, F., & Dengiz, B. (2011). Multi-objective optimization of a stochastic assembly line balancing: A hybrid simulated annealing algorithm. Computers & Industrial Engineering, 60(3), 376-384. doi: 10.1016/j.cie.2010.08.013

Chen, H. C., Lin, J. C., Yang, Y. K., & Tsai, C. H. (2010). Optimization of wire electrical discharge machining for pure tungsten using a neural network integrated simulated annealing approach. Expert Systems with Applications, 37(10), 7147-7153. doi: 10.1016/j.eswa.2010.04.020

Chen, S. M., & Chien, C. Y. (2011). Solving the traveling salesman problem based on the genetic simulated annealing ant colony system with particle swarm optimization techniques. Expert Systems with Applications, 38(12), 14439-14450. doi: 10.1016/j.eswa.2011.04.163

Chen, W. N., Zhang, J., Chung, H. S. H., Huang, R. Z., & Liu, O. (2010). Optimizing Discounted Cash Flows in Project Scheduling-An Ant Colony Optimization Approach. Ieee Transactions on Systems Man and Cybernetics Part C-Applications and Reviews, 40(1), 64-77. doi: 10.1109/tsmcc.2009.2027335

Chen, W. N., Zhang, J., Chung, H. S. H., Zhong, W. L., Wu, W. G., & Shi, Y. H. (2010).

A Novel Set-Based Particle Swarm Optimization Method for Discrete Optimization Problems. Ieee Transactions on Evolutionary Computation, 14(2), 278-300. doi: 10.1109/tevc.2009.2030331

Chen, Y. M., Miao, D. Q., & Wang, R. Z. (2010). A rough set approach to feature selection based on ant colony optimization. Pattern Recognition Letters, 31(3), 226-233. doi: 10.1016/j.patrec.2009.10.013

Cho, S.-B., et al. (2011). Integration of Swarm Intelligence and Artificial Neural Network. New Jersey, World Scientific.

Cho, S.-B., et al. (2011). Integration of Swarm Intelligence and Artificial Neural Network. New Jersey, World Scientific.

Cho, S.-B., et al. (2011). Integration of Swarm Intelligence and Artificial Neural Network. New Jersey, World Scientific.

Christodoulou, S. M. (2010). Scheduling Resource-Constrained Projects with Ant Colony Optimization Artificial Agents. Journal of Computing in Civil Engineering, 24(1), 45-55. doi: 10.1061/(asce)0887-3801(2010)24:1(45)

Coelho, L. D. (2010). Gaussian quantum-behaved particle swarm optimization approaches for constrained engineering design problems. Expert Systems with Applications, 37(2), 1676-1683. doi: 10.1016/j.eswa.2009.06.044

Collins, A. L., Zhang, Y., Walling, D. E., Grenfell, S. E., & Smith, P. (2010). Tracing sediment loss from eroding farm tracks using a geochemical fingerprinting procedure combining local and genetic algorithm optimisation. Science of the Total Environment, 408(22), 5461-5471. doi: 10.1016/j.scitotenv.2010.07.066

Dehuri, Satchidananda. Integration of Swarm Intelligence and Artificial Neural Network. SGP: World Scientific Publishing Co. 2011. p 69.

Dehuri, Satchidananda. Integration of Swarm Intelligence and Artificial Neural Network. SGP: World Scientific Publishing Co. 2011. p 69.

Dorigo, Marco; Stützle, Thomas. Ant Colony Optimization. Cambridge, MA, USA: MIT Press 2004. p 1.

Dorigo, Marco; Stützle, Thomas. Ant Colony Optimization. Cambridge, MA, USA: MIT Press 2004. p 33.

Ekren, O., & Ekren, B. Y. (2010). Size optimization of a PV/wind hybrid energy conversion system with battery storage using simulated annealing. Applied Energy, 87(2), 592-598. doi: 10.1016/j.apenergy.2009.05.022

Eslami, M., Shareef, H., Mohamed, A., & Khajehzadeh, M. (2010). Damping of Power System Oscillations Using Genetic Algorithm and Particle Swarm Optimization. International Review of Electrical Engineering-Iree, 5(6), 2745-2753.

Fernandez, M., Caballero, J., Fernandez, L., & Sarai, A. (2011). Genetic algorithm optimization in drug design QSAR: Bayesian-regularized genetic neural networks (BRGNN) and genetic algorithm-optimized support vectors machines (GA-SVM). Molecular Diversity, 15(1), 269-289. doi: 10.1007/s11030-010-9234-9

Garrett, Yevette. (2012). Optimization Algorithms Handbook. University Publications.

Geng, X. T., Chen, Z. H., Yang, W., Shi, D. Q., & Zhao, K. (2011). Solving the traveling salesman problem based on an adaptive simulated annealing algorithm with greedy search. Applied Soft Computing, 11(4), 3680-3689. doi: 10.1016/j.asoc.2011.01.039

Ghodratnama, A., Rabbani, M., Tavakkoli-Moghaddam, R., & Baboli, A. (2010). Solving a single-machine scheduling problem with maintenance, job deterioration and learning effect by simulated annealing. Journal of Manufacturing Systems, 29(1), 1-9. doi: 10.1016/j.jmsy.2010.06.004

Hatami, T., Meireles, M. A. A., & Zahedi, G. (2010). Mathematical modeling and genetic algorithm optimization of clove oil extraction with supercritical carbon dioxide. Journal of Supercritical Fluids, 51(3), 331-338. doi: 10.1016/j.supflu.2009.10.001

Hong, W. C. (2011). Traffic flow forecasting by seasonal SVR with chaotic simulated annealing algorithm. Neurocomputing, 74(12-13) 2096-2107. doi: 10.1016/j.neucom.2010.12.032

Ibe, Oliver C.. Fundamentals of Stochastic Networks. Hoboken, NJ, USA: Wiley 2011. p 184.

Ishaque, K., Salam, Z., Amjad, M., & Mekhilef, S. (2012). An Improved Particle Swarm Optimization (PSO)-Based MPPT for PV With Reduced Steady-State Oscillation. Ieee Transactions on Power Electronics, 27(8), 3627-3638. doi: 10.1109/tpel.2012.2185713

Jonker, R. and A. Volgenant (1987). "A shortest augmenting path algorithm for dense and sparse linear assignment problems." Computing **38**(4): 325-340.

Juang, C. F., & Chang, P. H. (2010). Designing Fuzzy-Rule-Based Systems Using Continuous Ant-Colony Optimization. Ieee Transactions on Fuzzy Systems, 18(1), 138-149. doi: 10.1109/tfuzz.2009.2038150

Kaveh, A., & Talatahari, S. (2010). An improved ant colony optimization for constrained engineering design problems. Engineering Computations, 27(1-2), 155-182. doi: 10.1108/02644401011008577

Kaveh, A., & Talatahari, S. (2010). An improved ant colony optimization for the design of planar steel frames. Engineering Structures, 32(3), 864-873. doi: 10.1016/j.engstruct.2009.12.012

Kaviri, A. G., Jaafar, M. N. M., & Lazim, T. M. (2012). Modeling and multi-objective exergy based optimization of a combined cycle power plant using a genetic algorithm. Energy Conversion and Management, 58, 94-103. doi: 10.1016/j.enconman.2012.01.002

Koshy, Thomas. Discrete Mathematics with Applications. Burlington, MA, USA: Academic Press 2003. p 746.

Kosmas, O. T., & Vlachos, D. S. (2012). Simulated annealing for optimal ship routing. Computers & Operations Research, 39(3), 576-581. doi: 10.1016/j.cor.2011.05.010

Kuo, R. J., & Han, Y. S. (2011). A hybrid of genetic algorithm and particle swarm optimization for solving bi-level linear programming problem - A case study on supply chain model. Applied Mathematical Modelling, 35(8), 3905-3917. doi: 10.1016/j.apm.2011.02.008

Kuo, Y. Y. (2010). Using simulated annealing to minimize fuel consumption for the time-dependent vehicle routing problem. Computers & Industrial Engineering, 59(1), 157-165. doi: 10.1016/j.cie.2010.03.012

Lee, C. Y., Lee, Z. J., Lin, S. W., & Ying, K. C. (2010). An enhanced ant colony optimization (EACO) applied to capacitated vehicle routing problem. Applied Intelligence, 32(1), 88-95. doi: 10.1007/s10489-008-0136-9

Lee, W. C., Chen, S. K., & Wu, C. C. (2010). Branch-and-bound and simulated annealing algorithms for a two-agent scheduling problem. Expert Systems with Applications, 37(9), 6594-6601. doi: 10.1016/j.eswa.2010.02.125

Leung, C. W., Wong, T. N., Mak, K. L., & Fung, R. Y. K. (2010). Integrated process planning and scheduling by an agent-based ant colony optimization. Computers & Industrial Engineering, 59(1), 166-180. doi: 10.1016/j.cie.2009.09.003

Li, X. D. (2010). Niching Without Niching Parameters: Particle Swarm Optimization Using a Ring Topology. Ieee Transactions on Evolutionary Computation, 14(1), 150-169. doi: 10.1109/tevc.2009.2026270

Lin, Y. T., Huang, Y. M., & Cheng, S. C. (2010). An automatic group composition system for composing collaborative learning groups using enhanced particle swarm optimization. Computers & Education, 55(4), 1483-1493. doi: 10.1016/j.compedu.2010.06.014

Liu, H., Cai, Z. X., & Wang, Y. (2010). Hybridizing particle swarm optimization with differential evolution for constrained numerical and engineering optimization. Applied Soft Computing, 10(2), 629-640. doi: 10.1016/j.asoc.2009.08.031

Magnier, L., & Haghighat, F. (2010). Multiobjective optimization of building design using TRNSYS simulations, genetic algorithm, and Artificial Neural Network. Building and Environment, 45(3), 739-746. doi: 10.1016/j.buildenv.2009.08.016

Meng, K., Wang, H. G., Dong, Z. Y., & Wong, K. P. (2010). Quantum-Inspired Particle Swarm Optimization for Valve-Point Economic Load Dispatch. Ieee Transactions on Power Systems, 25(1), 215-222. doi: 10.1109/tpwrs.2009.2030359

Mohan, B. C., & Baskaran, R. (2012). A survey: Ant Colony Optimization based recent research and implementation on several engineering domain. Expert Systems with Applications, 39(4), 4618-4627. doi: 10.1016/j.eswa.2011.09.076

Moradi, M. H., & Abedini, M. (2012). A combination of genetic algorithm and particle swarm optimization for optimal DG location and sizing in distribution systems. International Journal of Electrical Power & Energy Systems, 34(1), 66-74. doi: 10.1016/j.ijepes.2011.08.023

Moradi, M. H., & Abedini, M. (2012). A combination of genetic algorithm and particle swarm optimization for optimal DG location and sizing in distribution systems. International Journal of Electrical Power & Energy Systems, 34(1), 66-74. doi: 10.1016/j.ijepes.2011.08.023

Moslehi, G., & Mahnam, M. (2011). A Pareto approach to multi-objective flexible job-shop scheduling problem using particle swarm optimization and local search. International Journal of Production Economics, 129(1), 14-22. doi: 10.1016/j.ijpe.2010.08.004

Musa, R., Arnaout, J. P., & Jung, H. (2010). Ant colony optimization algorithm to solve for the transportation problem of cross-docking network. Computers & Industrial Engineering, 59(1), 85-92. doi: 10.1016/j.cie.2010.03.002

Naderi, B., Tavakkoli-Moghaddam, R., & Khalili, M. (2010). Electromagnetism-like mechanism and simulated annealing algorithms for flowshop scheduling problems minimizing the total weighted tardiness and makespan. Knowledge-Based Systems, 23(2), 77-85. doi: 10.1016/j.knosys.2009.06.002

Nazghelichi, T., Aghbashlo, M., & Kianmehr, M. H. (2011). Optimization of an artificial neural network topology using coupled response surface methodology and genetic algorithm for fluidized bed drying. Computers and Electronics in Agriculture, 75(1), 84-91. doi: 10.1016/j.compag.2010.09.014

Neumann, F., & Witt, C. (2010). Ant Colony Optimization and the minimum spanning tree problem. Theoretical Computer Science, 411(25), 2406-2413. doi: 10.1016/j.tcs.2010.02.012

Nickabadi, A., Ebadzadeh, M. M., & Safabakhsh, R. (2011). A novel particle swarm optimization algorithm with adaptive inertia weight. Applied Soft Computing, 11(4), 3658-3670. doi: 10.1016/j.asoc.2011.01.037

Niknam, T. (2010). A new fuzzy adaptive hybrid particle swarm optimization algorithm for non-linear, non-smooth and non-convex economic dispatch problem. Applied Energy, 87(1), 327-339. doi: 10.1016/j.apenergy.2009.05.016

Niknam, T., Mojarrad, H. D., & Nayeripour, M. (2010). A new fuzzy adaptive particle swarm optimization for non-smooth economic dispatch. Energy, 35(4), 1764-1778. doi: 10.1016/j.energy.2009.12.029

Niu, D. X., Wang, Y. L., & Wu, D. D. (2010). Power load forecasting using support vector machine and ant colony optimization. Expert Systems with Applications, 37(3), 2531-2539. doi: 10.1016/j.eswa.2009.08.019

Ozcan, U. (2010). Balancing stochastic two-sided assembly lines: A chance-constrained, piecewise-linear, mixed integer program and a simulated annealing algorithm. European Journal of Operational Research 205(1), 81-97. doi: 10.1016/j.ejor.2009.11.033

Park, J. B., Jeong, Y. W., Shin, J. R., & Lee, K. Y. (2010). An Improved Particle Swarm Optimization for Nonconvex Economic Dispatch Problems. Ieee Transactions on Power Systems, 25(1), 156-166. doi: 10.1109/tpwrs.2009.2030293

Pedemonte, M., Nesmachnow, S., & Cancela, H. (2011). A survey on parallel ant colony optimization. Applied Soft Computing, 11(8), 5181-5197. doi: 10.1016/j.asoc.2011.05.042

Pedersen, M. E. H., & Chipperfield, A. J. (2010). Simplifying Particle Swarm Optimization. Applied Soft Computing, 10(2), 618-628. doi: 10.1016/j.asoc.2009.08.029

Pishvaee, M. S., Kianfar, K., & Karimi, B. (2010). Reverse logistics network design using simulated annealing. International Journal of Advanced Manufacturing Technology, 47(1-4), 269-281. doi: 10.1007/s00170-009-2194-5

Poli, L., Rocca, P., Manica, L., & Massa, A. (2010). Handling Sideband Radiations in Time-Modulated Arrays Through Particle Swarm Optimization. Ieee Transactions on Antennas and Propagation, 58(4), 1408-1411. doi: 10.1109/tap.2010.2041165

Precup, R. E., David, R. C., Petriu, E. M., Preitl, S., & Radac, M. B. (2012). Fuzzy Control Systems With Reduced Parametric Sensitivity Based on Simulated Annealing. Ieee Transactions on Industrial Electronics, 59(8), 3049-3061. doi: 10.1109/tie.2011.2130493

Putha, R., Quadrifoglio, L., & Zechman, E. (2012). Comparing Ant Colony Optimization and Genetic Algorithm Approaches for Solving Traffic Signal Coordination under Oversaturation Conditions. Computer-Aided Civil and Infrastructure Engineering, 27(1), 14-28. doi: 10.1111/j.1467-8667.2010.00715.x

Putha, R., Quadrifoglio, L., & Zechman, E. (2012). Comparing Ant Colony Optimization and Genetic Algorithm Approaches for Solving Traffic Signal Coordination under Oversaturation Conditions. Computer-Aided Civil and Infrastructure Engineering, 27(1), 14-28. doi: 10.1111/j.1467-8667.2010.00715.x

Qin, X. S., Huang, G. H., & Liu, L. (2010). A Genetic-Algorithm-Aided Stochastic Optimization Model for Regional Air Quality Management under Uncertainty. Journal of the Air & Waste Management Association, 60(1), 63-71. doi: 10.3155/1047-3289.60.1.63

Queirolo, C. C., Silva, L., Bellon, O. R. P., & Segundo, M. P. (2010). 3D Face Recognition Using Simulated Annealing and the Surface Interpenetration Measure. Ieee Transactions on Pattern Analysis and Machine Intelligence, 32(2) 206-219. doi: 10.1109/tpami.2009.14

R. Venkata Rao; Vimal J. Savsani. (2012). Mechanical Design Optimization Using Advanced Optimization Techniques. Springer London

Rubinstein, Reuven Y.; Ridder, Ad; Vaisman, Radislav. Wiley Series in Probability and Statistics: Fast Sequential Monte Carlo Methods for Counting and Optimization. Somerset, NJ, USA: Wiley 2013. p 6.

Sanaye, S., & Hajabdollahi, H. (2010). Thermal-economic multi-objective optimization of plate fin heat exchanger using genetic algorithm. Applied Energy, 87(6), 1893-1902. doi: 10.1016/j.apenergy.2009.11.016

Sandou, Guillaume. FOCUS Series : Metaheuristic Optimization for the Design of Automatic Control Laws. Somerset, NJ, USA: Wiley 2013. p 13.

Sandou, Guillaume. FOCUS Series : Metaheuristic Optimization for the Design of Automatic Control Laws. Somerset, NJ, USA: Wiley 2013. p 9.

Santos, L., Coutinho-Rodrigues, J., & Current, J. R. (2010). An improved ant colony optimization based algorithm for the capacitated arc routing problem. Transportation Research Part B-Methodological, 44(2), 246-266. doi: 10.1016/j.trb.2009.07.004

Shankar, T. J., Sokhansanj, S., Bandyopadhyay, S., & Bawa, A. S. (2010). A Case Study on Optimization of Biomass Flow During Single-Screw Extrusion Cooking Using Genetic Algorithm (GA) and Response Surface Method (RSM). Food and Bioprocess Technology, 3(4), 498-510. doi: 10.1007/s11947-008-0172-9

Sharma, A. K., Son, K. H., Han, B. Y., & Sohn, K. S. (2010). Simultaneous Optimization of Luminance and Color Chromaticity of Phosphors Using a Nondominated Sorting Genetic Algorithm. Advanced Functional Materials 20(11), 1750-1755. doi: 10.1002/adfm.200902285

Shi, Y., Liu, H. C., Gao, L., & Zhang, G. H. (2011). Cellular particle swarm optimization. Information Sciences, 181(20), 4460-4493. doi: 10.1016/j.ins.2010.05.025

Singh, H. K., Ray, T., & Smith, W. (2010). C-PSA: Constrained Pareto simulated annealing for constrained multi-objective optimization. Information Sciences, 180(13), 2499-2513. doi: 10.1016/j.ins.2010.03.021

Sivapathasekaran, C., Mukherjee, S., Ray, A., Gupta, A., & Sen, R. (2010). Artificial neural network modeling and genetic algorithm based medium optimization for the improved production of marine biosurfactant. Bioresource Technology, 101(8), 2884-2887. doi: 10.1016/j.biortech.2009.09.093

Sousa, T., Morais, H., Vale, Z., Faria, P., & Soares, J. (2012). Intelligent Energy Resource Management Considering Vehicle-to-Grid: A Simulated Annealing Approach. Ieee Transactions on Smart Grid, 3(1), 535-542. doi: 10.1109/tsg.2011.2165303

Stivala, A. D., Stuckey, P. J., & Wirth, A. I. (2010). Fast and accurate protein substructure searching with simulated annealing and GPUs. Bmc Bioinformatics, 11. doi: 10.1186/1471-2105-11-446

Tatipamula, Mallikarjun; Oki, Eiji; Rojas-Cessa, Roberto. Advanced Internet Protocols, Services, and Applications. Hoboken, NJ, USA: Wiley 2012. p 48.

Tian, J., Ma, L. H., & Yu, W. Y. (2011). Ant colony optimization for wavelet-based image interpolation using a three-component exponential mixture model. Expert Systems with Applications, 38(10), 12514-12520. doi: 10.1016/j.eswa.2011.04.037

Tycko, R., & Hu, K. N. (2010). A Monte Carlo/simulated annealing algorithm for sequential resonance assignment in solid state NMR of uniformly labeled proteins with magic-angle spinning. Journal of Magnetic Resonance 205(2), 304-314. doi: 10.1016/j.jmr.2010.05.013

Unler, A., & Murat, A. (2010). A discrete particle swarm optimization method for feature selection in binary classification problems. European Journal of Operational Research 206(3), 528-539. doi: 10.1016/j.ejor.2010.02.032

Valdez, F., Melin, P., & Castillo, O. (2011). An improved evolutionary method with fuzzy logic for combining Particle Swarm Optimization and Genetic Algorithms. Applied Soft Computing, 11(2), 2625-2632. doi: 10.1016/j.asoc.2010.10.010

Wai, R. J., Lee, J. D., & Chuang, K. L. (2011). Real-Time PID Control Strategy for Maglev Transportation System via Particle Swarm Optimization. Ieee Transactions on Industrial Electronics, 58(2), 629-646. doi: 10.1109/tie.2010.2046004

Wang, J. F., Sun, Z. X., Dai, Y. P., & Ma, S. L. (2010). Parametric optimization design for supercritical CO_2 power cycle using genetic algorithm and artificial neural network. Applied Energy, 87(4), 1317-1324. doi: 10.1016/j.apenergy.2009.07.017

Wang, J. J., Jing, Y. Y., & Zhang, C. F. (2010). Optimization of capacity and operation for CCHP system by genetic algorithm. Applied Energy, 87(4), 1325-1335. doi: 10.1016/j.apenergy.2009.08.005

Wang, J. J., Zhai, Z. Q., Jing, Y. Y., & Zhang, C. F. (2010). Particle swarm optimization for redundant building cooling heating and power system. Applied Energy, 87(12), 3668-3679. doi: 10.1016/j.apenergy.2010.06.021

Wang, Y. C., Lv, J. A., Zhu, L., & Ma, Y. M. (2010). Crystal structure prediction via particle-swarm optimization. Physical Review B, 82(9). doi: 10.1103/PhysRevB.82.094116

Wang, Y., Li, B., Weise, T., Wang, J. Y., Yuan, B., & Tian, Q. J. (2011). Self-adaptive learning based particle swarm optimization. Information Sciences, 181(20), 4515-4538. doi: 10.1016/j.ins.2010.07.013

Wee, Hui-Ming. Management Science - Theory and Applications : Inventory Systems : Modeling and Research Methods. New York, NY, USA: Nova 2011. p 83.

Woodruff, David L. (Editor). Network Interdiction and Stochastic Integer Programming. Secaucus, NJ, USA: Kluwer Academic Publishers 2002. p 99.

Wu, D. F., Warwick, K., Ma, Z., Gasson, M. N., Burgess, J. G., Pan, S., & Aziz, T. Z. (2010). Prediction Of Parkinson's Disease Tremor Onset Using A Radial Basis Function Neural Network Based On Particle Swarm Optimization. International

Journal of Neural Systems 20(2), 109-116. doi: 10.1142/s0129065710002292

Wu, Q. (2010). A hybrid-forecasting model based on Gaussian support vector machine and chaotic particle swarm optimization. Expert Systems with Applications, 37(3), 2388-2394. doi: 10.1016/j.eswa.2009.07.057

Wu, W. Y., Simpson, A. R., & Maier, H. R. (2010). Accounting for Greenhouse Gas Emissions in Multiobjective Genetic Algorithm Optimization of Water Distribution Systems. Journal of Water Resources Planning and Management-Asce, 136(2), 146-155. doi: 10.1061/(asce)wr.1943-5452.0000020

Xavier-de-Souza, S., Suykens, J. A. K., Vandewalle, J., & Bolle, D. (2010). Coupled Simulated Annealing. Ieee Transactions on Systems Man and Cybernetics Part B-Cybernetics, 40(2), 320-335. doi: 10.1109/tsmcb.2009.2020435

Xing, L. N., Chen, Y. W., Wang, P., Zhao, Q. S., & Xiong, J. (2010). Knowledge-Based Ant Colony Optimization for Flexible Job Shop Scheduling Problems. Applied Soft Computing, 10(3), 888-896. doi: 10.1016/j.asoc.2009.10.006

Xu, T., Zuo, W. J., Xu, T. S., Song, G. C., & Li, R. C. (2010). An adaptive reanalysis method for genetic algorithm with application to fast truss optimization. Acta Mechanica Sinica, 26(2), 225-234. doi: 10.1007/s10409-009-0323-x

Yang, J. G., & Zhuang, Y. B. (2010). An improved ant colony optimization algorithm for solving a complex combinatorial optimization problem. Applied Soft Computing, 10(2), 653-660. doi: 10.1016/j.asoc.2009.08.040

Yang, Xin-She. Engineering Optimization: An Introduction with Metaheuristic Applications. Hoboken, NJ, USA: Wiley 2010. p 173-190.

Yang, Xin-She. Introduction to Mathematical Optimization: From Linear Programming to Metaheuristics. Cambridge, GBR: Cambridge International Science Publishing 2008. p 100-119.

Yu, B., & Yang, Z. Z. (2011). An ant colony optimization model: The period vehicle routing problem with time windows. Transportation Research Part E-Logistics and Transportation Review, 47(2), 166-181. doi: 10.1016/j.tre.2010.09.010

Yu, B., Yang, Z. Z., & Xie, J. X. (2011). A parallel improved ant colony optimization for multi-depot vehicle routing problem. Journal of the Operational Research Society, 62(1), 183-188. doi: 10.1057/jors.2009.161

Yu, V. F., Lin, S. W., Lee, W., & Ting, C. J. (2010). A simulated annealing heuristic for the capacitated location routing problem. Computers & Industrial Engineering, 58(2), 288-299. doi: 10.1016/j.cie.2009.10.007

Zhan, Z. H., Zhang, J., Li, Y., & Shi, Y. H. (2011). Orthogonal Learning Particle Swarm Optimization. Ieee Transactions on Evolutionary Computation, 15(6), 832-847. doi: 10.1109/tevc.2010.2052054

Zhang, B., Sun, X., Gao, L. R., & Yang, L. N. (2011). Endmember Extraction of Hyperspectral Remote Sensing Images Based on the Ant Colony Optimization (ACO) Algorithm. Ieee Transactions on Geoscience and Remote Sensing, 49(7), 2635-2646. doi: 10.1109/tgrs.2011.2108305

Zhang, R., & Wu, C. (2010). A hybrid immune simulated annealing algorithm for the job shop scheduling problem. Applied Soft Computing, 10(1), 79-89. doi: 10.1016/j.asoc.2009.06.008

Zhang, R., & Wu, C. (2011). A simulated annealing algorithm based on block properties for the job shop scheduling problem with total weighted tardiness objective. Computers & Operations Research, 38(5), 854-867. doi: 10.1016/j.cor.2010.09.014

Zhang, Y., Xu, J. L., Yuan, Z. H., Xu, H. J., & Yu, Q. (2010). Artificial neural network-genetic algorithm based optimization for the immobilization of cellulase on the smart polymer Eudragit L-100. Bioresource Technology, 101(9), 3153-3158. doi: 10.1016/j.biortech.2009.12.080

Part II

Chapter 4. Reducing the Kullback-Leibler Distance: The Cross-Entropy Method for Optimization

Abstract. This study presents a brief applied optimization technique, which utilizes the cross entropy (CE) method for finding minimum/maximum values of selected optimization problems. By utilizing the CE method, optimum solutions are achieved for sample optimization problems.

Keywords: Kullback-Leibler distance, cross-entropy method, optimization

1. Introduction

Rubinstein developed the Cross-Entropy (CE) method in 1997 and it is adapted for combinatorial optimization solutions (Rubinstein 1997, 1999, 2001; Rubinstein and Kroese 2004; Rubinstein and Melamed, 1998; Rubinstein and Shapiro 1993). The idea behind the CE method is to model an effective learning technique throughout the search process of the algorithm to solve combinatorial optimization problems. The method first produces a random sample from a pre-specified probability distribution function and then treats the sample to adjust the parameters of the probability distribution in order to generate a better sample in the next iteration.

The remainder of this study is organized as follows. Section 2 briefly reviews the literature on CE method. Section 3 introduces the method used for optimization. Section 4 presents some example functions for finding global minimum. The study is concluded in Section 5.

2. Literature review

The significance of the CE method is that it defines a precise mathematical framework for deriving fast, and in some sense "optimal" updating/learning rules, based on advanced simulation theory (de Boer *et al.* 2005).

While most of the stochastic algorithms for combinatorial optimization are based on local search (they employ local neighborhood structures), the cross-entropy method is a global random search procedure (Margolin 2005). The method consists of an iterative stochastic procedure that makes use of the importance sampling technique (Margolin 2005).

There are many examples for solving optimization problems using CE method. For instance, the literature about solving problems for vehicle routing (Chepuri and Homem-de-Mello, 2005), for max-cut and bipartition problems (Rubinstein, 2002), for project management (Cohen, Golany, and Shtub, 2005) and for scheduling

(Margolin, 2002, 2004). Some studies that use cross entropy method are highlighted in Table 1 through 5.

Table 1. The CE method literature (Mathematics and statistics)

An improved cross-entropy method applied to inverse problems (an *et al.* 2012).
The generalized cross entropy method, with applications to probability density estimation (Botev and Kroese 2011).
Improved cross-entropy method for estimation (Chan and Kroese 2012).
The cross-entropy method and its application to inverse problems (Ho and Yang 2010).
Multiobjective optimization of inverse problems using a vector cross entropy method (Ho and Yang 2012).

Table 2. The CE method literature (Computer science)

Design of a new interleaver using cross entropy method for turbo coding (Abderrahmane 2013).
A self-organization mechanism based on cross-entropy method for p2p-like applications (Chen *et al.* 2010).
Multi-objective buffer space allocation with the cross-entropy method (Bekker 2013).

Table 3. The CE method literature (Electronics and manufacturing)

Efficient capacity-based joint quantized precoding and transmit antenna selection using cross-entropy method for multiuser MIMO systems (Xhen *et al.* 2012).
Peak power reduction of OFDM systems through tone injection via parametric minimum cross-entropy method (Damavandi *et al.* 2013).
On the benefits of laplace samples in solving a rare event problem using cross-entropy method (Selvan *et al.* 2013).
Shaped beam synthesis of phased arrays using the cross entropy method (Weatherspoon *et al.* 2013).
Cross-entropy method for the optimization of optical alignment signals with diffractive effects (Chen *et al.* 2011).
Signal optimisation using the cross entropy method (Maher *et al.* 2013).
Sparse antenna array optimization with the cross-entropy method (Minvielle *et al.* 2011).
The cross-entropy method and its application to minimize the ripple of magnetic levitation forces of a maglev system (Zhang *et al.* 2010).

Table 4. The CE method literature (Optimization problems)

The cross-entropy method for combinatorial optimization problems of seaport logistics terminal (Yildiz and Yercan 2010).
Solving the multidimensional assignment problem by a cross-entropy method (Nguyen *et al.* 2014).
Simulation optimization using the cross-entropy method with optimal computing budget allocation (He *et al.* 2010).
The cross-entropy method in multi-objective optimisation: an assessment (Bekker and Aldrich 2011).

Table 5. The CE method literature (Others)

A suboptimal tone reservation algorithm based on cross-entropy method for PAPR reduction in OFDM systems (Chen *et al.* 2011).
Design of two-dimensional zero reference codes with cross-entropy method (Chen and Wen 2010).
Estimating change-points in biological sequences via the cross-entropy method (Evans *et al.* 2011).
A combined splitting-cross entropy method for rare-event probability estimation of queueing networks (Garvels 2011).
Optimal fuzzy control system using the cross-entropy method. A case study of a drilling process (Haber *et al.* 2010).
Cooperative cross-entropy method for generating entangled networks (Hui 2011).
The cross-entropy method with patching for rare-event simulation of large markov chains (Kaynar and Ridder 2010).
User selection method adopting cross-entropy method for a downlink multiuser MIMO system (Kim at al 2014).
The cross-entropy method for reliability assessment of cracked structures subjected to random markovian loads (Mattrand and Bourinet 2014).
Calibration of second order traffic models using continuous cross entropy method (Ngoduy and Maher 2012).
Stochastic inversion of ocean color data using the cross-entropy method (Salama and Shen 2010).
Enhanced cross-entropy method for dynamic economic dispatch with valve-point effects (Selvakumar 2011).
Fast reconstruction of computerized tomography images based on the cross-entropy method (Wang *et al.* 2011).

3. The CE Method

The global maximum of function $S(x)$ is represented by,

$$S(x^*) = \gamma^* = \max_{x \in X}(S(x))$$

The probability that the score function $S(x)$ evaluated at a particular state x is close to γ^* is classified as rare-event.

$$P_v(S(x) \geq \gamma) = E_v I_{\{S(x) \geq \gamma\}}$$

The important element of *maximum likelihood estimation* (MLE) is that there is a definable probability function that could be used to generate the likelihood of the observed event. It is shown as,

$$\hat{v}^* = \arg\max_v \left((1/N_s) \sum_{i=1}^{N_s} I_{\{S(x) \geq \gamma\}} \ln \varphi(x_i, v) \right)$$

The above statement comes from the definition of Kullback-Leibler (KL) distance. The Kullback-Leibler distance or cross entropy is a measure of the distance between two probability distributions $D_{KL}(P, Q)$. The stochastic optimization problem is solved by identifying the optimal *importance sampling* (IS) density that minimizes KL distance regarding the original density function. KL distance is the cross entropy between the original density function and the importance sampling density function. The distance $D_{KL}(g,h)$ is determined as a particular suitable criterion between densities of g and h. The KL distance (cross-entropy) is

$$D_{KL}(P,Q) = E_P \ln \frac{P(x)}{Q(x)} = \int P(x) \ln P(x) dx - \int P(x) \ln Q(x) dx$$

For discrete variables, the Kullback-Leibler (KL) divergence of Q from P is,

$$D_{KL}(P,Q) = \sum_x P(x) \log(\frac{P(x)}{Q(x)}) = \sum_x P(x) \log(P(x)) - \sum_x P(x) \log(Q(x))$$

For continuous variables, the Kullback-Leibler (KL) divergence of Q from P is,

$$D_{KL}(P,Q) = \int_{-\infty}^{+\infty} P(x) \ln(\frac{P(x)}{Q(x)}) dx = \int_{-\infty}^{+\infty} P(x) \ln(P(x)) dx - \int_{-\infty}^{+\infty} P(x) \ln(Q(x)) dx$$

$$where \; D_{KL}(P,Q) = H(P,Q) - H(P)$$

$H(P,Q)$ is named as the cross-entropy between P and Q. $H(P)$ is the entropy of P. The minimization of the KL distance (cross-entropy) provides definition for the parameters of the density functions and generations of enhanced feasible vectors. The method aborts when it comes together into a solution in the feasible region.

$$D_{KL}(P,Q) = E_P \ln \frac{P(x)}{Q(x)}$$

$$D_{KL}(P,Q) = \int P(x) \ln P(x) dx - \int P(x) \ln Q(x) dx$$

Minimizing the KL divergence between $P(x)$ and $Q(x)$ depends on the second term of the above statement, where $Q(x) = \varphi(x,v)$, then the equivalent maximization statement is,

$$\min_v \left(-\int P^*(x) \ln \varphi(x,v) dx \right) = \max_v \left(\int P^*(x) \ln \varphi(x,v) dx \right)$$

$$P^*(x) = \frac{I_{\{S(x) \geq \gamma\}} \varphi(x,u)}{l}$$

By substituting $P^*(x)$ in to the statement,

$$\max_v \left(\int \frac{I_{\{S(x) \geq \gamma\}} \varphi(x,u)}{l} \ln \varphi(x,v) dx \right)$$

The equivalent statement with an expectation E operator is,

$$\max_{v} D(v) = \max_{v} \left(E_u I_{\{S(x) \geq \gamma\}} \ln \varphi(x, v) dx \right)$$

3.1 The continuous CE optimization

The normal distribution $N(\mu, \sigma^2)$ function $\varphi(x)$ is,

$$\varphi(x) = \frac{1}{\sqrt{2\pi\sigma^2}} e^{-\frac{1}{2}\left(\frac{x-\mu}{\sigma}\right)^2}, \quad x \in R$$

By taking natural logarithms of both sides of the distribution function,

$$\ln\left(\varphi(x)\right) = \ln\left(\frac{1}{\sqrt{2\pi\sigma^2}} e^{-\frac{1}{2}\left(\frac{x-\mu}{\sigma}\right)^2} \right)$$

Thus, the statement becomes,

$$\ln\left(\varphi(x)\right) = -\left(\frac{1}{2}\ln(2\pi) + \frac{1}{2}\ln(\sigma^2) + \frac{1}{2}\left(\frac{x-\mu}{\sigma}\right)^2 \right)$$

the maximum log-likelihood estimator is,

$$\hat{v}^* = \arg\max_{v} \left((1/N_s) \sum_{i=1}^{N_s} I_{\{S(x) \geq \gamma\}} \ln \varphi(x_i, v) \right)$$

By substituting the function $\varphi(x)$ into the maximum log-likelihood estimator,

$$\hat{v}^* = \arg\max_{v} \left((1/N_s) \sum_{i=1}^{N_s} I_{\{S(x) \geq \gamma\}} \left[-\left(\frac{1}{2}\ln(2\pi) + \frac{1}{2}\ln(\sigma^2) + \frac{1}{2}\left(\frac{x_i-\mu}{\sigma}\right)^2 \right) \right] \right)$$

The maximization problem becomes a minimization, since from inspection of the above statement,

$$-\max(-f(x)) = \min(f(x))$$

Then,

$$\hat{\tau}^* = \min_{\tau} \left((1/2N_s) \sum_{i=1}^{N_s} I_{\{S(x) \geq \gamma\}} \left[\left(\ln(2\pi) + \ln(\sigma^2) + \left(\frac{x-\mu}{\sigma} \right)^2 \right) \right] \right)$$

By eliminating the non-variable parts, the minimization can be simplified as,

$$\min \left(\ln(\sigma^2) \sum_{i=1}^{N_s} I_i + \left(\frac{1}{\sigma^2} \right) \sum_{i=1}^{N_s} I_i (x_i - \mu)^2 \right)$$

By taking partial derivatives with respect to μ,

$$\frac{\partial}{\partial \mu} \left(\ln(\sigma^2) \sum_{i=1}^{N_s} I_i + \left(\frac{1}{\sigma^2} \right) \sum_{i=1}^{N_s} I_i (x_i - \mu)^2 \right) =$$

$$\frac{2}{\sigma^2} \left(\sum_{i=1}^{N_s} I_i \mu - \sum_{i=1}^{N_s} I_i x_i \right) = 0$$

$$\frac{2}{\sigma^2} \sum_{i=1}^{N_s} I_i \mu = \frac{2}{\sigma^2} \sum_{i=1}^{N_s} I_i x$$

Moreover, placing μ on the left hand side, the following estimate is obtained,

$$\hat{\mu} = \frac{\sum_{i=1}^{N_s} I_i x}{\sum_{i=1}^{N_s} I_i}$$

By taking partial derivatives with respect to σ^2,

$$\frac{\partial}{\partial \sigma^2} \left(\ln(\sigma^2) \sum_{i=1}^{N_s} I_i + \left(\frac{1}{\sigma^2} \right) \sum_{i=1}^{N_s} I_i (x_i - \mu)^2 \right) =$$

$$\frac{1}{\sigma^2} \sum_{i=1}^{N_s} I_i - \frac{1}{(\sigma^2)^2} \sum_{i=1}^{N_s} I_i (x_i - \mu)^2 = 0$$

$$\sum_{i=1}^{N_s} I_i = \frac{1}{(\sigma^2)} \sum_{i=1}^{N_s} I_i (x_i - \mu)^2$$

By placing σ^2 on the left hand side, the following estimate is obtained,

$$\hat{\sigma}^2 = \frac{\sum_{i=1}^{N_s} I_i (x_i - \mu)^2}{\sum_{i=1}^{N_s} I_i}$$

3.2 The combinatorial CE optimization

The random vector is $X = (X_1, ..., X_n) \sim Ber(p)$. Density function φ on X parameterized by a vector $p \in [0,1]^n$. Bernoulli density function, under the following probability density function (pdf) is,

$$\varphi(x, p) = \prod_{i=1}^{n} (p_i)^{x_i} (1 - p_i)^{1-x_i} \quad ,$$

$$x \in \{0, 1\}$$

which is,

$$\varphi(x; p) = \begin{cases} p & \text{if } x = 1, \\ 1 - p & \text{if } x = 0, \\ 0 & \text{otherwise} \end{cases}$$

$$P_v(S(x) \geq \gamma) = E_v I_{\{S(x) \geq \gamma\}}$$

By taking natural logarithms of both sides of the distribution function $\varphi(x; p)$,

$$\ln \varphi(x; p) = \ln \left(p^x (1 - p)^{1-x} \right)$$

yields,

$$\ln \varphi(x; p) = x \ln p + (1 - x) \ln(1 - p)$$

By taking partial derivatives with respect to p,

$$\frac{\partial}{\partial p} \ln \varphi(x; p) = \frac{x_i}{p} - \frac{1 - x_i}{1 - p}$$

The statement becomes,

$$\frac{\partial}{\partial p} \ln \varphi(x; p) = \frac{x_i - p}{p(1 - p)}$$

The maximum log-likelihood is,

$$\hat{v}^* = \arg\max_{v} \left((1/N_s) \sum_{i=1}^{N_s} I_{\{S(x)\geq\gamma\}} \ln\varphi(x_i,v) \right)$$

By substituting,

$$\frac{\partial}{\partial p} \left(\sum_{i=1}^{N_s} I_{\{S(x)\geq\gamma\}} \ln\varphi(x_i,v) \right) = \sum_{i=1}^{N_s} I_{\{S(x)\geq\gamma\}} \frac{x_i - p}{p(1-p)}$$

$$= (1/p(1-p)) \sum_{i=1}^{N_s} I_{\{S(x)\geq\gamma\}} (x_i - p)$$

In addition, placing p on the left hand side, the following estimate is obtained,

$$\hat{p} = \frac{\displaystyle\sum_{i=1}^{N_s} I_{\{S(x)\geq\gamma\}} x_i}{\displaystyle\sum_{i=1}^{N_s} I_{\{S(x)\geq\gamma\}}}$$

The CE method involves an iterative procedure where each iteration can be broken down into two phases (de Boer *et al.* 2005):

1. Generate a random data sample (trajectories, vectors, etc.) according to a specified mechanism.

2. Update the parameters of the random mechanism based on the data to produce a "better" sample in the next iteration.

The general cross-entropy method consists of three main steps (Margolin 2005):

1. Choosing a probability family. Initializing of the distribution parameters and the method parameters. Generating feasible solutions according to the chosen distribution.

2. Updating the distribution parameters, based on the Kullback-Leibler cross-entropy.

3. Checking the stopping rule. Updating the method parameters in the case the stopping rule fails.

A general view of the pseudo-code for the cross-entropy algorithm is (Connor 2008):

1. Initialize parameters: Set initial parameter $\hat{}$, a small value of ρ , set population size K, smoothing constant α and set iteration counter t=1

2. Update $\gamma^{(t)}$: Given $\hat{}$, let $\gamma^{(t)}$ be the $(1-\rho)$- quantile of $Z(x)$ satisfying

$$P_{v^{(t-1)}}(Z(x) \geq \gamma^{(t)}) \geq \rho$$

$$P_{v^{(t-1)}}(Z(x) \leq \gamma^{(t)}) \geq 1-\rho$$

with x sampled from $f(\cdot, \hat{})$. Then, the estimate of $\gamma^{(t)}$ is calculated computed as $\gamma^{(t)} = Z_{(\lceil 1-\rho K \rceil)}$, where $\lceil \cdot \rceil$ rounds $(1-\rho)K$ towards infinity.

3. Update $\hat{}$: Given $\hat{}$, determine $\hat{}$ by solving the CE program

$$\hat{} \quad \frac{1}{N_s} \sum_{i=1}^{N_s} I_{\{Z(x_i \geq \gamma^{(t)})\}} \ln f(x_i, v)$$

v

4. Optional step: (Smooth update of $\hat{}$) To decrease the probability of the CE procedure converging too quickly to a suboptimal solution, a smoothed update of $\hat{}$ can be computed.

$$\hat{} \qquad \hat{}$$

where $\check{v}^{(t)}$ is the estimate of the parameter vector computed with (3), $\hat{}$ is the parameter estimate from the previous iteration and $\alpha (for\ 0 < \alpha \leq 1)$ is a constant smoothing coefficient. By setting $\alpha = 1$, the update will not be smoothed.

5. If convergence is reached then stop; otherwise, Set t=t+1 and reiterate from step 2 to 4, until the stopping criteria is satisfied.

4. Examples

In this section, some test functions are used to evaluate the characteristics of the CE algorithm's search for the global minimum. These test functions are the Giunta, Ackley, Rastrigin, and Hölder table. Each subsection demonstrates the CE method's search for the global minimum.

4.1 Finding the minimum value of the Giunta test function

In this subsection, a sample optimization is carried out on the Giunta test function (Giunta 1997) by utilizing the CE method. The Giunta test function is defined in the search domain $x_1, x_2 \in [-1,1]$ as,

$$f(x) = \frac{3}{5} + \sum_{i=1}^{2} [\sin(\frac{16}{15}x_i - 1) + \sin^2(\frac{16}{15}x_i - 1) + \frac{1}{50}\sin 4((\frac{16}{15}x_i - 1))]$$

The Giunta function is known that it has minimum for x_1, x_2 both having values of

0.45834282 (see Figure 1).

$$f_{\min}(x_1, x_2) = 0.0602472184$$

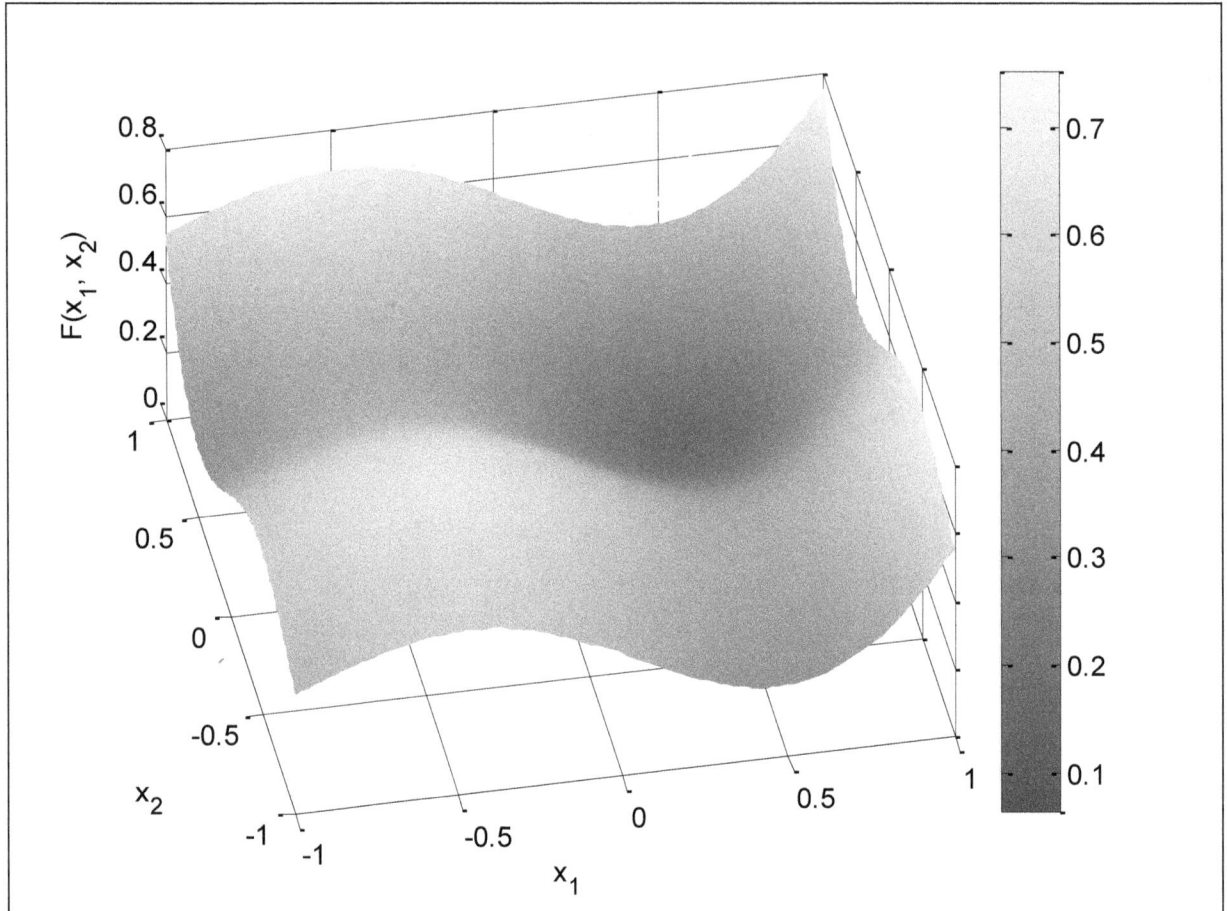

Figure 1. The plot of the Giunta function in 3D

The search for the optimum value is shown in the Figure 2. At each iteration, better vectors are created and each of these vectors are used to generate better values. Algorithm will stop when it converges to a global optimum value. The global minimum is shown with a cross symbol in the Figure 2.

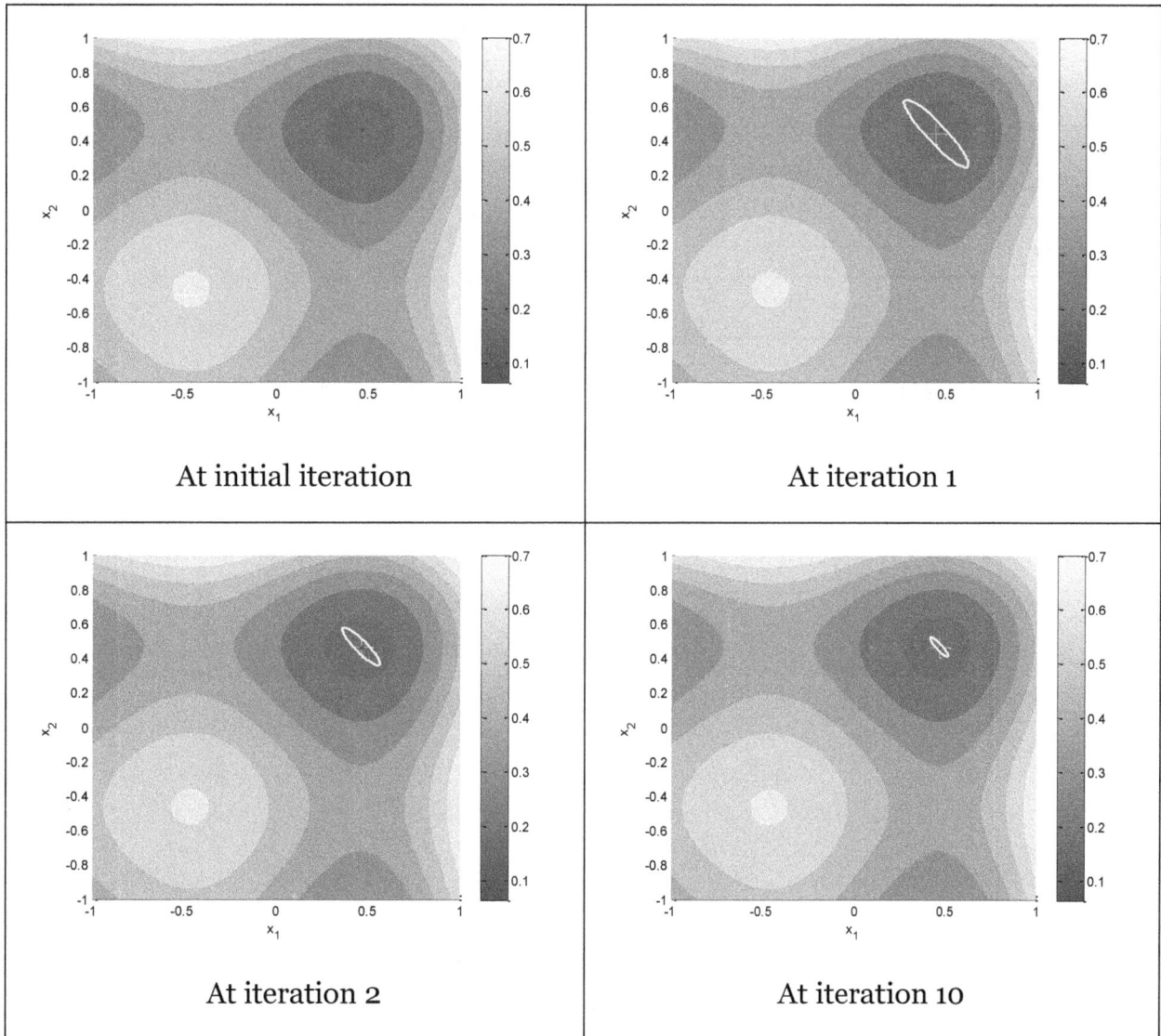

Figure 2. The contour plot of the Giunta test function – The search of a global minimum at each iteration.

In this example, with the author's own limitation to 10 iterations, the optimum value is found (see Table 6). At each iteration, algorithm tries better (x_1, x_2) values to obtain $f_{\min}(x_1, x_2)$. At the 10th iteration, minimum value is

$$f_{\min}(0.4673, 0.4673) = 0.0645$$

113

Table 6. CE algorithm's search for the optimum values

Iteration number	$\gamma^{(t)}$	$\hat{\gamma}$	x_1	x_2
1	0.0645	0.1043	0.3997	0.3956
2	0.0645	0.0666	0.4444	0.4441
3	0.0645	0.0645	0.4657	0.4653
4	0.0645	0.0645	0.4674	0.4674
5	0.0645	0.0645	0.4676	0.4673
6	0.0645	0.0645	0.4674	0.4673
7	0.0645	0.0645	0.4674	0.4673
8	0.0645	0.0645	0.4672	0.4674
9	0.0645	0.0645	0.4674	0.4673
10	0.0645	0.0645	0.4673	0.4673

4.2 Finding the minimum value of the Ackley test function

The Ackley test function f is defined as,

$$F(x1, x2) = 20 - 20\ e^{\left(-0.2\sqrt{0.5\ x1^2 + 0.5\ x2^2}\right)} - e^{(0.5\ \cos(2\ \pi\ x1) + 0.5\ \cos(2\ \pi\ x2))} + e$$

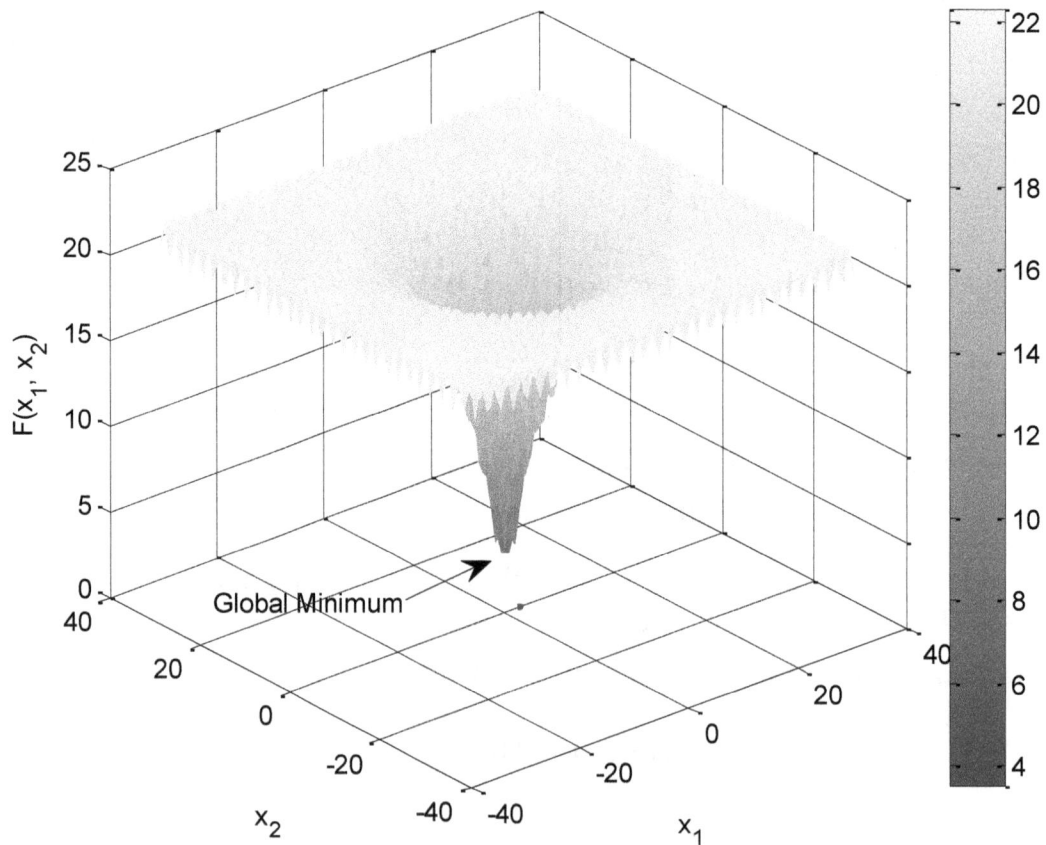

Figure 3. The plot of the Ackley function in 3D.

114

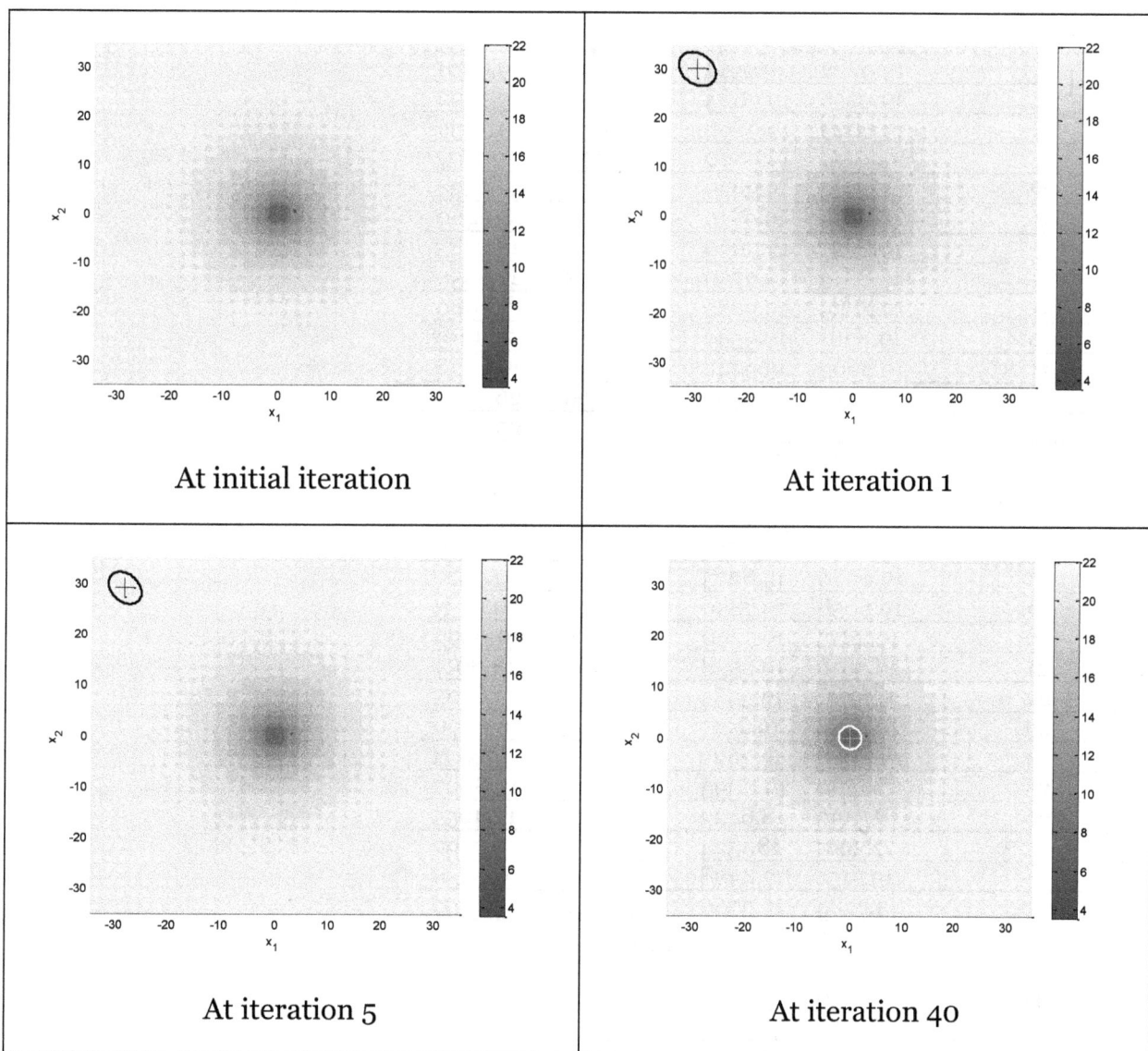

Figure 4. The contour plot of the Ackley test function – The search of a global minimum at each iteration.

Table 7. CE algorithm's search for the optimum values

Iteration number	$\tilde{\gamma}$	$\hat{\gamma}$	x_1	x_2
1	19.9222	19.9914	-29.7293	29.8061
2	19.9256	19.9884	-29.3937	29.5592
3	19.9029	19.9868	-29.1167	29.3436
4	19.8990	19.9820	-28.8572	29.1877
5	19.9092	19.9839	-28.5431	28.9334
6	19.9014	19.9810	-28.2489	28.6460
7	19.8937	19.9690	-27.9809	28.2785
8	19.8819	19.9730	-27.6325	27.8550
9	19.8829	19.9636	-27.2615	27.4680
10	19.8381	19.9542	-26.8110	27.1919
11	19.8603	19.9446	-26.2334	26.6838
12	19.8458	19.9400	-25.7706	26.2210
13	19.8364	19.9291	-25.2951	25.6356
14	19.8117	19.9131	-24.7197	24.9332
15	19.7725	19.8996	-24.1052	24.2560
16	19.7552	19.8755	-23.4588	23.4716
17	19.7017	19.8531	-22.6867	22.8329
18	19.6688	19.8273	-21.8956	21.9577
19	19.6058	19.7850	-20.9415	20.9773
20	19.4875	19.7290	-20.0486	20.0599
21	19.3816	19.6538	-19.0600	18.8917
22	19.1975	19.5585	-17.9663	17.8725
23	19.0518	19.4317	-16.7615	16.6490
24	18.8028	19.2627	-15.5357	15.3171
25	18.4400	19.0093	-14.1977	13.8807
26	18.0130	18.6423	-12.8371	12.4055
27	17.2961	18.1524	-11.3578	10.7792
28	15.5152	17.4308	-9.7410	9.2858
29	14.7284	16.3760	-7.9840	7.4752
30	12.5507	14.5835	-5.8932	5.6041
31	8.7427	11.6643	-3.8726	3.4926
32	3.5871	7.2291	-1.7185	1.5046
33	0.0294	2.1498	-0.1155	0.0991
34	0.0249	0.5409	-0.0064	0.0072
35	0.0053	0.5255	-0.0060	0.0011
36	0.0088	0.5036	-0.0041	0.0010
37	0.0347	0.5199	0.0013	-0.0001
38	0.0301	0.5014	-0.0000	-0.0013
39	0.0251	0.4653	-0.0001	-0.0039
40	0.0229	0.4697	-0.0007	-0.0030

4.3 Finding the minimum value of the Rastrigin test function

The Rastrigin test function F is defined as

$$F(x1, x2) = x1^2 + x2^2 - 10 \cos(2 \pi x1) - 10 \cos(2 \pi x2) + 20$$

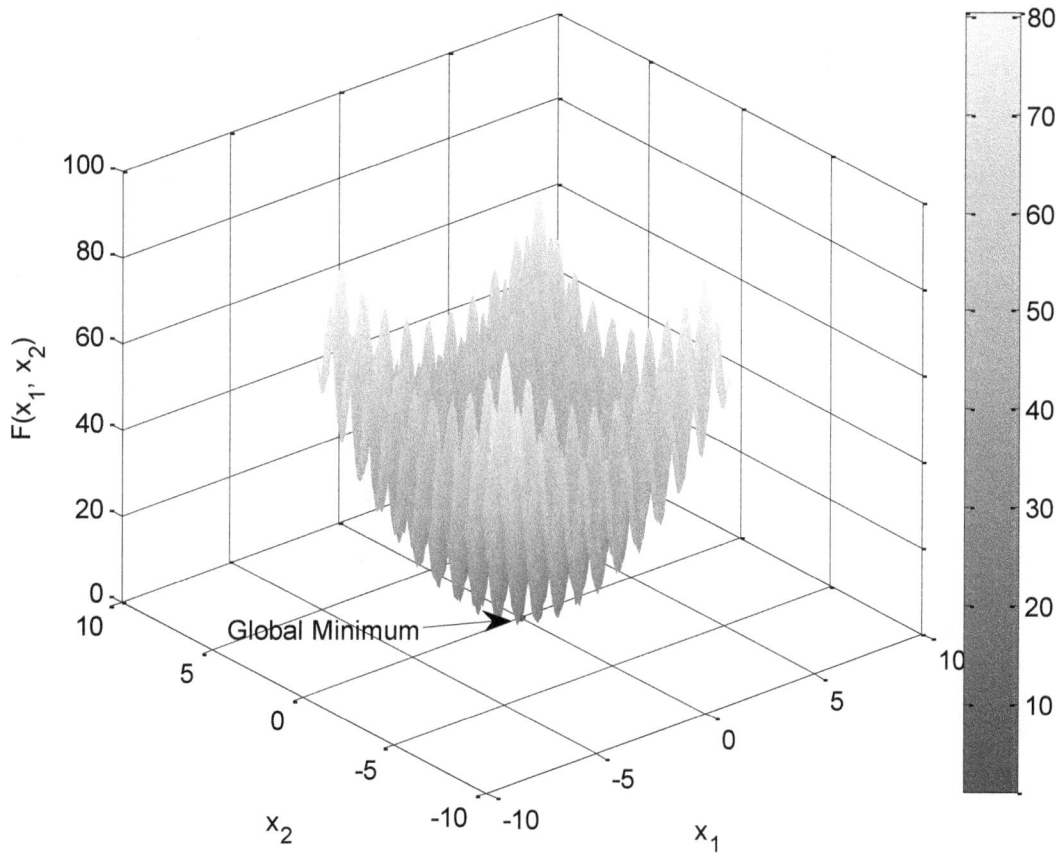

Figure 5. The plot of the Rastrigin test function in 3D.

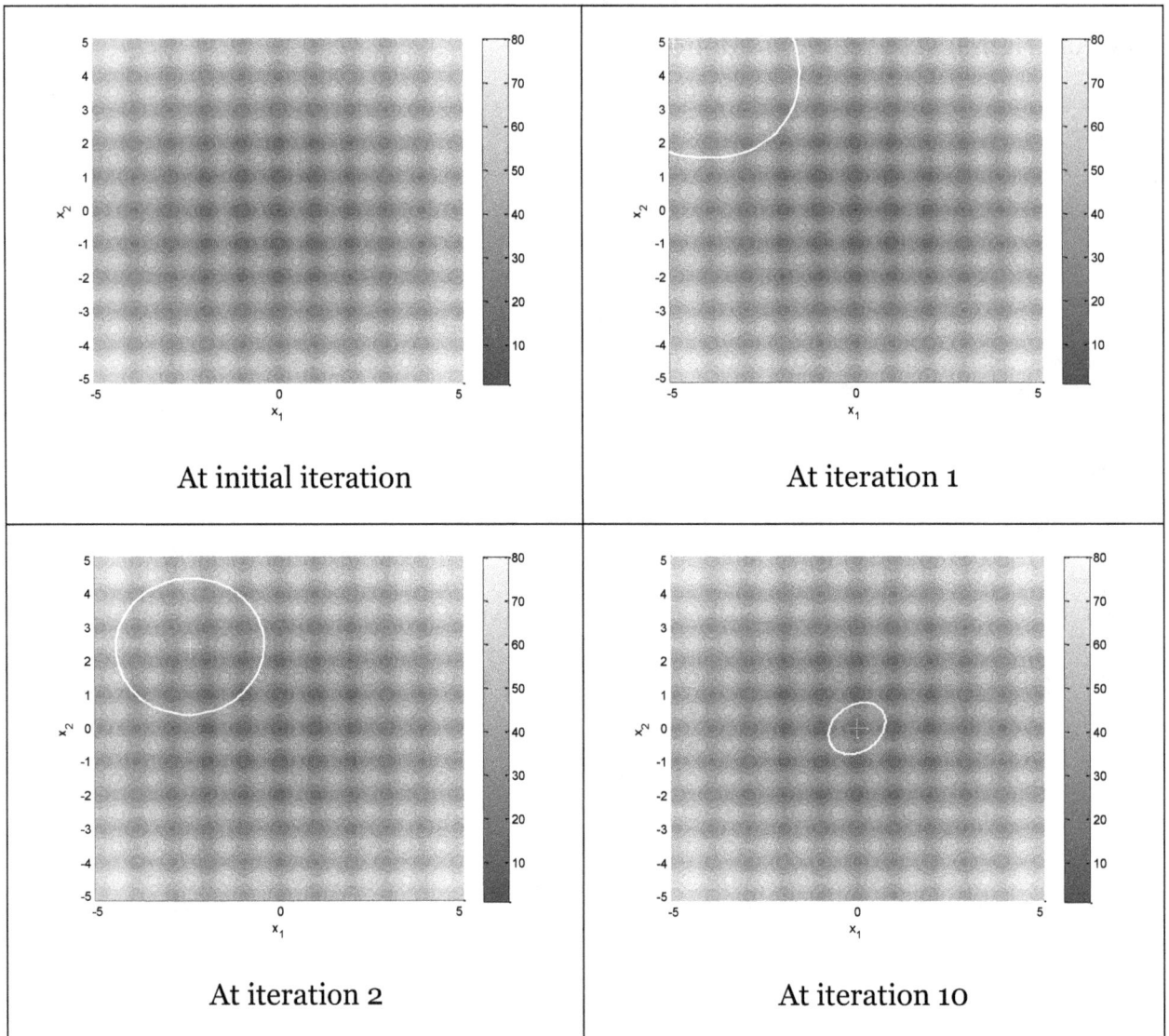

Figure 6. The contour plot of the Rastrigin test function – The search of a global minimum at each iteration.

Table 8. CE algorithm's search for the optimum values

Iteration number	$\gamma^{(t)}$	$\hat{}$	x_1	x_2
1	6.5395	21.4974	-2.4731	2.4337
2	1.6262	7.8246	-1.4369	1.3483
3	0.9443	2.9052	-0.8404	0.8250
4	0.0003	1.6696	-0.4264	0.3515
5	0.0003	1.0270	-0.0684	0.0540
6	0.0018	0.5191	-0.0031	0.0027
7	0.0043	0.3716	0.0007	0.0006
8	0.0021	0.2854	0.0012	-0.0019
9	0.0016	0.2287	0.0009	-0.0020
10	0.0008	0.1880	0.0004	0.0020

4.4 *Finding the minimum value of the Hölder table test function*

The Hölder table test function f is defined as,

$$F(x1, x2) = -\left| \sin(x1)\cos(x2)\, \mathbf{e}^{\left| 1 - \frac{\sqrt{x1^2 + x2^2}}{\pi} \right|} \right|$$

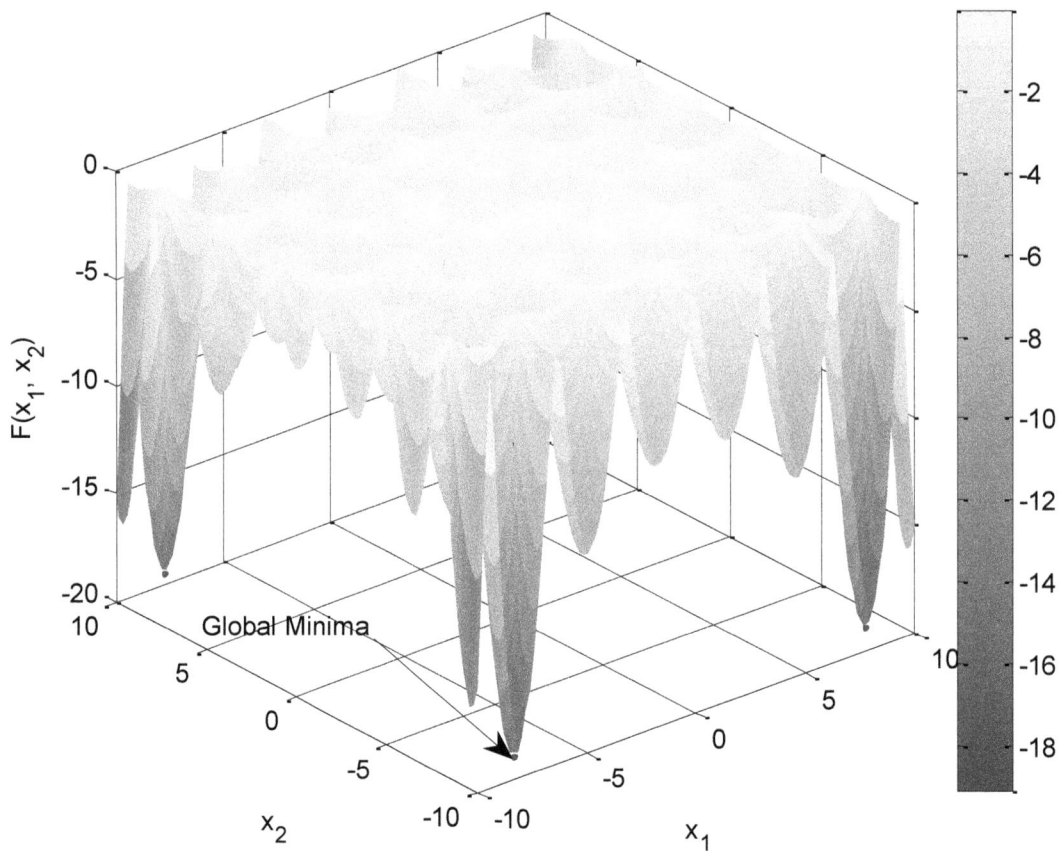

Figure 7. The plot of the Hölder table function in 3D.

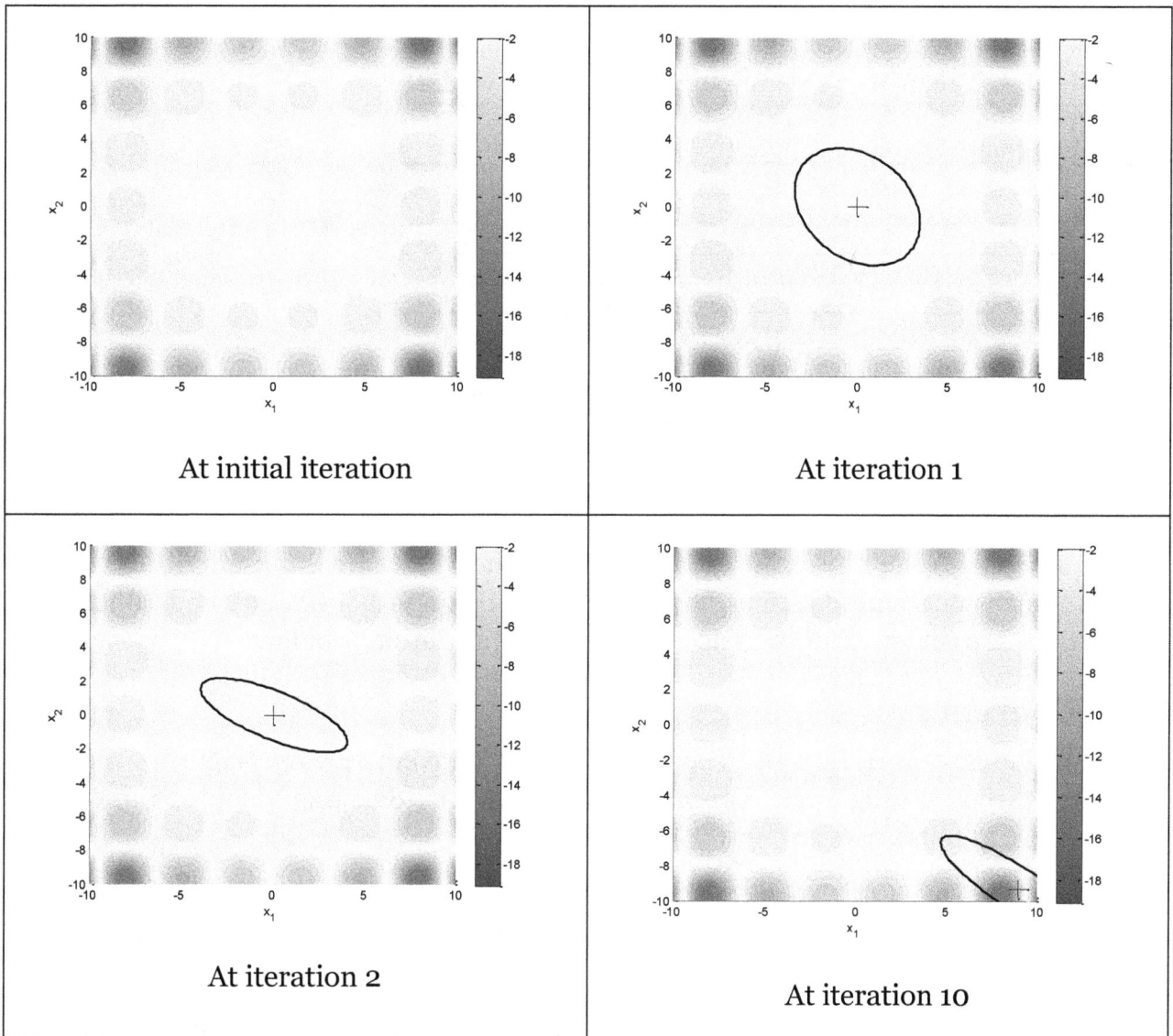

Figure 8. The contour plot of the Hölder table test function – The search of a global minimum at each iteration.

121

Table 9. CE algorithm's search for the optimum values

Iteration number	$\gamma^{(t)}$	$\hat{\gamma}$	x_1	x_2
1	-2.4783	-1.7217	0.2467	-0.0682
2	-5.1073	-1.7292	1.0025	-0.5533
3	-9.3766	-3.6231	5.8590	-3.2323
4	-9.4566	-7.6462	8.3307	-6.0783
5	-13.8899	-9.3822	8.1698	-6.5472
6	-14.4456	-9.4426	8.2758	-6.7798
7	-14.9241	-9.4609	8.4256	-7.0645
8	-16.2653	-9.4812	9.1023	-8.2246
9	-18.8213	-14.5481	9.0045	-9.3178
10	-19.2075	-18.8066	8.1178	-9.6154

5. Conclusion

This study presented the cross entropy (CE) method for finding minimum/maximum values of some optimization problems. By utilizing the CE method, optimum solutions are achieved for the selected sample optimization problems.

6. References

Abderrahmane, L. H. (2013). Design of a new interleaver using cross entropy method for turbo coding. *Iet Communications*, 7(9), 828-835. doi: 10.1049/iet-com.2012.0599

An, S. G., Yang, S. Y., Ho, S. L., & Ni, P. H. (2012). An Improved Cross-Entropy Method Applied to Inverse Problems. *Ieee Transactions on Magnetics*, 48(2), 327-330. doi: 10.1109/tmag.2011.2173303

Bekker, J. (2013). Multi-Objective Buffer Space Allocation With The Cross-Entropy Method. *International Journal of Simulation Modelling*, 12(1), 50-61. doi: 10.2507/ijsimm12(1)5.228

Bekker, J., & Aldrich, C. (2011). The cross-entropy method in multi-objective optimisation: An assessment. *European Journal of Operational Research*, 211(1), 112-121. doi: 10.1016/j.ejor.2010.10.028

Botev, Z. I., & Kroese, D. P. (2011). The Generalized Cross Entropy Method, with Applications to Probability Density Estimation. *Methodology and Computing in Applied Probability*, 13(1), 1-27. doi: 10.1007/s11009-009-9133-7

Chan, J. C. C., & Kroese, D. P. (2012). Improved cross-entropy method for estimation. *Statistics and Computing*, 22(5), 1031-1040. doi: 10.1007/s11222-011-9275-7

Chen, G., Sarrafzadeh, A., Low, C. P., & Zhang, L. A. (2010). A Self-Organization Mechanism Based on Cross-Entropy Method for P2P-Like Applications. *Acm Transactions on Autonomous and Adaptive Systems*, 5(4). doi: 10.1145/1867713.1867716

Chen, J. C., & Wen, C. K. (2010). Design of two-dimensional zero reference codes with cross-entropy method. *Applied Optics*, 49(18), 3560-3565.

Chen, J. C., Chiu, M. H., Yang, Y. S., & Li, C. P. (2011). A Suboptimal Tone Reservation Algorithm Based on Cross-Entropy Method for PAPR Reduction in OFDM Systems. *Ieee Transactions on Broadcasting*, 57(3), 752-756. doi: 10.1109/tbc.2011.2127590

Chen, J. C., Chiu, M. H., Yang, Y. S., Liao, K. Y., & Li, C. P. (2012). Efficient Capacity-Based Joint Quantized Precoding and Transmit Antenna Selection Using Cross-Entropy Method for Multiuser MIMO Systems. *International Journal of Antennas and Propagation*. doi: 10.1155/2012/965834

Chen, J. C., Wang, S. H., Lee, M. K., & Li, C. P. (2011). Cross-Entropy Method for the Optimization of Optical Alignment Signals With Diffractive Effects. *Journal of Lightwave Technology*, 29(18), 2706-2714. doi: 10.1109/jlt.2011.2163182

Chepuri, K. and T. Homem-de-Mello. (2005). "Solving the Vehicle Routing Problem with Stochastic Demands using the Cross-Entropy Method." *Annals of Operations Research,* 134, 153–181

Cohen, I., B. Golany, and A. Shtub. (2005). "Managing Stochastic Finite Capacity Multi-Project Systems Through the Cross-Entropy Method." *Annals of Operations Research,* 134, 183–199

Connor, J. D. (2008). Antenna array synthesis using the cross entropy method, Ph.D. Thesis, The Florida State University.

Damavandi, M. G., Abbasfar, A., & Michelson, D. G. (2013). Peak Power Reduction of OFDM Systems Through Tone Injection via Parametric Minimum Cross-Entropy Method. *Ieee Transactions on Vehicular Technology*, 62(4), 1838-1843. doi: 10.1109/tvt.2012.2233507

de Boer, P.-T., *et al.* (2005). "A Tutorial on the Cross-Entropy Method." *Annals of Operations Research,* 134(1): 19-67.

Evans, G. E., Sofronov, G. Y., Keith, J. M., & Kroese, D. P. (2011). Estimating change-points in biological sequences via the cross-entropy method. *Annals of Operations Research*, 189(1), 155-165. doi: 10.1007/s10479-010-0687-0

Garvels, M. J. J. (2011). A combined splitting-cross entropy method for rare-event probability estimation of queueing networks. *Annals of Operations Research*, 189(1), 167-185. doi: 10.1007/s10479-009-0608-2

Giunta, A. A. (1997). Aircraft Multidisciplinary Design Optimization Using Design of Experiments Theory and Response Surface Modeling Methods, Ph. D. Thesis, Virginia Polytechnic Institute and State University.

Haber, R. E., del Toro, R. M., & Gajate, A. (2010). Optimal fuzzy control system using the cross-entropy method. A case study of a drilling process. *Information Sciences*, 180(14), 2777-2792. doi: 10.1016/j.ins.2010.03.030

He, D. H., Lee, L. H., Chen, C. H., Fu, M. C., & Wasserkrug, S. (2010). Simulation Optimization Using the Cross-Entropy Method with Optimal Computing Budget Allocation. *Acm Transactions on Modeling and Computer Simulation*, 20(1). doi: 10.1145/1667072.1667076

Ho, S. L., & Yang, S. Y. (2010). The Cross-Entropy Method and Its Application to Inverse Problems. *Ieee Transactions on Magnetics*, 46(8), 3401-3404. doi: 10.1109/tmag.2010.2044380

Ho, S. L., & Yang, S. Y. (2012). Multiobjective Optimization of Inverse Problems Using a Vector Cross Entropy Method. *Ieee Transactions on Magnetics*, 48(2), 247-250. doi: 10.1109/tmag.2011.2175437

Hui, K. P. (2011). Cooperative Cross-Entropy method for generating entangled networks. *Annals of Operations Research*, 189(1), 205-214. doi: 10.1007/s10479-009-0589-1

Kaynar, B., & Ridder, A. (2010). The cross-entropy method with patching for rare-event simulation of large Markov chains. *European Journal of Operational Research*, 207(3), 1380-1397. doi: 10.1016/j.ejor.2010.07.002

Kim, K., Yang, H., & Choi, S. (2014). User Selection Method Adopting Cross-Entropy Method for a Downlink Multiuser MIMO System. *Wireless Personal Communications*, 74(2), 789-802. doi: 10.1007/s11277-013-1321-7

Maher, M., Liu, R. H., & Ngoduy, D. (2013). Signal optimisation using the cross entropy method. *Transportation Research Part C-Emerging Technologies*, 27, 76-88. doi: 10.1016/j.trc.2011.05.018

Margolin, L. (2002). "Application of the Cross-Entropy Method to Scheduling Problems." Master's thesis, Technion, Industrial Engineering.

Margolin, L. (2004). "The Cross-Entropy Method for the Single Machine Total

Weighted Tardiness Problem." Unpublished.

Margolin, L. (2005). "On the Convergence of the Cross-Entropy Method." *Annals of Operations Research,* 134(1): 201-214.

Mattrand, C., & Bourinet, J. M. (2014). The cross-entropy method for reliability assessment of cracked structures subjected to random Markovian loads. *Reliability Engineering & System Safety,* 123, 171-182. doi: 10.1016/j.ress.2013.10.009

Minvielle, P., Tantar, E., Tantar, A. A., & Berisset, P. (2011). Sparse Antenna Array Optimization With the Cross-Entropy Method. *Ieee Transactions on Antennas and Propagation,* 59(8), 2862-2871. doi: 10.1109/tap.2011.2158941

Ngoduy, D., & Maher, M. J. (2012). Calibration of second order traffic models using continuous cross entropy method. *Transportation Research Part C-Emerging Technologies,* 24, 102-121. doi: 10.1016/j.trc.2012.02.007

Nguyen, D. M., Thi, H. A. L., & Dinh, T. P. (2014). Solving the Multidimensional Assignment Problem by a Cross-Entropy method. *Journal of Combinatorial Optimization,* 27(4), 808-823. doi: 10.1007/s10878-012-9554-z

Rubinstein, R.Y. & Kroese, D.P. (2004). The Cross-Entropy Method: A Unified Approach to Combinatorial Optimization, Monte-Carlo Simulation and Machine Learning. Springer-Verlag, New York.

Rubinstein, R.Y. & Melamed, B. (1998). Modern simulation and modeling. Wiley series in probability and Statistics.

Rubinstein, R.Y. & Shapiro, A. (1993). Discrete Event Systems: Sensitivity Analysis and Stochastic Optimization via the score function method. Wiley.

Rubinstein, R.Y. (1997). Optimization of computer simulation models with rare events. *European Journal of Operations Research,* pp. 89-112.

Rubinstein, R.Y. (1999). The cross-entropy method for combinatorial and continuous optimization. *Methodology and Computing in Applied Probability,* pp. 127-190.

Rubinstein, R.Y. (2001). Combinatorial optimization, cross-entropy, ants and rare events. In S. Uryasev and P. M. Pardalos, editors, Stochastic Optimization: Algorithms and Applications, Kluwer, pp. 304-358.

Rubinstein, R.Y. (2002). "Cross-Entropy and Rare-Events for Maximal Cut and Bipartition Problems." *ACM Transactions on Modeling and Computer Simulation,* 27–53

Salama, M. S., & Shen, F. (2010). Stochastic inversion of ocean color data using the cross-entropy method. *Optics Express*, 18(2), 479-499.

Selvakumar, A. I. (2011). Enhanced cross-entropy method for dynamic economic dispatch with valve-point effects. *International Journal of Electrical Power & Energy Systems*, 33(3), 783-790. doi: 10.1016/j.ijepes.2011.01.001

Selvan, S. E., Subathra, M. S. P., Christinal, A. H., & Amato, U. (2013). On the benefits of Laplace samples in solving a rare event problem using cross-entropy method. *Applied Mathematics and Computation*, 225, 843-859. doi: 10.1016/j.amc.2013.10.011

Wang, Q., Wang, H. X., & Yan, Y. (2011). Fast reconstruction of computerized tomography images based on the cross-entropy method. *Flow Measurement and Instrumentation*, 22(4), 295-302. doi: 10.1016/j.flowmeasinst.2011.03.010

Weatherspoon, M. H., Connor, J. D., & Foo, S. Y. (2013). Shaped beam synthesis of phased arrays using the cross entropy method. *International Journal of Numerical Modelling-Electronic Networks Devices and Fields*, 26(6), 630-642. doi: 10.1002/jnm.1893

Yildiz, T., & Yercan, F. (2010). The Cross-Entropy Method for Combinatorial Optimization Problems Of Seaport Logistics Terminal. *Transport*, 25(4), 411-422. doi: 10.3846/transport.2010.51

Zhang, W., Yang, S. Y., Bai, Y. A., & Machado, J. M. (2010). The cross-entropy method and its application to minimize the ripple of magnetic levitation forces of a maglev system. *International Journal of Applied Electromagnetics and Mechanics*, 33(3-4), 1063-1068. doi: 10.3233/jae-2010-1221

Chapter 5. Logistics systems and optimization strategies under uncertain operational environment

Abstract. Transportation and logistics systems are characterized by their highly dynamic structures along with numerous interconnected processes. The natures of these systems involve various levels of resource allocation decisions where usually it is not always possible to execute these decisions in the field on time at the best possible way because of the unpredictable factors in plans. By considering the uncertain operational environment, this study explores the uncertainty issue within operational systems and deals with the problem of allocating resources to maximize expected total profit and minimize inefficiencies under uncertainty. The aim is to design, develop, visualize, and effectively deal with a more realistic model to satisfy uncertain demand nodes by leaving minimal or none unsatisfied zones within an operational environment at seaports, transportation, logistics and supply chain systems. A representative optimization model, which is developed to address the uncertainty issue, has been solved by using an optimization algorithm. The results show that operational plans without the utilization of uncertainty models could have negative impacts, including increased emissions, negative environmental effects, along with higher costs to organizations.

Keywords: Uncertainty, optimization algorithm, resource allocation problem, transportation, logistics, environment

1. Introduction

Logistics systems have been at the center of attention of enterprises for over a decade with today's heightened expectation of customers and harsh competitive environment in markets. Enterprises are continuously seeking the best development and enterprise-level best solution strategies for their logistics systems to meet the present and possible future expectations and to stay beyond their counterparts. With the advancement of information, communication systems, and optimization techniques, it became possible to model and to obtain solutions for large-scale complex problems.

Optimization is a well-established field due to comprehensive research conducted over past decades. In decision science, optimization is an essential tool. As stated by Nocedal (1999), optimization is an important tool in decision science and in the analysis of physical systems. In order to benefit from optimization, it is necessary to identify some objectives and then some quantitative measures of the system.

The objective of a system could be maximizing profit, minimizing total time for completing a task or a group of tasks, or maximizing/minimizing any quantity or composition of quantities that can be represented by mathematical models. The objective depends on certain characteristics of the system, called variables or unknowns. The goal of optimization is to find correct values for the variables that optimize the objective. The variables are usually restricted, or constrained, in some way (Nocedal 1999). Many types of optimization problems have been addressed and various types of algorithms have been researched. Several methodologies for optimizing objectives of systems have been used in various practical applications and the range of applications is continuously growing (Arora 2007).

For example, the objective function might be linear or nonlinear, differentiable or nondifferentiable, concave or convex, etc. The decision variables might be continuous or discrete. The feasible region might be convex or nonconvex. These differences each impact how the model can be solved, and thus optimization models are classified according to these differences (Sarker 2002).

Global optimization is made even more difficult because supply chains need to be designed for and operated in uncertain environments (Simchi-Levi 2004). Operations research uses quantitative models to analyze and predict the behavior of systems, and to provide information for decision makers. Two key concepts in operations research are optimization and uncertainty. Uncertainty is emphasized in operations research that could be called "stochastic operations research" in which uncertainty is described by stochastic models. The typical models in stochastic operations research are queuing models, inventory models, financial engineering models, reliability models, and simulation models (Dohi 2009).

This paper is organized as follows. Section 2 reviews the literature. Section 3 provides general background information on mathematical programming with uncertainty components and recourse variables. The study is concluded in Section 4.

2. Literature review

Hall (2003) stated that transportation science covers research from many fields such as geography, economics, and location theory. Methodologies of transportation science come from physics, operations research, probability, and control theory; it is fundamentally a quantitative discipline, relying on mathematical models and optimization algorithms to explain the phenomena of transportation (Hall 2003). Frazelle (2002) explained the overall goal of transportation as it should be: to connect sourcing locations with customers at the lowest possible transportation cost within the constraints of the customer service policy. Thus, the transportation optimization equation could be expressed as follows: minimizing total transportation costs subject to customer service policy constraints (Frazelle 2002). Frazelle (2002) also emphasized that transportation expenses are rising quickly versus other logistics costs, with smaller, more frequent orders, increasing international trade and global logistics, rising fuel charges, labor shortages, decreased carrier competition due to carrier mergers and acquisitions, and increased union penetration in the labor market.

Additionally, Simchi-Levi (2004) acknowledged that it is challenging to design and operate a supply chain so that total system wide costs are minimized and system wide service levels are maintained. Indeed, it is frequently difficult to operate a single facility so that costs are minimized and service level is maintained. The difficulty increases exponentially when an entire system is being considered. The process of finding the best system wide strategy is known as global optimization (Simchi-Levi 2004). The process of identifying objectives, variables, and constraints for a given problem is known as modeling.

Construction of an appropriate model is the first step— sometimes the most important step— in the optimization process. If the model is too simplistic, it will not give useful insights into the practical problem, but if it is too complex, it may become too difficult to solve (Nocedal 1999). Before solving an optimization model, it is important to consider the form and mathematical properties of the objective function, constraints, and decision variables.

There are a large number of optimization problems in organized systems, such as in industrial systems, business systems, transportation and logistics systems, where at strategic, tactical, and operational levels, planners, analysts, strategists, and engineers are confronted with uncertainty. Many optimization problems arising from these systems have deterministic (certain) parts along with uncertain components, which planners could disambiguated based on the predictable and probabilistic information. Along with the other various factors, Pardalos (2004) pointed out the crucial issue of developing efficient methods of analyzing this information in order to understand the internal structure of the market and make effective strategic decisions for the successful operation of a business.

In addition, referring to the efficiency of the transportation infrastructure, Hall (2003) expressed that planners and engineers need to forecast the demand of transportation to make informed transportation infrastructure planning decisions. Arora (2007) stated for any engineering system that the uncertainties in system characteristics and demand prevent assurance from being given with absolute certainty. For supply chain systems, Simchi-Levi (2004) emphasized that it is necessary to design systems that eliminate as much uncertainty as possible and it is necessary to deal effectively with the uncertainty that remains.

Much research has been devoted to tackling optimization problems under uncertainty. Researchers have investigated various systems and have dealt with the uncertainty issue. Table 1 highlights recent and leading studies on the applications of stochastic programming with uncertainty depending on the fields studied.

Table 1a. Studies on stochastic programming with uncertainty (Optimization problems).

A multi-objective robust stochastic programming model for disaster relief logistics under uncertainty (Bozorgi-Amiri *et al.* 2013).
A multistage stochastic programming approach for capital budgeting problems under uncertainty (Beraldi *et al.* 2013).
Integration of Scheduling and Dynamic Optimization of Batch Processes under Uncertainty: Two-Stage Stochastic Programming Approach and Enhanced Generalized Benders Decomposition Algorithm (Chu and You 2013).
Scheduling jobs sharing multiple resources under uncertainty: A stochastic programming approach (Keller and Bayraksan 2010).
Determining supply requirement in the sales-and-operations-planning (S&OP) process under demand uncertainty: a stochastic programming formulation and a spreadsheet implementation (Sodhi and Tang 2011).
A multi-stage stochastic programming approach for production planning with uncertainty in the quality of raw materials and demand (Zanjani *et al.* 2010).
Capacities-based supply chain network design considering demand uncertainty using two-stage stochastic programming (Singh *et al.* 2013).
Location of cross-docking centers and vehicle routing scheduling under uncertainty: A fuzzy possibilistic-stochastic programming model (Mousavi *et al.* 2014).
A stochastic programming winner determination model for truckload procurement under shipment uncertainty (Ma *et al.* 2010).
Stochastic programming approach to re-designing a warehouse network under uncertainty (Kiya and Davoudpour 2012).
Solution strategies for multistage stochastic programming with endogenous uncertainties (Gupta and Grossmann 2011).
A stochastic programming approach for optimal microgrid economic operation under uncertainty using 2m+1 point estimate method (Mohammadi *et al.* 2013).

Table 1b. Studies on stochastic programming with uncertainty (Industry).

Design under uncertainty of hydrocarbon biorefinery supply chains: Multiobjective stochastic programming models, decomposition algorithm, and a Comparison between CVaR and downside risk (Gebreslassie *et al.* 2012).
Electric sector investments under technological and policy-related uncertainties: a stochastic programming approach (Bistline and Weyant 2013).
Sustainable development and planning of coal industry under uncertainty using system dynamic and stochastic programming (Xu and Wu 2010).

Table 1c. Studies on stochastic programming with uncertainty (Environment).

An interval-fuzzy two-stage stochastic programming model for planning carbon dioxide trading under uncertainty (Li *et al.* 2011).
Energy and environmental systems planning under uncertainty-An inexact fuzzy-stochastic programming approach (Li *et al.* 2010).
Two-Stage Stochastic Programming Model for Planning CO_2 Utilization and Disposal Infrastructure Considering the Uncertainty in the CO_2 Emission (Han and Lee 2011).
Developing a Two-Stage Stochastic Programming Model for CO_2 Disposal Planning under Uncertainty (Han *et al.* 2012).
A two-stage inexact-stochastic programming model for planning carbon dioxide emission trading under uncertainty (Chen *et al.* 2010).

Table 1d. Studies on stochastic programming with uncertainty (water resources management).

Inexact Fuzzy-Stochastic Programming for Water Resources Management Under Multiple Uncertainties (Guo *et al.* 2010).
Inexact joint-probabilistic stochastic programming for water resources management under uncertainty (Li and Huang 2010).
An inexact fuzzy parameter two-stage stochastic programming model for irrigation water allocation under uncertainty (Li *et al.* 2013).

Table 1e. Studies on stochastic programming with uncertainty (Waste management).

An interval-parameter mean-CVaR two-stage stochastic programming approach for waste management under uncertainty (Dai *et al.* 2014).
A Superiority-Inferiority-Based Inexact Fuzzy Stochastic Programming Approach for Solid Waste Management Under Uncertainty (Tan *et al.* 2010).

Table 1f. Studies on stochastic programming with uncertainty (Energy systems).

Development of an interval multi-stage stochastic programming model for regional energy systems planning and GHG emission control under uncertainty (Li *et al.* 2012).
Decomposition Based Stochastic Programming Approach for Polygeneration Energy Systems Design under Uncertainty (Liu *et al.* 2010).
An interval fixed-mix stochastic programming method for greenhouse gas mitigation in energy systems under uncertainty (Xie *et al.* 2010).

Moreover, while taking into consideration efficient systems design under uncertainty, critical awareness about the environment is important. The transport sector is one of the few sectors that have ever-increasing greenhouse gas (GHG) emissions. According to Intergovernmental Panel on Climate Change (IPCC) reports, transport accounts for 13.1% of GHG emissions, 23% of global carbon dioxide (CO_2) emissions, 26% of total world energy use, and the transport sector is one of the constantly growing sectors. Thus, it is also critical to take seriously into consideration inefficient transportation and logistics system designs that can have negative effects on the environment.

A report by Pronello and Andrè (2000) proposed that pollution caused by a set of vehicle routes can be unreliable. However, it is obvious that reducing the total distance and thus putting to practical use of optimal routes by algorithms will provide environmental benefits as a result of the reduction in fuel consumption and the consequent pollutants (Sbihi and Eglese 2007). Palmer (2004) also analyzed a study about the connection between efficient network design and thus reducing environmental pollutants.

3. A Background: Mathematical programming

This section presents a brief general background about mathematical programming with *second-stage* or *recourse* variables. As stated by Sahinidis (2004), second-stage variables can be interpreted as corrective measures or recourse against any infeasibilities arising due to a particular realization of uncertainty. The most common programming models are stochastic linear/non-linear/integer programming, probabilistic or chance-constrained programming, robust stochastic programming, stochastic dynamic programming, and fuzzy programming. The next section will present brief fundamentals of mathematical programming with uncertainty. Further details of mathematical programming concepts have been well documented in the textbooks of Bertsekas and Tsitsiklis (1996), Birge and Louveaux (1997), Kall and Wallace (1994), Prékopa (1995), and Zimmermann (1991). Early works involving uncertainty studies were included research by Beale (1955), Bellman (1957), Bellman and Zadeh (1970), Charnes and Cooper (1959), Dantzig (1955), and Tintner (1955).

3.1 Stochastic linear programming with recourse

A stochastic programming can be stated as

$$\min \ f(X) = C^t X = \sum_{j=1}^{n} c_j x_j$$

subject to

$$A_i^T X = \sum_{j=1}^{n} a_{ij} x_j \leq b, \ i = 1, 2, 3, ..., m$$

$$x_j \geq 0, \ j = 1, 2, 3, ..., n$$

Based on the above formulation, the general formulation for two-stage (*recourse*) stochastic linear program (Birge and Louveaux 1997; Infanger 1994; Kall and Wallace 1994 and Prékopa 1995) is

The first stage of the problem,

$$\min \ c^t x + E_{\omega \in \Omega}[Q(x, \omega)]$$
$$s.t. \ Ax \leq b, \ x \geq 0, \ x \in X$$

and the second stage of the problem,

$$Q(x, \omega) = \min f(\omega)^t y$$

Subject to

$$D(\omega)y \geq h(\omega) + T(\omega)x, \quad y \in Y$$

ω is a random variable from a probability space (Ω, F, P) and within the interval

132

$\Omega = \left[\omega^{L.B.}, \omega^{U.B.} \right]$. The expected value function $E_{\omega \in \Omega}[Q(x,\omega)]$ is

$$E_{\omega \in \Omega}[Q(x,\omega)] = \int_{\omega^{L.B.}}^{\omega^{U.B.}} Q(x,\omega)d(P)$$

The first stage decision variables are represented by vector x and the second stage variables are represented by vector y.

A is a first stage ($m \times n$) constraint matrix. b is a first stage (m component) right hand side vector. c is a first stage (n component) cost vector. x is a vector of first stage vector variables. y is a vector of second stage variables. Ω is the all possible variables of the random variable ω. D is a second stage recourse matrix. $T(\omega)$ is a second stage matrix. $h(\omega)$ is a second stage right hand side vector. $Q(\cdot)$ is a cost function of the second stage. P is the probability measure of the variable ω. The variables $\omega^{U.B.}, \omega^{L.B.}$ are upper and lower bounds of the variable ω.

Probabilistic/Chance-constrained programming:

$$\min \ c'x$$

$$s.t. \ P(Tx \geq \xi) \geq p$$

$$Ax = b, \ x \geq 0$$

p is the fixed probability. $P(Tx \geq \xi) \geq p$ is the probabilistic constraint.

Stochastic non-linear programming with recourse:

The first stage of the problem,

$$\min \ f(x) + E_{\omega \in \Omega}[Q(x,\omega)]$$

subject to

$$g(x) \leq 0$$

with the second stage of the problem,

$$Q(x,\omega) = \min F(\omega, x, y)$$

subject to

$$G(\omega, x, y) \leq 0, \ y \in Y$$

133

where ω is a random variable from a probability space (Ω, F, P).

3.2 A sample model with an uncertainty component

The expected cost φ is the total cost and plus the expected revenue lost due to the unassigned cargo spaces/seats.

The objective functions $\varphi_{1,2}$ are

$$Min. \; \varphi_1 = \sum_{i=1}^{\mu}\sum_{j=1}^{v} c_{ij}x_{ij} + \left[\sum_{j=1}^{v} k_j \omega_j - \sum_{j=1}^{v} k_j \sum_{h=1}^{\rho} \gamma_{hj}\psi_{hj} \right]$$

$$Min. \; \varphi_2 = \sum_{i=1}^{\mu}\sum_{j=1}^{v} c_{ij}x_{ij} + \left[\sum_{jh} k_j \lambda_{jh}\beta_{jh} \right]$$

subject to:

$$\psi_{jh} \leq \sum_{i} \pi_{ij}x_{ij}$$

$$\beta_{jh} = \delta_{jh} - \psi_{jh}$$

Vehicle balance:

$$\sum_{j} x_{ij} \leq \upsilon_i$$

Demand balance:

$$\sum_{i} \pi_{ij}x_{ij} \geq \sum_{h\$\Delta b_{jh}} \psi_{jh}$$

where

$i, j, h \in \Re_1$

$i = 1,2,3,...,\mu-1$ is vehicle types and unassigned cargo container/passenger. $i = \mu$, if unassigned. $j = 1,2,3,...,v-1$ is the type of route to which a vehicle is assigned. $j = v$, if unassigned. $h = 1,2,3,...,\rho$ is the demand state. c_{ij} is the cost per vehicle type i on route j. π_{ij} is the capacity of vehicle type i on route j. υ_i is the vehicle availability. λ_{jh} is the probability of demand state h on route j. δ_{jh} is the demand distribution on route j based on the demand state h. γ_{hj} is the probability of exceeding demand increment h on route j. ω_j is the expected demand on route j. k_j is the revenue lost. x_{ij} is the vehicle type i assigned to route j. ψ_{hj} is the load actually carried on route j based on the demand state h.

β_{hj} is the load not carried on route j based on the demand state h. $\sum_j x_{ij}$ is the total available vehicle type i on route j. The first term of the objective function $\sum_{i=1}^{\mu} \sum_{j=1}^{v} c_{ij} x_{ij}$ is the operating cost. $\sum_{j=1}^{v} k_j$ is the total revenue lost and $\sum_{j=1}^{v} \omega_j$ is the total expected demand, thus $\sum_{j=1}^{v} k_j \omega_j$ is the expected revenue, in case all allocations are made sufficiently. The second terms of the objective functions $\left[\sum_{j=1}^{v} k_j \omega_j - \sum_{j=1}^{v} k_j \sum_{h=1}^{\rho} \gamma_{hj} \psi_{hj} \right]$ and $\left[\sum_{jh} k_j \lambda_{jh} \beta_{jh} \right]$ are the expected revenue lost due to unassigned spaces. For more information about the mathematical model, refer to a slightly different model from Dantzig (1963).

3.3 Initialization with data and simulation

This sub-section initializes the sample model described in the previous section with data to represent a realistic problem. Sample data initialization includes few steps as shown in Figure 1. To solve the optimization problem with the initialized data, IpOpt algorithm is used. The IpOpt algorithm implements an interior point line search filter method that aims to find a local solution of the objective function subject to constraints. The mathematical details of the algorithm are in various publications (Waechter 2002; Waechter and Biegler 2005, 2006; Nocedal *et al.* 2005). The algorithm is intended to solve non-linear problems; however, it is also capable of providing solutions for various other types of problems as well as linear problems.

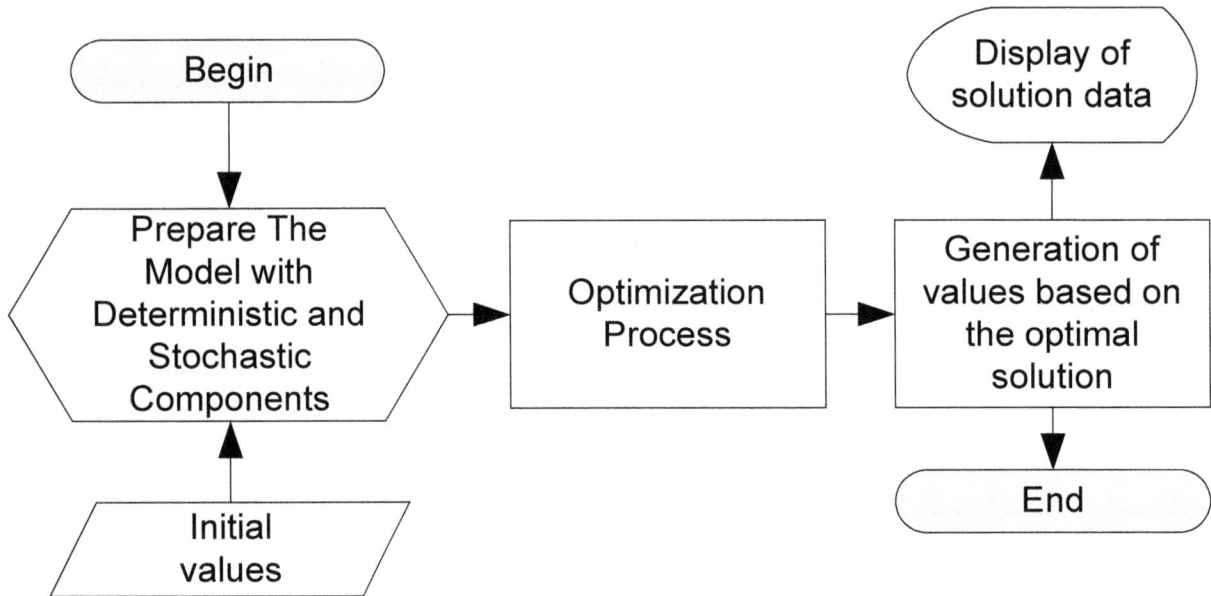

Figure 1. The flowchart of the optimization process

Waechter and Biegler (2006) stated that the interest in efficient optimization methods led to the development of interior-point or barrier methods for large-scale nonlinear programming. In addition, Waechter and Biegler (2006) further emphasized the fact that in particular, these methods provide an attractive alternative to active set strategies in handling problems with large numbers of inequality constraints. For the convergence properties of interior-point methods (Forsgren 2002) over the past years there has been a better understanding and efficient algorithms with desirable global and local convergence properties have been developed (Waechter and Biegler 2006). Waechter and Biegler (2006) also stated that to allow convergence from poor starting points, interior-point methods, in both trust region and line-search frameworks, have been developed that use exact penalty merit functions to enforce progress toward the solution (Byrd *et al.* 2000; Tits *et al.* 2003; Yamashita 1998; Waechter and Biegler 2006).

The next part shows a solution to the optimization problem under uncertainty having 100 vehicles and 25 different routes. In this case, the total number of variables is 3250 (See Table 2). Table 2 shows the optimization process.

Table 2. The optimization process (IpOpt algorithm)

Iter. No.	Objective	Inf_pr	Inf_du	Lg(mu)	||d||	Lg(rg)	Alpha_du	Alpha_pr	Ls
0	2.0000000e-002	7.64e+004	1.19e+000	0.0	0.00e+000	-	0.00e+000	0.00e+000	0
1	2.2453154e-002	7.32e+004	2.83e+001	-5.3	2.74e+001	-	1.40e-003	4.29e-002h	1
2	3.5510215e-002	6.59e+004	2.61e+001	-1.2	1.94e+002	-	1.66e-003	9.94e-002h	1
3	3.4950269e-002	6.42e+004	2.54e+001	-5.3	1.35e+003	-	1.41e-002	2.62e-002h	1
4	4.0035380e-002	6.19e+004	2.45e+001	-1.2	6.87e+002	-	6.64e-002	3.45e-002h	1
5	3.6059254e-002	4.93e+004	1.95e+001	-5.3	6.43e+002	-	1.70e-002	2.05e-001h	1
6	3.5044461e-002	4.36e+004	1.72e+001	-5.3	5.15e+002	-	8.76e-002	1.15e-001h	1
7	3.5879943e-002	3.91e+004	1.55e+001	-5.4	4.64e+002	-	5.95e-002	1.03e-001h	1
8	4.0438994e-002	3.58e+004	1.42e+001	-5.4	4.33e+002	-	1.07e-001	8.31e-002h	1
9	7.0272037e-002	3.20e+004	1.27e+001	-5.5	4.23e+002	-	4.96e-002	1.07e-001h	1
10	2.2892969e-001	2.94e+004	1.35e+001	-5.5	3.94e+002	-	3.01e-002	8.25e-002h	1
11	1.1670179e+002	8.78e+003	2.11e+001	-1.6	2.98e+002	-	1.44e-003	7.01e-001f	1
12	1.0487235e+002	2.12e+003	4.99e+000	-5.5	9.57e+001	-	1.36e-002	7.59e-001h	1
13	9.4358075e+001	5.93e+002	1.36e+000	-5.5	2.74e+001	-	3.91e-001	7.20e-001h	1
14	7.7650402e+001	8.49e+001	2.20e-001	-5.7	1.95e+001	-	4.39e-001	8.57e-001h	1
15	5.4069763e+001	2.51e+000	5.78e-003	-6.0	2.43e+001	-	9.80e-001	9.70e-001f	1
16	2.7939077e+000	2.10e-002	9.09e-005	-7.4	5.17e+001	-	9.79e-001	9.92e-001f	1
17	3.0061904e-002	2.06e-004	9.32e-007	-8.9	2.79e+000	-	9.90e-001	9.90e-001f	1
18	4.8234800e-005	3.42e-007	1.50e-009	-10.9	3.01e-002	-	9.98e-001	9.98e-001f	1
19	3.8114012e-008	7.56e-009	4.81e-005	-9.5	6.72e+000	-	1.00e+000	9.79e-001h	1
20	8.4073551e-011	6.92e-009	3.43e+000	-9.5	1.09e-005	-4.0	1.00e+000	3.86e-002h	1
21	4.3991391e-008	1.71e-009	1.18e+000	-9.3	4.98e-005	-4.5	1.00e+000	7.75e-001h	1
22	2.4409058e-008	7.95e-010	4.02e-001	-9.3	1.49e-004	-5.0	1.00e+000	5.03e-001f	1
23	1.8460108e-009	1.35e-010	6.91e-011	-9.3	6.08e+000	-	1.00e+000	1.00e+000f	1

Table 2 and Figure 2 show the optimization process. Table 3 and Table 4 show the final output of the process. The column "*iter*" is the iteration counter and the column "*objective*" is the current value of the objective function. Other values of the columns are primal infeasibility (*inf_pr*), dual infeasibility (*inf_du*), logarithm of current barrier parameter (*lg(mu)*), max-norm of the primal search direction ($\|d\|$), logarithm of Hessian perturbation (*lg(rg)*), dual step size (*alpha_du*), primal step size (*alpha_pr*), and number of backtracking steps (*ls*).

Table 3. About the optimized model.

Number of nonzeros in equality constraint Jacobian	3218
Number of nonzeros in inequality constraint Jacobian	5709
Number of nonzeros in Lagrangian Hessian	0
Total number of variables	3250
variables with only lower bounds	3250
variables with lower and upper bounds	0
variables with only upper bounds	0
Total number of equality constraints	2
Total number of inequality constraints	125
inequality constraints with only lower bounds	25
inequality constraints with lower and upper bounds	0
inequality constraints with only upper bounds	100

Ipopt shows information (see Tables 2, 3, 4) of the optimization procedure and closes with statistical information about the computational effort. In the Table 2, the first column "*iter*" is the iteration counter. It is not reset after an update of the barrier parameter or when the algorithm switched between the restoration phase and the regular algorithm. The next two columns "*objective*" and "*inf_pr*" indicate the value of the objective function. The fourth column "*inf_du*" is a measure of optimality; as the Ipopt aims to find a point satisfying the optimality conditions. The last column "*ls*" is an indication about how many trial points needed to be evaluated. Additionally, as the value of the barrier parameter is going to zero, there is a decrease in the number in column "*lg(mu)*". Furthermore, the larger the step sizes of columns "*alpha_du*" and "*alpha_pr*", the better is the usually the progress. (For more details about the algorithm, refer to https://projects.coin-or.org/Ipopt).

Table 4. The final output of the optimization process.

	(unscaled)	(scaled)
Objective	1.8460108023708330e-009	1.8460108023708330e-009
Dual infeasibility	6.9146319266848223e-011	6.9146319266848223e-011
Constraint violation	1.3451350211095411e-010	1.6007106751203540e-010
Complementarity	7.5684299981862521e-010	7.5684299981862521e-010
Overall NLP error	7.5684299981862521e-010	7.5684299981862521e-010

The process provides an optimal solution for the sample model having over a hundred thousand variables/constraints (See Table 2). The return of an optimal solution value takes less than few minutes on a computer having an Intel® Core™2 Quad Q6600 at 2.4GHz CPU. Many commercial optimization algorithms (i.e. IBM® Cplex, and more) return solutions in less than a few seconds even with millions of variables and constraints.

4. Conclusion

While there is no single effective solution that can be applied to various other types of real-world problems having some uncertainty components, the IpOpt optimization method presented on this paper and theoretically applied based on a typical scenario. In brief, this paper has introduced optimization issues under uncertainty along with brief fundamentals of mathematical programming concepts that address various possible uncertainty situations. It then presented a sample optimization model with uncertainty variables. Finally, it showed optimal solutions found at the last iteration of the optimization algorithm. The optimal solution found at the last iteration provides critical information, which addresses the best system configuration and conditions under changing operational environment.

The sample optimization model put the IpOpt optimization algorithm to practical use; this model was developed to address the uncertainty issue. The results indicate that operational plans without taking into consideration of uncertainty models could have negative impacts that lead to non-optimal operational design and thus, higher costs to organizations along with possible harmful environmental effects (i.e. increased emissions, pollutants, and more). This study solved a sample model having more than a hundred thousand variables/constraints. The aim was to visualize an optimal solution under uncertainty for a sample logistics problem. Many other solvers and optimization algorithms are capable of solving highly complex problems with millions of variables in a short period depending on the CPU power of the computer.

5. References

Arora, J.S. (Ed.). (2007) Optimization of Structural and Mechanical Systems. River

Edge, NJ, USA: World Scientific. p 1-271.

Beale, E.M.L. (1955) On minimizing a convex function subject to linear inequalities. *Journal of the Royal Statistical Society*, 17B, pp. 173–184.

Bellman, R. E. (1957) Dynamic programming. Princeton, PA: Princeton University Press.

Bellman, R., Zadeh, L.A. (1970) Decision-making in a fuzzy environment. *Management Science*, 17, pp. 141–161.

Beraldi, P., Violi, A., De Simone, F., Costabile, M., Massabo, I., & Russo, E. (2013). A multistage stochastic programming approach for capital budgeting problems under uncertainty. *Ima Journal of Management Mathematics*, 24(1), 89-110. doi: 10.1093/imaman/dps018

Bertsekas, D. P., Tsitsiklis, J. N. (1996) Neuro-dynamic programming. Belmont, MA: Athena Scientific.

Birge, J. R., Louveaux, F. V. (1997) Introduction to stochastic programming. New York, NY: Springer.

Bistline, J. E., & Weyant, J. P. (2013). Electric sector investments under technological and policy-related uncertainties: a stochastic programming approach. *Climatic Change*, 121(2), 143-160. doi: 10.1007/s10584-013-0859-4

Bozorgi-Amiri, A., Jabalameli, M. S., & Al-e-Hashem, S. (2013). A multi-objective robust stochastic programming model for disaster relief logistics under uncertainty. *Or Spectrum*, 35(4), 905-933. doi: 10.1007/s00291-011-0268-x

Byrd, R. H., Gilbert, J. Ch., Nocedal, J. (2000) A trust region method based on interior point techniques for nonlinear programming. *Mathematical Programming*, 89, 149–185

Charnes, A., Cooper, W.W. (1959) Chance-constrained programming. *Management Science*, 6, pp. 73–79.

Chen, W. T., Li, Y. P., Huang, G. H., Chen, X., & Li, Y. F. (2010). A two-stage inexact-stochastic programming model for planning carbon dioxide emission trading under uncertainty. *Applied Energy*, 87(3), 1033-1047. doi: 10.1016/j.apenergy.2009.09.016

Chu, Y. F., & You, F. Q. (2013). Integration of Scheduling and Dynamic Optimization of Batch Processes under Uncertainty: Two-Stage Stochastic Programming Approach and Enhanced Generalized Benders Decomposition Algorithm.

Industrial & Engineering Chemistry Research, 52(47), 16851-16869. doi: 10.1021/ie402621t

Dai, C., Cai, X. H., Cai, Y. P., Huo, Q., Lv, Y., & Huang, G. H. (2014). An interval-parameter mean-CVaR two-stage stochastic programming approach for waste management under uncertainty. *Stochastic Environmental Research and Risk Assessment*, 28(2), 167-187. doi: 10.1007/s00477-013-0738-6

Dantzig, G.B. (1955) Linear programming under uncertainty. Management Science 1, pp. 197–206.

Dantzig, G.B. (1963) Chapter 28. In Linear Programming and Extensions. Princeton University Press, Princeton, New Jersey.

Dohi, T. (Ed.). (2009) Recent Advances in Stochastic Operations Rsearch II. SGP: World Scientific, p 6.

Forsgren, A., Gill, P. E., Wright, M. H. (2002) Interior methods for nonlinear optimization. *SIAM Review*, 44 (4), 525–597

Frazelle, E. H. (2002) Supply Chain Strategy.Blacklick, OH, USA: McGraw-Hill Education Group, 2002. p 169-174.

Gebreslassie, B. H., Yao, Y., & You, F. Q. (2012). Design under uncertainty of hydrocarbon biorefinery supply chains: Multiobjective stochastic programming models, decomposition algorithm, and a Comparison between CVaR and downside risk. *Aiche Journal*, 58(7), 2155-2179. doi: 10.1002/aic.13844

Guo, P., Huang, G. H., & Li, Y. P. (2010). Inexact Fuzzy-Stochastic Programming for Water Resources Management Under Multiple Uncertainties. *Environmental Modeling & Assessment*, 15(2), 111-124. doi: 10.1007/s10666-009-9194-6

Gupta, V., & Grossmann, I. E. (2011). Solution strategies for multistage stochastic programming with endogenous uncertainties. *Computers & Chemical Engineering*, 35(11), 2235-2247. doi: 10.1016/j.compchemeng.2010.11.013

Hall, R.W.(Ed.). (2003) Handbook of Transportation Science, Second Edition. Secaucus, NJ, USA: Kluwer Academic Publishers. p 2-39.

Han, J. H., & Lee, I. B. (2011). Two-Stage Stochastic Programming Model for Planning CO_2 Utilization and Disposal Infrastructure Considering the Uncertainty in the CO_2 Emission. *Industrial & Engineering Chemistry Research*, 50(23), 13435-13443. doi: 10.1021/ie200362y

Han, J. H., Ryu, J. H., & Lee, I. B. (2012). Developing a Two-Stage Stochastic

Programming Model for CO2 Disposal Planning under Uncertainty. *Industrial & Engineering Chemistry Research*, 51(8), 3368-3380. doi: 10.1021/ie201148x

Infanger, G. (1994) Planning under uncertainty: Solving large scale stochastic linear programs. Danvers, MA: Boyd and Fraser Publishing Co.

Kall, P., Wallace, S. W. (1994) Stochastic programming. New York, NY: Wiley.

Keller, B., & Bayraksan, G. N. (2010). Scheduling jobs sharing multiple resources under uncertainty: A stochastic programming approach. *Iie Transactions*, 42(1), 16-30. doi: 10.1080/07408170902942683

Kiya, F., & Davoudpour, H. (2012). Stochastic programming approach to re-designing a warehouse network under uncertainty. *Transportation Research Part E-Logistics and Transportation Review*, 48(5), 919-936. doi: 10.1016/j.tre.2012.04.005

Li, G. C., Huang, G. H., Lin, Q. G., Cai, Y. P., Chen, Y. M., & Zhang, X. D. (2012). Development of an interval multi-stage stochastic programming model for regional energy systems planning and GHG emission control under uncertainty. *International Journal of Energy Research*, 36(12), 1161-1174. doi: 10.1002/er.1867

Li, M. W., Li, Y. P., & Huang, G. H. (2011). An interval-fuzzy two-stage stochastic programming model for planning carbon dioxide trading under uncertainty. *Energy*, 36(9), 5677-5689. doi: 10.1016/j.energy.2011.06.058

Li, M., Guo, P., Fang, S. Q., & Zhang, L. D. (2013). An inexact fuzzy parameter two-stage stochastic programming model for irrigation water allocation under uncertainty. *Stochastic Environmental Research and Risk Assessment*, 27(6), 1441-1452. doi: 10.1007/s00477-012-0681-y

Li, Y. F., Li, Y. P., Huang, G. H., & Chen, X. (2010). Energy and environmental systems planning under uncertainty-An inexact fuzzy-stochastic programming approach. *Applied Energy*, 87(10), 3189-3211. doi: 10.1016/j.apenergy.2010.02.030

Li, Y. P., & Huang, G. H. (2010). Inexact joint-probabilistic stochastic programming for water resources management under uncertainty. *Engineering Optimization*, 42(11), 1023-1037. doi: 10.1080/03052151003622539

Liu, P., Pistikopoulos, E. N., & Li, Z. (2010). Decomposition Based Stochastic Programming Approach for Polygeneration Energy Systems Design under Uncertainty. *Industrial & Engineering Chemistry Research*, 49(7), 3295-3305. doi: 10.1021/ie901490g

Ma, Z., Kwon, R. H., & Lee, C. G. (2010). A stochastic programming winner determination model for truckload procurement under shipment uncertainty. *Transportation Research Part E-Logistics and Transportation Review*, 46(1), 49-60. doi: 10.1016/j.tre.2009.02.002

Mohammadi, S., Mozafari, B., & Soleymani, S. (2013). A stochastic programming approach for optimal microgrid economic operation under uncertainty using 2m+1 point estimate method. *Journal of Renewable and Sustainable Energy*, 5(3). doi: 10.1063/1.4808039

Mousavi, S. M., Vandani, B., Tavakkoli-Moghaddam, R., & Hashemi, H. (2014). Location of cross-docking centers and vehicle routing scheduling under uncertainty: A fuzzy possibilistic-stochastic programming model. *Applied Mathematical Modelling*, 38(7-8), 2249-2264. doi: 10.1016/j.apm.2013.10.029

Nocedal, J. (1999) Numerical Optimization.Secaucus, NJ, USA: Springer, p 23.

Nocedal, J., Waechter, A., Waltz, R.A. (2005) Adaptive barrier strategies for nonlinear interior methods. Technical Report RC 23563, IBM T.J. Watson Research Center, Yorktown Heights, USA, March 2005.

Palmer, A. (2004) The environmental implications of grocery home delivery, ELA doctorate workshop, Centre for Logistics and Supply Chain Management, Cranfield University

Pardalos, P.M. (Ed.). (2002) Combinatorial and Global Optimization. River Edge, NJ, USA: World Scientific. p 10pq.

Pardalos, P.M. (Ed.). (2004) Supply Chain and Finance. Singapore: World Scientific Publishing Company, Incorporated. p v.

Prékopa, A. (1995) Stochastic programming. Dordrecht, The Netherlands: Kluwer Academic Publishers.

Pronello, C., André, M. (2000) Pollutant emissions estimation in road transport models. INRETS-LTE report, vol. 2007

Sahinidis, N.V. (2004) Optimization under uncertainty: state-of-the-art and opportunities, *Computers & Chemical Engineering*, 28(6-7): 971-983. doi: 10.1016/j.compchemeng.2003.09.017

Sarker, R. (Ed.). (2002) Evolutionary Optimization.Secaucus, NJ, USA: Kluwer Academic Publishers, p 4.

Sbihi, A., Eglese, R.W. (2007) Combinatorial optimization and Green logistics. *4OR:*

A Quarterly Journal of Operations Research, 5(2), pp. 99–116.

Simchi-Levi, D. (2004) Managing the Supply Chain. Blacklick, OH, USA: McGraw-Hill Professional Publishing. p 2-3.

Singh, A. R., Jain, R., & Mishra, P. K. (2013). Capacities-based supply chain network design considering demand uncertainty using two-stage stochastic programming. *International Journal of Advanced Manufacturing Technology*, 69(1-4), 555-562. doi: 10.1007/s00170-013-5054-2

Sodhi, M. S., & Tang, C. S. (2011). Determining supply requirement in the sales-and-operations-planning (S&OP) process under demand uncertainty: a stochastic programming formulation and a spreadsheet implementation. *Journal of the Operational Research Society*, 62(3), 526-536. doi: 10.1057/jors.2010.93

Tan, Q. A., Huang, G. H., & Cai, Y. P. (2010). A Superiority-Inferiority-Based Inexact Fuzzy Stochastic Programming Approach for Solid Waste Management Under Uncertainty. *Environmental Modeling & Assessment*, 15(5), 381-396. doi: 10.1007/s10666-009-9214-6

Tintner, G. (1955) Stochastic linear programming with applications to agricultural economics. In H. A. Antosiewicz (Ed.), Proceedings of the Second Symposium in Linear Programming (pp. 197–228), National Bureau of Standards, Washington, DC.

Tits, A. L., Wächter, A., Bakhtiari, S., Urban, T. J., Lawrence, C. T. (2003) A primal-dual interior-point method for nonlinear programming with strong global and local convergence properties. *SIAM Journal on Optimization*, 14 (1), 173–199

Waechter A., Biegler, L.T. (2005) Line search filter methods for nonlinear programming: Local convergence. *SIAM Journal on Optimization*, 16(1):32-48

Waechter A., Biegler, L.T. (2006) On the implementation of a primal-dual interior point filter line search algorithm for large-scale nonlinear programming. *Mathematical Programming*, 106(1):25-57

Waechter, A. (2002) An Interior Point Algorithm for Large-Scale Nonlinear Optimization with Applications in Process Engineering. Ph.D. thesis, Carnegie Mellon University, Pittsburgh, PA, USA, January 2002.

Waechter, A., Biegler, L.T. (2005) Line search filter methods for nonlinear programming: Motivation and global convergence. *SIAM Journal on Optimization*, 16(1):1-31

Xie, Y. L., Li, Y. P., Huang, G. H., & Li, Y. F. (2010). An interval fixed-mix stochastic programming method for greenhouse gas mitigation in energy systems under uncertainty. *Energy, 35*(12), 4627-4644. doi: 10.1016/j.energy.2010.09.045

Xu, J. P., & Wu, D. D. (2010). Sustainable development and planning of coal industry under uncertainty using system dynamic and stochastic programming. *International Journal of Environment and Pollution, 42*(4), 371-387.

Yamashita, H. (1998) A globally convergent primal-dual interior-point method for constrained optimization. *Optimization Methods and Software, 10*, 443–469

Zanjani, M. K., Nourelfath, M., & Ait-Kadi, D. (2010). A multi-stage stochastic programming approach for production planning with uncertainty in the quality of raw materials and demand. *International Journal of Production Research, 48*(16), 4701-4723. doi: 10.1080/00207540903055727

Zimmermann, H.-J. (1991) Fuzzy set theory and its application (2nd ed.). Boston: Kluwer Academic Publishers.

Chapter 6. Transportation Network Design by Heuristic Methods

Abstract. This study suggests and employs an optimization method; the Genetic Algorithm, on the problems of the selected transportation systems, by taking into consideration the substantial environmental effects of these transportation systems. Several environmental impacts and logistic system costs can be reduced by applying optimal solutions and putting into application the heuristic methods. Computational solutions disclose that applied methods are effective, flexible, and easy to enforce in solving problems. This thence reduce ecological effects on particular logistical actions resulting from inefficient network designs. This study establishes that heuristics methods have the ability to provide optimal solutions for the problems associated with logistic systems with the aim of reducing environmental impacts on particular logistical actions with ineffective network designs.

Keywords: genetic algorithm, transportation costs, environmental effects

1. Introduction

Transport sector is one of the few sectors, which have ever-increasing greenhouse gas (GHG) emissions. According to IPCC reports, the transport sector accounts for 13.1% of GHG emissions, 23% of global carbon dioxide (CO_2) emissions, and 26% of total world energy use. The transport sector is one of the constantly growing sectors.

The growth of the transport sector heavily depends on fossil fuels and imports. Thus, with the growing of transport sector, the demand on oil is expected to increase in the years to come. Consumption data of fuel is one of the reliable elements used to calculate emission levels.

On the other hand, from the global scale based on the estimates of the years between 2000 and 2050, the International Energy Agency (IEA) warns that there is a possibility of 50% increase in CO_2 emissions originating from transport sector.

The next section of this paper covers the general literature. The third section looks at the fundamental combinatorial optimization background as well as details of genetic algorithm (GA) method and fundamental routing problems are exhibited with designated constraints. In the same section, scenario based travelling salesmen (TSP) path problems with randomly assigned locations are used and solved by the heuristic methods and a fundamental reason for choosing the GA method for problems are briefly given. Finally, this paper is summarized in section 4.

2. Literature review

Practices correlated with environmental issues and performances constitute both the internal and external activities (Cousins 2006). For instance, the more products are stored or shipped in a given cubic capacity, the more the associated unit costs. This implies that the environmental impact may be cut down. Packaging in the form of condensed containers, such as pallets, or recyclable containers, will often necessitate returning them to the point of origin for them to be reused.

For logisticians, the problem manifests itself in the form of reverse logistics. Waste packaging requires to be taken back up the supply chain (Rushton 2006). According to Stock (2002), to reduce environmental impacts, some typical measures in logistics activities are, but not limited to:

- miles per gallon/liters per kilometer of fuel used,
- per centum of fleet using less polluting fuels,
- utilization of vehicle load space expressed as a percentage,
- per centum of empty miles or kilometers run by vehicles,
- targets for reducing waste packaging,
- targets for reducing noise levels

Kutz (2003) states that in the design process, the main emphasis lies on defining a network that connects and/or links up selected nodes with a certain quality. It is taken for granted that the accessibility of these nodes is thereby assured. Of adequate importance is the quality of the designed network as well as a corroboration of other sustainability criteria, such as:

- The environment and livability,
- Traffic safety,
- Network accessibility,
- Economic accessibility, and
- The Costs.

Environment and livability effects are emissions of harmful substances to man, fauna, and flora; noise nuisance; and fragmenting the landscape. Traffic emits a number of harmful substances such as CO, CO_2, CxHy, NOx, Pb, and particulate matter. The size of these emissions depends on various factors, such as fuel usage, type of fuel, speed, driving cycle, and gradient. One has to employ a highly complex method of calculation in order to take account of all these factors (Kutz 2003). Critical cognizance and responsiveness to the situation is crucial.

However, environmental effects tend to be deemed secondary objectives besides the primary criteria of the cost, quality, and delivery at supply chains and logistics units of some organizations. Pressure to consider an environmental prospect from legislation is not the only business concern (Stock 2002). A "green" image has become an important marketing opportunity. Thus, it is imperative for businesses to revisit their supply chain performance metrics in response to developing external institutional pressure as well as the clients' environmental requirements (Stock 2002).

Grounded on the sort of decision variables, the most noteworthy network design problems can be separated into distinct and continuous models according to Gallo *et al.* (2010) and Beltran *et al.* (2009).

- Continuous variable models were developed and formulated in the papers by Dantzig *et al.* (1979), Abdulaal and Le Blanc (1979), Marcotte (1983), Harker and Friesz (1984), Le Blanc and Boyce (1986), Suwansirikul *et al.* (1987), Friesz *et al.* (1992), Davis (1994), Cho and Lo (1999), Meng *et al.* (2001), Meng and Yang (2002), and Chiou (2005).
- Discrete variable models were developed and formulated in the papers by Billheimer and Gray (1973), Le Blanc (1975), Los (1979), Boyce and Janson (1980), Foulds (1981), Los and Lardinois (1982), Poorzahedy and Turnquist (1982), Chen and Alfa (1991), Herrmann *et al.* (1996), Solanki *et al.* (1998), Cruz *et al.* (1999), Drezner and Wesolowsky (2003), Gao *et al.* (2005), Poorzahedy and Abulghasemi (2005), Poorzahedy and Rouhani (2007) and Ukkusuri *et al.* (2007).

For network design problems, poly-criteria proficiency for urban networks with the usage of genetic algorithm proposed by Pattnaik *et al.* (1998), Dhingra *et al.* (2000), Ngamchai and Lovell (2003), Cantarella and Vitetta (2006), Russo and Vitetta (2006). Cantarella (2006) also suggests other methods, such as Simulated Annealing (SA), Tabu Search (TS), Path Relinking, Climbing, and Genetic Algorithms. On account of the non-convexity of the transportation network design problem as reported by Newell (1979), the most beneficial and effective solution methods are based on heuristic procedures. For the other most noteworthy works about network design, it is possible to mention Baaj and Mahmassani (1992, 1995), Ceder and Israeli (1993) and Carrese and Gori (2002).

Table 1. Some highlighted literature about transportation network design

A bilevel flow model for hazmat transportation network design (Bianco *et al.* 2009).
Model for Microcirculation Transportation Network Design (Chen and Shi 2012).
A review of urban transportation network design problems (Farahani *et al.* 2013).
Multimodal Freight Transportation Network Design Problem for Reduction of Greenhouse Gas Emissions (Kim *et al.* 2013).
Global optimization method for mixed transportation network design problem: A mixed-integer linear programming approach (Luathep *et al.* 2011).
System-optimal stochastic transportation network design (Patil and Ukkusuri 2007).
Mixed Transportation Network Design under a Sustainable Development Perspective (Qin *et al.* 2013).
Approximation Techniques for Transportation Network Design Problem under Demand Uncertainty (Sharma *et al.* 2011).
Pareto Optimal Multiobjective Optimization for Robust Transportation Network Design Problem (Sharma *et al.* 2009).
Robust transportation network design under demand uncertainty (Ukkusuri *et al.* 2007).
Multi-period transportation network design under demand uncertainty (Ukkusuri and Patil 2009).
Freight Transportation Network Design Problem for Maximizing Throughput under Uncertainty (Unnikrishnan and Waller 2009).
Transportation Network Design considering Morning and Evening Peak-Hour Demands (Wang *et al.* 2014).
Hazmats Transportation Network Design Model with Emergency Response under Complex Fuzzy Environment (Xu *et al.* 2013).
Bilevel programming model and solution method for mixed transportation network design problem (Zhang and Gao 2009).

3. Data and methods

A combinatorial optimization problem can be written as

$$x^* = \min_{x \in D \subseteq X} f(x), \quad (1)$$

where the objective is to find $x^* \in D \subseteq X$. X is bounded by a finite space and $D \subseteq X$ is the subspace of feasible solutions. $f : X \to R^1$ is the objective function.

To obtain solutions for the types of problems, as shown in (1), there are several approaches.

Routing problem I: The problem is to ascertain the operation plan satisfying the demand at various zones at minimum cost.

Objective function is

$$\text{Minimize } f_{obj} = \sum_{i=1}^{G} \sum_{j=1}^{Z} \sum_{k=1}^{F} C_{ijk} x_{ijk}$$

Subject to

$$\sum_{i=1}^{G}\sum_{k=1}^{F} L_k x_{ijk} \geq D_j \qquad \forall j$$

$$\sum_{j=1}^{Z}\sum_{k=1}^{F} L_k x_{ijk} \leq S_j \qquad \forall i$$

$$\sum_{j=1}^{Z} L_k x_{ijk} \leq U_{ki} \qquad \forall k,i$$

$$x_{ijk} \geq 0 \qquad \forall i,j,k$$

where parameters are

G = Number of source locations (index i)

Z = Number of receiving nodes for containers (index j)

F = Number of trailers available (index k)

L_k = Load capacity of trailer k

S_i = Quantity of available containers for transportation from location i

D_j = Quantity of containers required by zone j

C_{ijk} = Unit cost of transporting from location i to zone j by trailer k

U_{ik} = Maximum allowable containers that can be transported from location i by trailer k in a given period

and variables

x_{ijk} = the number of trips required by trailer k from location i to zone j

Routing Problem II: A generic model that practitioners encounter in many planning and decision processes. For instance, the delivery and collection of containers/cargos, etc.

Objective function is

$$\text{Minimize Z} = \sum_{k=1}^{K}\sum_{(i,j)\in A} C_{ij} x_{kij}$$

Subject to

$$\sum_{i=1}^{n} y_{ij} = 1, \qquad j = 2, 3, ..., n$$

$$\sum_{j=1}^{n} y_{ij} = 1, \qquad i = 2, 3, ..., n$$

$$\sum_{j=1}^{n} y_{1j} = K$$

$$\sum_{j=1}^{n} y_{i1} = K$$

$$\sum_{i=1}^{n} \sum_{j=2}^{n} D_j x_{kij} \leq U, \qquad k = 1, 2, ..., K$$

$$\sum_{k=1}^{K} x_{kij} = y_{ij} \qquad \forall i, j$$

$$\sum_{(i,j) \in S \times S} y_{ij} \leq |S| - 1, \qquad \text{for all subsets } S \text{ of } \{2, 3, ..., n\}$$

$$x_{kij} = 0 \ or \ 1 \quad \forall (i, j) \in A \ and \ \forall k$$

$$y_{ij} = 0 \ or \ 1 \quad \forall (i, j) \in A$$

- A fleet of M capacitated vehicles located in a depot (i=1)
- A set of target zones (of size N-1), each having a demand D_j (j=2,...,N)
- A cost C_{ij} of traveling from location i to location j
- The problem is to find a set of routes for delivering / picking up goods to/from the target zones at minimum possible cost.

The vehicle fleet is homogeneous and that each vehicle has a capacity of U units.

and variables:

x_{kij} = 1 if the vehicle k travels on the arc i to j, 0 otherwise

y_{ij} = 1 if any vehicle travels on the arc (i,j), 0 otherwise

3.1 Various scenarios and network configurations – Optimal solutions by heuristics methods

In this section, several potential scenarios are considered while designing a transport network. On arbitrarily distributed locations, how heuristics methods can be employed to meet particular transport network design needs are considered. Developed scenarios are based on a theoretical assumption that city locations are all accessible from every direction without geographical limitation, other constraints, and route affecting factors. Therefore, founded on the supposition, five scenarios are introduced. Scenarios 1 to 4 are formulated using genetic algorithm (See Figure 1). GA network designs are generated and simulated by utilizing Matlab® codes.

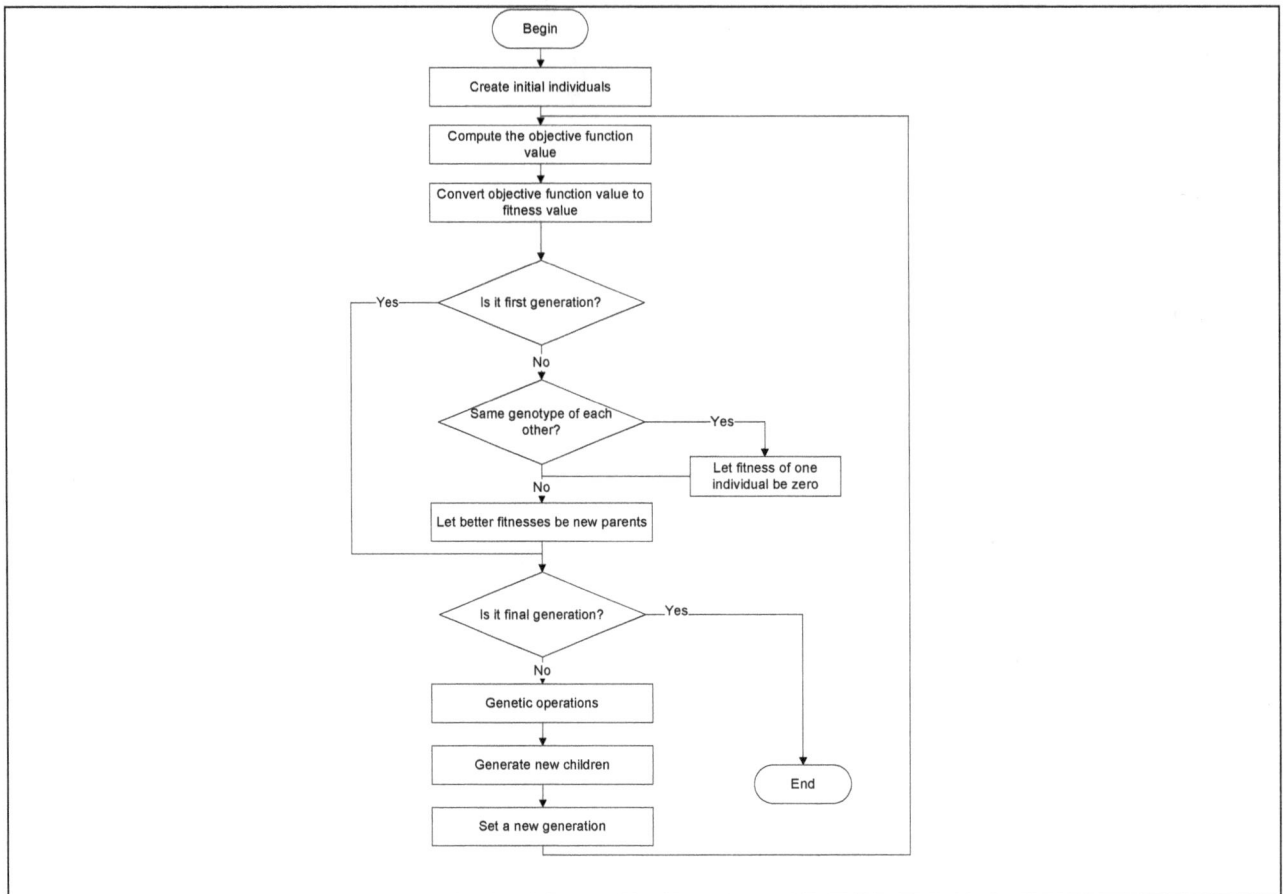

Figure 1. Flowchart of the genetic algorithm with pseudo-code for genetic algorithm

Pseudo-code for genetic algorithm:

1. begin **Genetic_Algorithm**()
2. g:=0 //set counter
3. **Initialize population** P(g) ;
4. **Evaluate population** P(g) ; //compute fitness values
5. while not done do
6.   g:=g+1
7.   Select P(g) from P(g-1)
8.   **Crossover** P(g) ;
9.   **Mutate** P(g) ;
10.   **Evaluate** P(g) ;
11. end while
12. end **Genetic_Algorithm**()

Scenario 1

In this scenario, closed loop transportation networks for 60 city locations with three trucks are taken into consideration. An arbitrarily distributed city locations on an XY plane and fixed number of vehicles, e.g. 3 trucks, are given a task to visit all of their assigned locations as such that the total distance travelled by these vehicles will be minimum.

To downplay the total distance, GA method is applied to find out the assigned locations for each vehicle and to keep the travelled distance for these vehicles at minimum. Based on the algorithm's outcome, minimum total distance for three vehicles can only be achieved by the network configuration shown at Figure 6a. The solution network design (Figure 2a) is based on the randomly selected and predetermined city locations (see top left of the Figure 2b). The best solution history (bottom right of Figure 2b) is the gradually reduction of the total distance at each iteration.

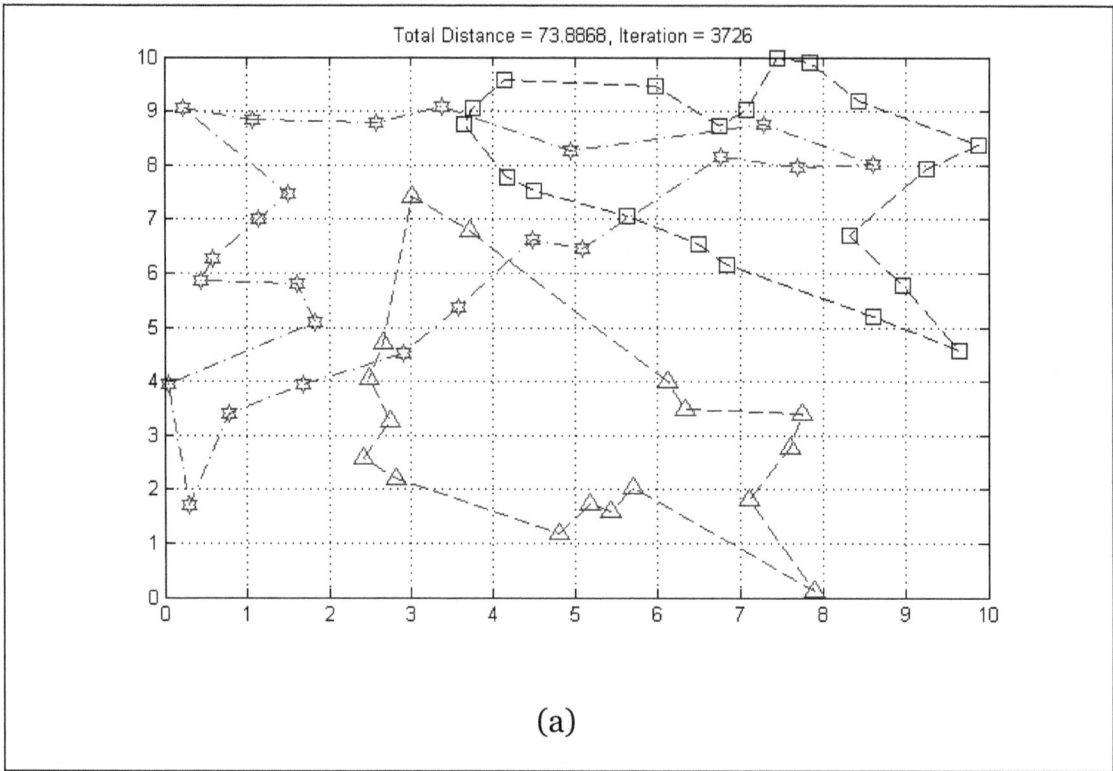

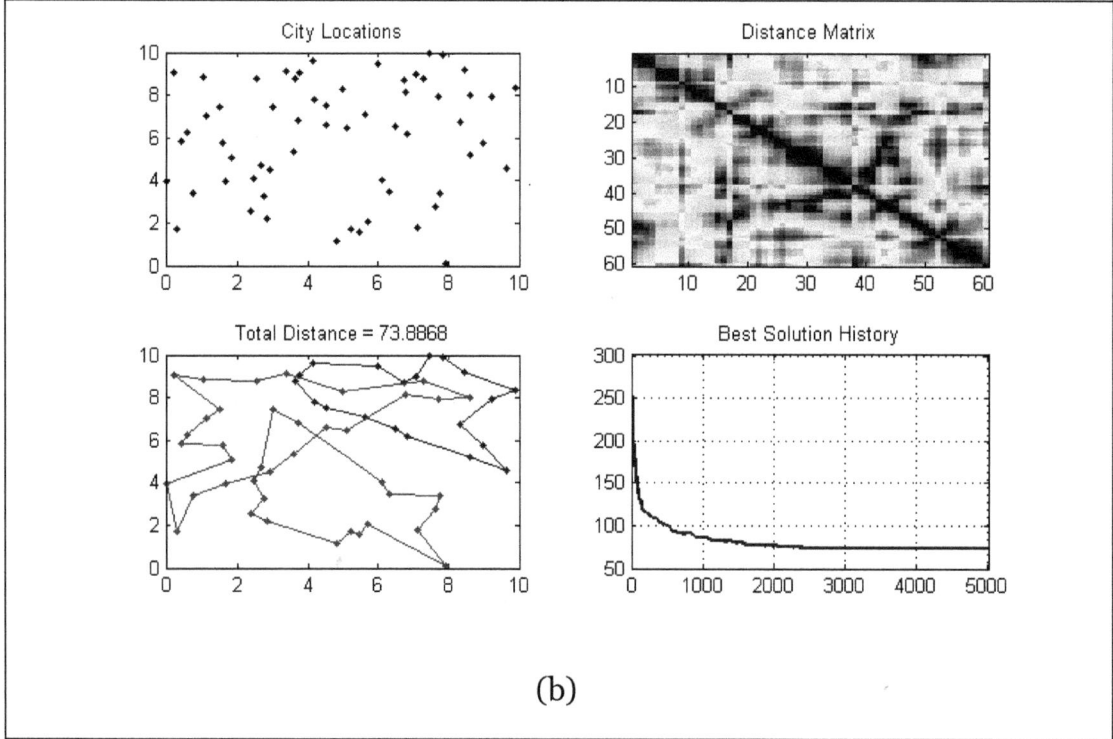

Figure 2. Three trucks travel at minimum possible total distance. All locations visited.

Scenario 2

In this scenario, a transportation network for 60 city locations with three trucks are considered with a fixed start node. On a randomly distributed city locations, vehicles are assigned to specific city locations by the GA algorithm to minimize the total travelling distance of the vehicles.

A closed loop network for each vehicle has been generated. The total distance can only be minimized by the configuration shown at Figure 3b.At each iteration of the algorithm, the total distance reduces to a global minimum as shown at the best solution history (see bottom right of the Figure 3b).

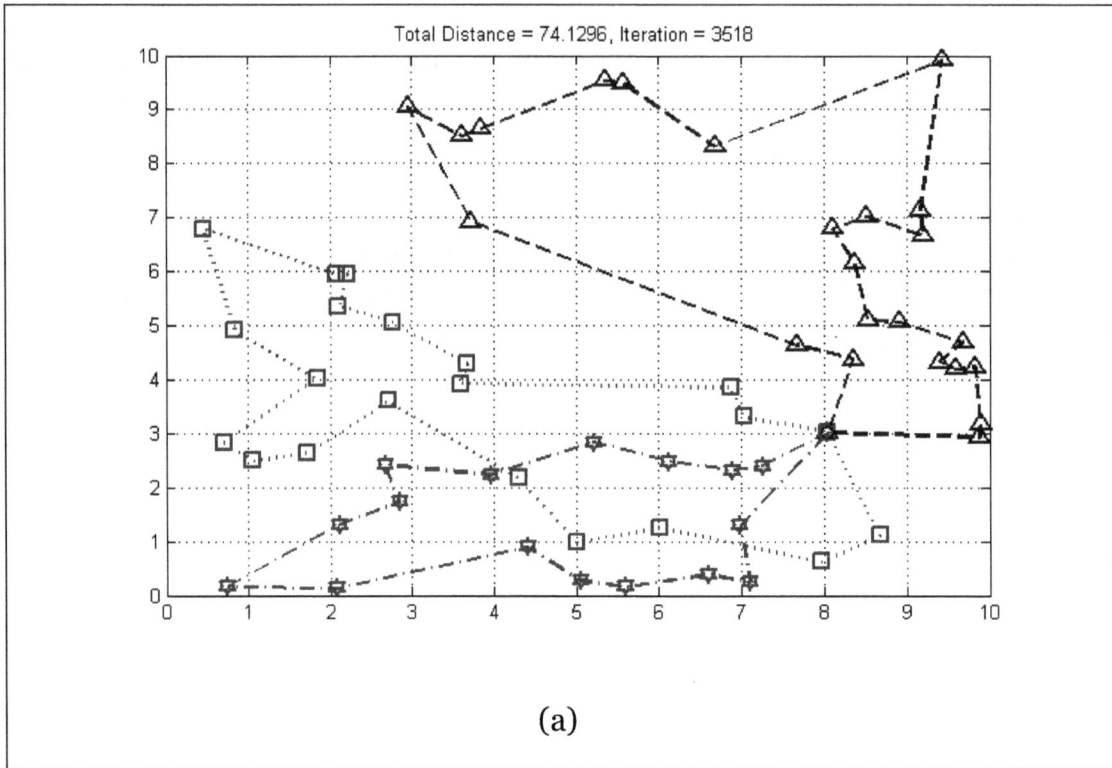

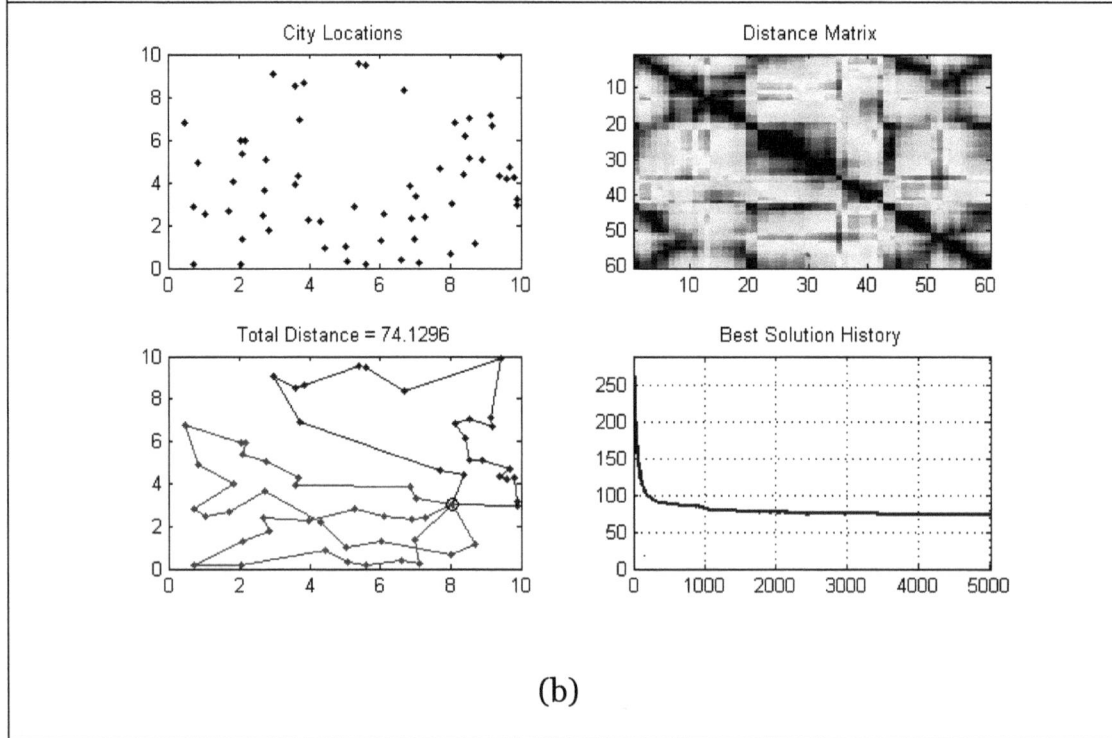

Figure 3. Three trucks, starting from a fixed location, travel at minimum possible total distance. All locations visited.

Scenario 3

A scenario of a fixed start point with open ends for three trucks of 60 randomly distributed city locations is considered. All locations are covered on the transportation network. Minimum total distance can only be achieved by the network configuration shown at the Figure 4a.

"Fixed start-open ends" configuration shown at the Figure 4a provides the minimum possible total distance for all the vehicles. Other details of the configuration can be seen at the Figure 4b.

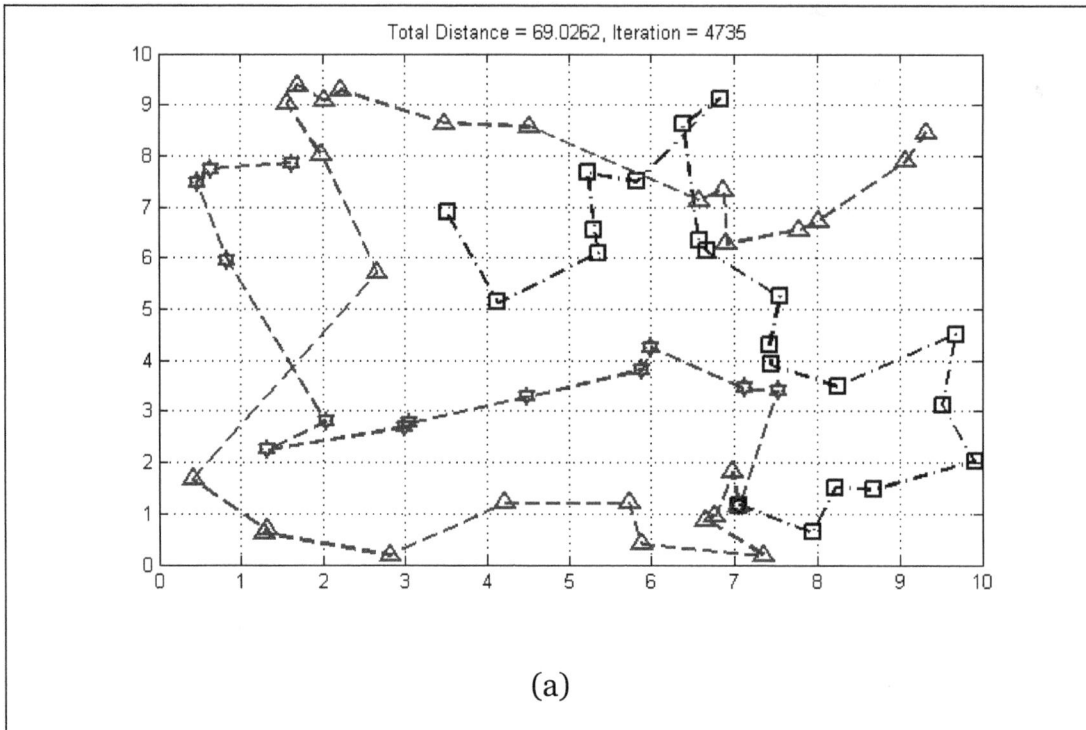

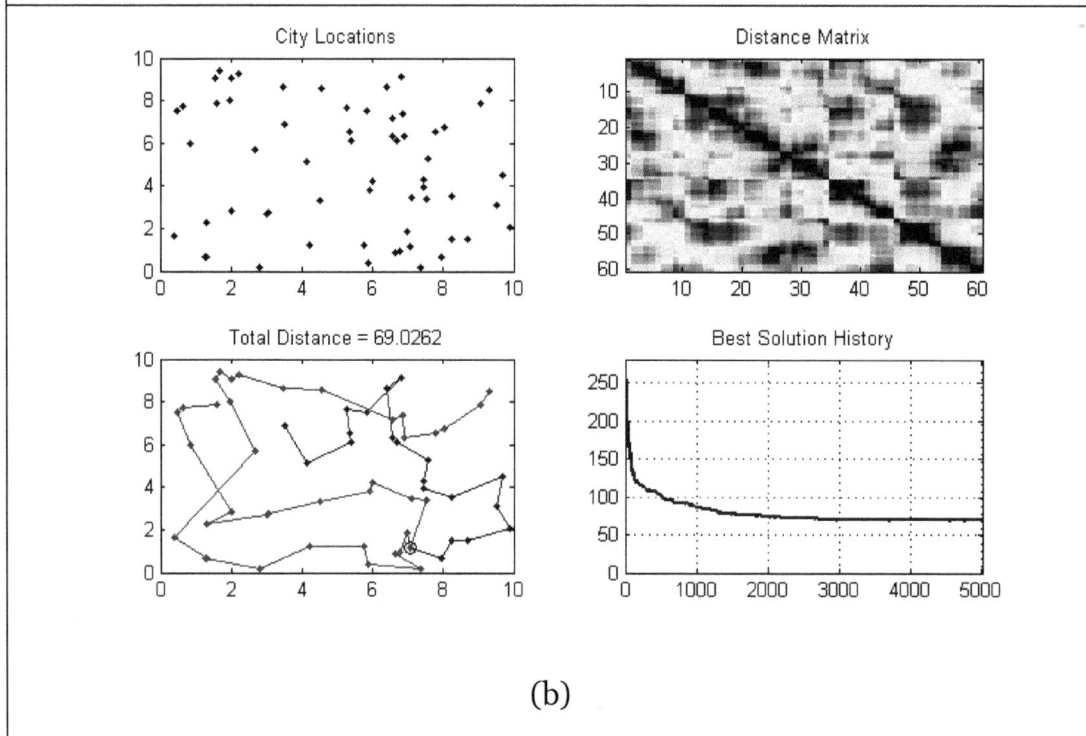

Figure 4. Three trucks, starting from a fixed location, travel at minimum possible total distance. All locations visited.

Scenario 4

This scenario considers the basic minimum total distance for *fixed start* and *fixed end* location for a travelling vehicle. Minimum total distance by visiting all the locations can only be achieved by following the route shown at the Figure 5a.

All locations are visited with a minimum total distance. Other details of the configuration can be seen at the Figure 5b.

In addition, based on all the above assumptions, comparing the algorithms for network design performance from different theoretical and empirical categories is a complicated task. Owing to the fact that there is not a specific empirical baseline, that enables unbiased comparison among algorithms.

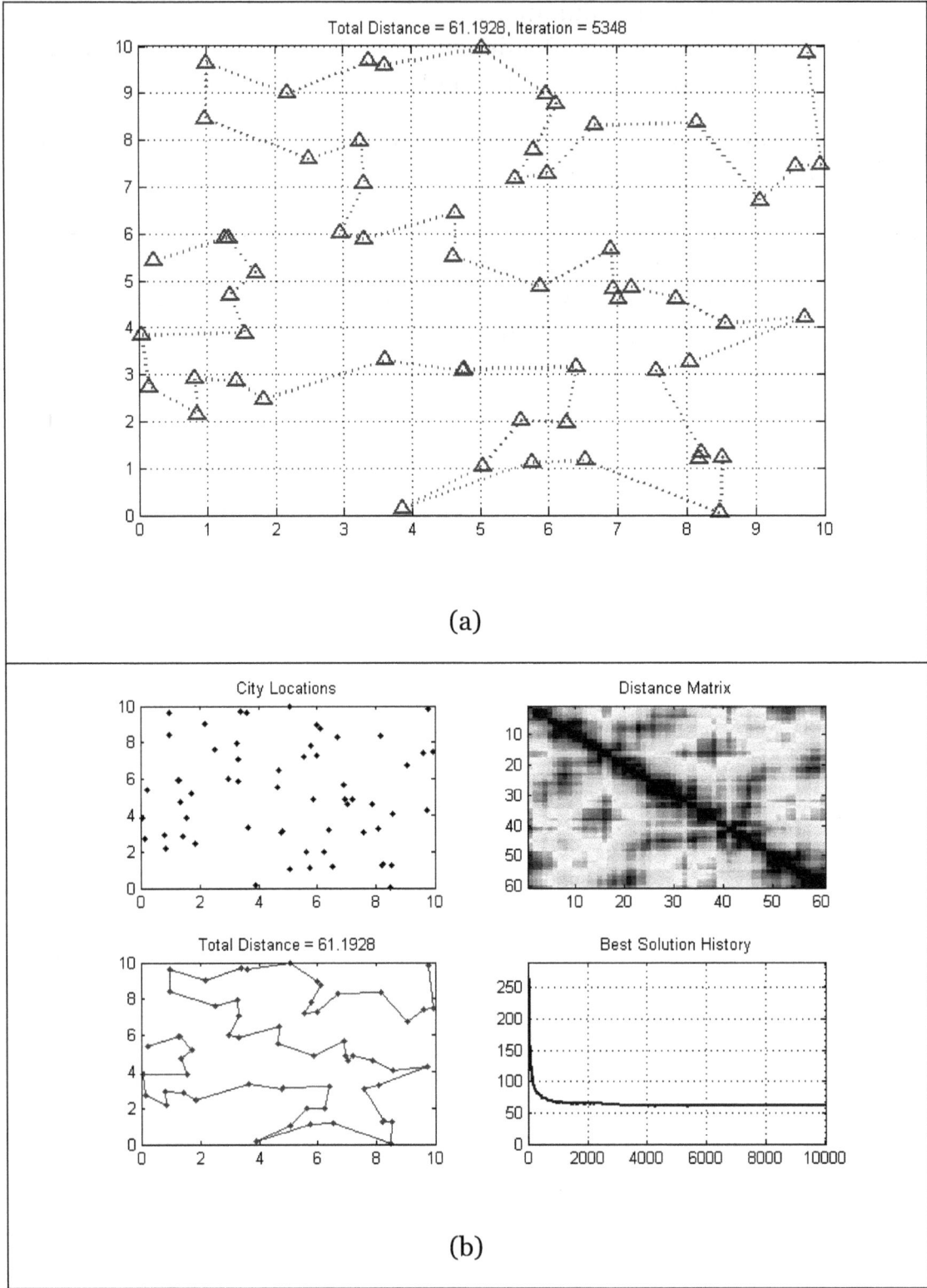

Figure 5. Minimum possible total distance by covering all locations on the route. 1 vehicle with a fixed start and fixed end locations.

4. Research findings and discussions

Additional meta-heuristic algorithms such as Simulated Annealing (SA), Tabu Search (TS), Ant Colony Optimization (ACO), Particle Swarm Optimization (PSO), Memetic Algorithms (MA), etc. are rather mutual in solving several forms of problems. On the other hand, there are important differences among them. These differences originate from theoretical and empirical grounds of algorithms (Aarts and Korst 1989; Goldberg 1989; Dorigo *et al.* 1999; Ehrgott 2002).

Measuring and quantifying environmental pollutants based on the design of network has not been stressed in literature. It is therefore a difficult and even practically inconceivable task to achieve. In a report by Pronello and Andrè (2000), it is indicated that pollution induced by a set of vehicle routes can be treacherous. However, it is unmistakable that shortening the total distance and thence utilizing optimal routes by heuristics algorithms will provide environmental benefits due to the reduction in fuel consumption and the consequent pollutants (Sbihi and Eglese 2007).

5. Conclusion

While there is no single efficacious solution that conforms to all transport emission reduction problems, the optimization method presented on this paper is satisfactorily proposed as route optimization methods that are theoretically applied. By applying optimal solutions and putting into practice the heuristics algorithm method such as Genetic Algorithm, numerous environmental impacts as well as an entire logistics system cost can be cut down. Computational results disclose that the Applied Method is efficient, versatile, and easy to use in solving problems and reducing ecological effects of particular logistical actions with inefficient network designs.

Generally, based on the number and complexity of transportation networks, obtaining optimal solutions with heuristic methods is a non-deterministic polynomial-time (NP) hard problem and computational time exponentially increases contingent to the number of resources involved in the problem. It is stated in particular with detail on this paper that heuristics algorithm's approach gives stable solutions by finding out optimal values. By utilizing and running the proposed high performing algorithms for the problems of transportation network designs, it is apparent that there will be significant reduction in emissions along with the total operational costs.

Lastly, it has been demonstrated that heuristics methods are capable of providing optimal solutions to the problems of logistics systems with the concerns of reducing environmental influences of particular logistical actions with inefficient network designs. As green issues in logistical and supply-chain systems have been receiving a growing concentration from the last decade, the proposed solutions by using heuristic methods to the environmental issues will find encouraging areas on network design problems.

6. References

Aarts, E., Korst, J., (1989) Simulated Annealing and Boltzmann Machines. Wiley. 284 p.

Abdulaal, M., Le Blanc, L., (1979) Continuous equilibrium network design models, *Transportation Research Part B*, 13 (1), 19–32.

Baaj, H., Mahmassani, H.S., (1992) TRUST: a lisp program for the analysis of transit route configurations, *Transportation Research Record*, 1283, 125–135.

Baaj, H., Mahmassani, H.S., (1995) Hybrid route generation heuristic algorithm for the design of transit networks, *Transportation Research Part C*, 3 (1), 31–50.

Beltran, B., Carrese, S., Cipriani, E., Petrelli, M. (2009) Transit network design with allocation of green vehicles: A genetic algorithm approach, *Transportation Research Part C: Emerging Technologies*, 17 (5), 475-483.

Bianco, L., Caramia, M., & Giordani, S. (2009). A bilevel flow model for hazmat transportation network design. *Transportation Research Part C-Emerging Technologies*, 17(2), 175-196. doi: 10.1016/j.trc.2008.10.001

Billheimer, J.W., Gray, P., (1973) Network design with fixed and variable cost elements, *Transportation Science*, 7 (1), 49–74.

Boyce, D.E., Janson, B.N., (1980) A discrete transportation network design problem with combined trip distribution and assignment, *Transportation Research Part B*, 14 (1-2), 147–154.

Cantarella, G.E., Pavone, G., Vitetta, A., (2006) Heuristics for urban road network design: Lane layout and signal settings, *European Journal of Operational Research*, 175 (3), 1682–1695.

Cantarella, G.E., Vitetta, A., (2006) The multi-criteria road network design problem in an urban area, *Transportation*, 33 (6), 567–588.

Carrese, S., Gori, S., (2002) An urban bus network design procedure. In: Patriksson, M., Labbè, M., (Eds.), Transportation Planning: State of the Art, 177–196.

Ceder, A., Israeli, Y., (1993) Design and evaluation of transit routes in urban networks. In: Proceedings of the 3rd International Conference on Competition and Ownership in Surface Passenger Transport, Ontario, Canada.

Chen, M. and Alfa, A.S., (1991) A network design algorithm using a stochastic incremental traffic assignment approach, *Transportation Science*, 25 (3), 215–224.

Chen, Q., & Shi, F. (2012). Model for Microcirculation Transportation Network Design. *Mathematical Problems in Engineering*. doi: 10.1155/2012/379867

Chiou, S.W., (2005) Bilevel programming for the continuous transport network design problem, *Transportation Research Part B*, 39 (4), 361–383.

Cho, H.J., Lo, S.C., (1999) Solving bilevel network design problem using linear reaction function without nondegeneracy assumption, *Transportation Research Record*, 1667, 96–106.

Claudio B. Cunha, Silva M.R., (2007) A genetic algorithm for the problem of configuring a hub-and-spoke network for a LTL trucking company in Brazil, *European Journal of Operational Research*, Volume 179, Issue 3, 16, Pages 747-758

Colorni, A., Dorigo, M., Maffioli, F., Maniezzo, V., Righini, G., Trubian, M., (1996) Heuristics from nature for hard combinatorial problems. *International Transactions in Operational Research*, 3 (1), 1–21.

Cousins, Paul D. (Ed.), (2006) Supply Chain Management Theory and Practice : The Emergence of an Academic Discipline? Bradford, GBR: Emerald Group Publishing Limited, p 797.

Cruz, F.R.B., Smith, J.M.G., Mateus, G.R., (1999) Algorithms for a multi-level network optimization problem, *European Journal of Operational Research*, 118 (1), 164–180.

Dantzig, G.B., Harvey, R.P., Lansdowne, Z.F., Robinson, D.W., Maier, S.F., (1979) Formulating and solving the network design problem by decomposition, *Transportation Research Part B*, 13 (1), 5–17.

Davis, G.A., (1994) Exact local solution of the continuous network design problem via stochastic user equilibrium assignment, *Transportation Research Part B*, 28 (1), 61–75.

Dhingra, S.L., Muralidhar S. & Krishna Rao K.V., (2000) Public transport routing and scheduling using genetic algorithms. In: Proceedings Presented at the CASPT 8th International Conference, Berlin, Germany.

Dorigo, M., Di Caro, G., Gambardella, L. M. (1999) Ant algorithms for discrete optimization. *Artificial life*, 5 (2): 137–172.

Drezner, Z., Wesolowsky, G.O., (2003) Network design: Selection and design of links and facility location, *Transportation Research Part A*, 37 (3), 241–256.

Ehrgott, M., Gandibleux, X. (eds.), (2002) Multiple Criteria Optimization: State of the Art Annotated Bibliographic Surveys. Secaucus, NJ, USA: Kluwer Academic Publishers. 1st edition. Springer. 520 p.

Farahani, R. Z., Miandoabchi, E., Szeto, W. Y., & Rashidi, H. (2013). A review of urban transportation network design problems. *European Journal of Operational Research*, 229(2), 281-302. doi: 10.1016/j.ejor.2013.01.001

Foulds, R.L., (1981) A multicommodity flow network design problem, *Transportation Research Part B*, 15 (4), 273–283.

Friesz, T.L., Cho, Hsun-Jung, Mehta, Nihal J., Tobin, Roger L., Anandalingam, G., (1992) A simulated annealing approach to the network design problem with variational inequality constraints, *Transportation Science*, 26 (1), 18–26.

Gallo, M., D'Acierno, L., Montella, B., (2010) A meta-heuristic approach for solving the Urban Network Design Problem, *European Journal of Operational Research*, 201 (1), 144-157

Gao, Z., Wu, J., Sun, H., (2005) Solution algorithm for the bi-level discrete network design problem, *Transportation Research Part B*, 39 (6), 479–495.

Garey and Johnson, (1979) Computer and Intractability,Freeman, San Fransico, CA.

Glover, Fred (Ed.), (2002) Handbook of Metaheuristics. Secaucus, NJ, USA: Kluwer Academic Publishers, p 55.

Goldberg, D., (1989) Genetic Algorithms in Search, Optimization and Machine Learning. Addison Wesley.

Harker, T.P., Friesz, T.L., (1984) Bounding the solution of the continuous equilibrium network design problem. In: Volmuller, J., Hamerslag, R. (Eds.), Proceedings of 9th International Symposium on Transportation and Traffic Theory. VNU Science Press, Utrecht, The Netherlands, 233–252.

Herrmann, J.W., Ioannou, G., Minis, I., Proth, J.M. (1996) A dual ascent approach to the fixed-charge capacitated network design problem, *European Journal of Operational Research*, 95 (3), 476–490.

IEA, (2001) Saving Oil and Reducing CO2 Emissions in Transport. International Energy Agency, OECD, 194 pp.

IEA, (2002a) Transportation Projections in OECD Regions – Detailed report.

International Energy Agency, 164 pp.

IEA, (2002b) Bus Systems for the Future: Achieving Sustainable Transport Worldwide. International Energy Agency, 188 pp.

IEA, (2003) Transport Technologies and policies for energy security and CO2 Reductions. Energy technology policy and collaboration papers, International Energy Agency, ETPC paper no 02/2003.

IEA, (2004a) World Energy Outlook 2004. International Energy Agency, 570 pp.

IEA, (2004b) Energy Technologies for a Sustainable Future: Transport. International Energy Agency, Technology Brief, 40 pp.

IEA, (2004c) Biofuels for Transport: An International Perspective. International Energy Agency, Paris, 210 pp.

IEA, (2004d) Reducing Oil Consumption in Transport - Combining Three Approaches.

IEA, (2004d) Reducing Oil Consumption in Transport - Combining Three Approaches. IEA/EET working paper by L. Fulton, International Energy Agency, Paris, 24 pp. Jones, M. Tim, 2003. AI Application Programming. Herndon, VA, USA: Charles River Media, p 115.

IEA, (2005) Prospects for Hydrogen and Fuel Cells. International Energy Agency, Paris, 253 pp.

IEA, (2006a) Energy Technology Perspectives 2006; Scenarios & Strategies to 2050. International Energy Agency, Paris, 479 pp. 383

Kim, S., Park, M., & Lee, C. (2013). Multimodal Freight Transportation Network Design Problem for Reduction of Greenhouse Gas Emissions. *Transportation Research Record*, (2340), 74-83. doi: 10.3141/2340-09

Kutz, Myer. (2003) Handbook of Transportation Engineering. New York, NY, USA: McGraw-Hill. p 69-70.

Le Blanc, L.J., (1975) An algorithm for the discrete network design problem, *Transportation Science*, 9 (3), 183–199.

Le Blanc, L.J., Boyce, D.E., (1986) A bilevel programming algorithm for exact solution of the network design problem with user optimal flows, *Transportation Research Part B*, 20, 259–265.

Los, M. and Lardinois, C., (1982) Combinatorial programming, statistical optimization and the optimal transportation network problem, *Transportation Research Part B*, 16 (2), 89–124.

Los, M., (1979) A discrete-convex programming approach to the simultaneous optimization of land use and transportation, *Transportation Research Part B*, 13 (1), 33–48.

Luathep, P., Sumalee, A., Lam, W. H. K., Li, Z. C., & Lo, H. K. (2011). Global optimization method for mixed transportation network design problem: A mixed-integer linear programming approach. *Transportation Research Part B-Methodological*, 45(5), 808-827. doi: 10.1016/j.trb.2011.02.002

Meng, Q, Yang, H., (2002) Benefit distribution and equity in road network design, *Transportation Research Part B*, 36, 19–35.

Meng, Q., Yang, H., Bell, M.G.H., (2001) An equivalent continuously differentiable model and a locally convergent algorithm for the continuous network design problem, *Transportation Research Part B*, 35 (1), 83–105.

Newell, G.F., (1979) Some issue relating to the optimal design of bus lines, *Transportation Science*, 13 (1), 20–35.

Ngamchai, S., Lovell, D.J., (2003) Optimal time transfer in bus transit route network design using a genetic algorithm, *Journal of Transportation Engineering*, 129 (5), 510–521.

Olivier, J.G.J. *et al.*, (2005) Recent trends in global greenhouse gas emissions: regional trends 1970-2000 and spatial distribution of key sources in 2000. *Environmental Science*, 2(2-3), pp. 81-99.

Olivier, J.G.J., *et al.*, (2006) Part III: Greenhouse gas emissions: 1. Shares and trends in greenhouse gas emissions; 2. Sources and Methods; Greenhouse gas emissions for 1990, 1995 and 2000. In CO2 emissions from fuel combustion 1971-2004, 2006 Edition, pp. III.1-III.41. International Energy Agency (IEA), Paris

Palmer, A., (2004) The environmental implications of grocery home delivery, ELA doctorate workshop, Centre for Logistics and Supply Chain Management, Cranfield University

Pardalos, P.M.(Ed.)., (2002) Combinatorial and Global Optimization. River Edge, NJ, USA: World Scientific. p 10pq.

Patil, G. R., & Ukkusuri, S. V. (2007). System-optimal stochastic transportation network design. *Transportation Research Record*, (2029), 80-86. doi: 10.3141/2029-09

Pattnaik, S.B., Mohan, S., Tom, V.M., (1998) Urban bus transit network design using genetic algorithm, *Journal of Transportation Engineering*, 124 (4), 368–375.

Poorzahedy, H., Abulghasemi, F., (2005) Application of ant system to network design problem, *Transportation*, 32 (3), 251–273.

Poorzahedy, H., Rouhani, O.M., (2007) Hybrid meta-heuristic algorithms for solving network design problem, *European Journal of Operational Research*, 182 (2), 578–596.

Poorzahedy, H., Turnquist, M.A., (1982) Approximate algorithms for the discrete network design problem, *Transportation Research Part B*, 16 (1), 45–55.

Pronello C., André M., (2000) Pollutant emissions estimation in road transport models. INRETS-LTE report, vol 2007

Qin, J., Ni, L. L., & Shi, F. (2013). Mixed Transportation Network Design under a Sustainable Development Perspective. *Scientific World Journal*. doi: 10.1155/2013/549735

Reeves, Colin R., (2002) Genetic Algorithms - Principles and Perspectives : A Guide to GA Theory. Secaucus, NJ, USA: Kluwer Academic Publishers, p1.

Rushton, Alan, (2006) Handbook of Logistics and Distribution Management (3rd Edition). London, GBR: Kogan Page, Limited, p 579.

Russo, F., Vitetta, A., (2006) A topological method to choose optimal solutions after solving the multi-criteria urban network design problem, *Transportation*, 33, 347–370.

Sarker R., (2008) Optimization Modelling: a practical approach. CRC Press

Sarker, R. (Ed.), (2002) Evolutionary Optimization. Secaucus, NJ, USA: Kluwer Academic Publishers, p 29.

Sbihi, A., Eglese, R.W., (2007) Combinatorial optimization and Green logistics. *4OR: A Quarterly Journal of Operations Research*, 5 (2), 99–116.

Sergienko, I.V., Hulianytskyi, L.F., Sirenko, S.I., (2009) Classification of applied methods of combinatorial optimization. *Cybernetics and Systems Analysis*, 45 (5), 732-741

Sharma, S., Mathew, T. V., & Ukkusuri, S. V. (2011). Approximation Techniques for Transportation Network Design Problem under Demand Uncertainty. *Journal of Computing in Civil Engineering*, 25(4), 316-329. doi: 10.1061/(asce)cp.1943-5487.0000091

Sharma, S., Ukkusuri, S. V., & Mathew, T. V. (2009). Pareto Optimal Multiobjective Optimization for Robust Transportation Network Design Problem. *Transportation Research Record*, (2090), 95-104. doi: 10.3141/2090-11

Sivanandam S.N. and Deepa S.N., (2008) Introduction to Genetic Algorithms, Terminologies and Operators of GA, Springer, pp 39-81

Solanki, R.S., Gorti, J.K., Southworth, F. (1998) The highway network design problem, *Transportation Research Part B*, 32, 127–140.

Stock, James R.(Ed.), (2002) Qualitative methods and approaches in logistics: part 2. Bradford, GBR: Emerald Group Publishing Limited, p 24.

Suwansirikul, C., Friesz, T.L., Tobin, V. (1987) Equilibrium decomposed optimization: A heuristic for the continuous equilibrium network design problem, *Transportation Science,* 21, pp. 254–263.

Ukkusuri, S. V., & Patil, G. (2009). Multi-period transportation network design under demand uncertainty. *Transportation Research Part B-Methodological*, 43(6), 625-642. doi: 10.1016/j.trb.2009.01.004

Ukkusuri, S. V., Mathew, T. V., & Waller, S. T. (2007). Robust transportation network design under demand uncertainty. *Computer-Aided Civil and Infrastructure Engineering*, 22(1), 6-18. doi: 10.1111/j.1467-8667.2006.00465.x

Ukkusuri, S.V., Mathew, T.V., Waller, S.T., (2007) Robust transportation network design under demand uncertainty, *Computer-Aided Civil and Infrastructure Engineering*, 22 (1), 6–18.

Unnikrishnan, A., & Waller, S. T. (2009). Freight Transportation Network Design Problem for Maximizing Throughput Under Uncertainty. *Transportation Research Record*, (2090), 105-114. doi: 10.3141/2090-12

Wang, H., Xiao, G. Y., Zhang, L. Y., & Ji, Y. B. B. (2014). Transportation Network Design considering Morning and Evening Peak-Hour Demands. *Mathematical Problems in Engineering*. doi: 10.1155/2014/806916

WBCSD, (2002) Mobility 2001: World Mobility at the End of the Twentieth Century, and its Sustainability. World Business Council for Sustainable Development

WBCSD, (2004a) Mobility 2030: Meeting the Challenges to Sustainability. <http://www.wbcsd.ch/> accessed 30/05/07.

WBSCD, (2004b) IEA/SMP Model Documentation and Reference Projection. Fulton, L. and G. Eads, <http://www.wbcsd.org/web/publications/mobility/smp-model-

document.pdf> accessed 30/05/07

Xu, J. P., Gang, J., & Lei, X. (2013). Hazmats Transportation Network Design Model with Emergency Response under Complex Fuzzy Environment. *Mathematical Problems in Engineering*. doi: 10.1155/2013/517372

Yildiz, T., Yercan, F., (2010) The cross-entropy method for combinatorial optimization problems of seaport logistics terminal. *Journal Transport*, 25 (4), 411-422.

Zhang, H. Z., & Gao, Z. Y. (2009). Bilevel programming model and solution method for mixed transportation network design problem. *Journal of Systems Science & Complexity*, 22(3), 446-459. doi: 10.1007/s11424-009-9177-3

Chapter 7. A closed-loop supply chain system: A multi-echelon, multi-period, multi-product system

Abstract. Many business activities take place within a supply chain. Therefore, in a typical supply chain system, numerous business processes interact. On the other hand, supply chain optimization studies are dedicated to some various smaller parts of a whole business system; often the big picture is left unseen. Moreover, many studies focus relatively small theoretical optimization models using various methods. However, there is a huge potential of opportunities; as such, reducing the overall costs and waste, when the supply chain system is viewed as a complete "closed-loop" structure. The term "closed-loop" refers to both integrated systems of a forward and a reverse chain. In this regard, by taking into account of the practical outcomes of both forward and reverse chain system optimization; this study presents a model of a more realistic supply chain system with an optimization source code.

Keywords: closed-loop supply chain, forward chain, reverse chain, logistics, GAMS modeling

1. Introduction

The closed-loop supply chain consists of both a forward and a reverse chain. These both chains are considered as integrated chains in the closed-loop supply chain system. Therefore, the investigation of the opportunities brought by considering the supply chain system as a whole system of elements in a real world scenario requires vast knowledge about the interaction of all the elements within the chain (See Figure 1 and Table 1). More specifically, the optimization of such an integrated system requires inclusion of all the elements and the processes into the mathematical formulation to better view and to evaluate the results.

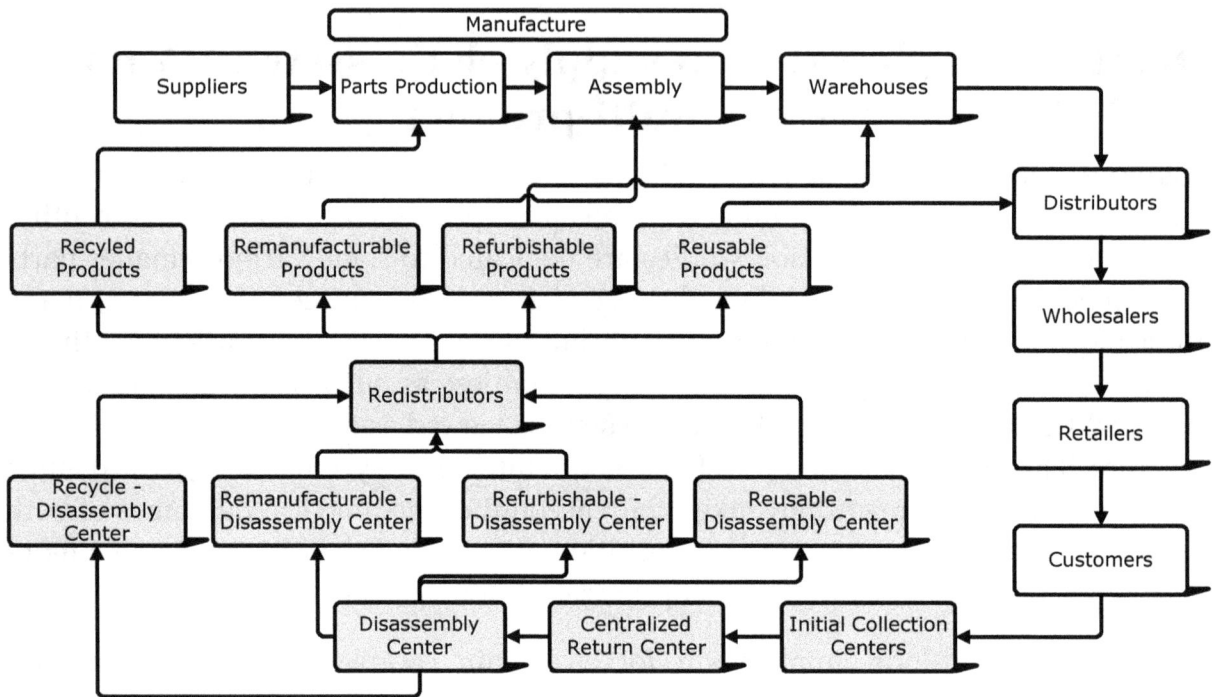

Figure 1. A closed-loop supply chain: the integrated model of forward and reverse chains. Reverse chain processes are shown in gray, while the forward chain processes are shown in white.

Table 1. The activities in a generic CLSC

Process (From)	Process (To)	Product Status	Time	Main Activity
Suppliers	Parts Production	Raw Material	Time	Raw Material acquisition cost
Parts Production	Assembly	Semi Product	Time	Semi Product manufacturing cost
Assembly	Warehouses	Final Product	Time	Final Product storage cost
Warehouses	Distributors	Final Product	Time	Final Product Price at warehouse
Distributors	Wholesalers	Final Product	Time	Wholesaler Final Product acquisition price
Wholesalers	Retailers	Final Product	Time	Retailer Final Product acquisition price
Retailers	Customers	Final Product	Time	End Customer price
Customers	Initial Collection Centers	Used Product	Time	Used Product return ratio
Initial Collection Centers	Centralized Return Center	Used Product	Time	Used Product collection cost
Centralized Return Center	Disassembly Center	Used Product	Time	Used Product transferring cost
Disassembly Center	Redistributors	Recycled Product	Time	Recycled Product processing cost
Disassembly Center	Redistributors	Remanufacturable Product	Time	Remanufacturable Product processing cost
Disassembly Center	Redistributors	Refurbishable Product	Time	Refurbishable Product processing cost
Disassembly Center	Redistributors	Reusable Product	Time	Reusable Product processing cost
Recycled Product	Parts Production	Raw Material	Time	Recycled - acquisition cost
Remanufacturable Product	Assembly	Semi Product	Time	Remanufacturable - acquisition cost
Refurbished Product	Warehouses	Final Product	Time	Refurbished - acquisition cost
Reusable Product	Distributors	Final Product	Time	Reusable - acquisition cost

This study considers the main elements of a Closed-Loop Supply Chain (here after CLSC) system and states an optimization model for reducing the total costs of the system. In this model, strategic and tactical management levels are taken into consideration. Based on the foregoing, the present study develops a closed loop supply chain system to visualize a more realistic scenario. This study adopts a CLSC model and develops a General Algebraic Modeling System (here after GAMS) model of a more realistic supply chain system. The GAMS is a high-level modeling system for mathematical programming and optimization. It consists of a source code compiler and some integrated solvers. GAMS is used for complex, large scale modeling applications, and allows for building small or large maintainable models that can be adapted to changing situations (See gams.com website for more information).

The remainder of this paper is organized as follows. Section 2 reviews the literature on closed-loop supply chains and logistics. Section 3 describes the details of the CLSC model. Section 4 presents initialization of sample data and simulation of the CLSC system. The paper concludes in Section 5.

2. Literature review

Studies of CLSC have been made by many researchers. The most recent research reflects the very significant amount of international literature in the area based on numerous different viewpoints and approaches.

Inventory related researches of CLSC have been made by many researchers. For instance, Mitra (2012) developed an inventory management in a two-echelon closed-loop supply chain with correlated demands and returns. They conclude the research by providing managerial implications and directions for future research. Zhou and Disney (2006) explored bullwhip and inventory variance in a closed loop supply chain. Chung *et al.* (2008) worked on the optimal policy for a closed-loop supply chain inventory system with remanufacturing. A multi-echelon inventory system with remanufacturing capability is proposed. Akcali and Cetinkaya (2011) proposed quantitative models for inventory and production planning in closed-loop supply chains. They provided a comprehensive exposition of the state-of-art in quantitative models for CLSC systems. Yuan *et al.* (2015) examined inventory decision-making models for a closed-loop supply chain system with different decision-making structures. Yuan and Gao (2010) examined inventory decision-making models for a closed-loop supply chain system. Diabat *et al.* (2015) studied a closed-loop location-inventory problem with spare parts consideration. They worked on a closed-loop location-inventory problem. The problem is formulated using a mixed integer nonlinear location allocation model. Yang *et al.* (2010) performed a sequential and global optimization for a closed-loop deteriorating inventory supply chain. Their research considers out of date as a type of deterioration and generalizes the some cases. The analytical results of their study show that a significant increase in the joint profit will result when the integrated policy is adopted. Wu and Ryan (2014) performed the joint optimization of asset and inventory management in a product-service system. A numerical example is provided to illustrate the computational procedures.

Recycling related researches of CLSC have been made by many researchers. For instance, Chen *et al.* (2015) developed an integrated closed-loop supply chain model with location allocation problem and product recycling decisions. In their paper, a comprehensive closed-loop supply chain (CLSC) model is established. Chern *et al.* (2014) focused on solving a multi-objective master planning problem with substitution and a recycling process for a capacitated multi-commodity supply chain network. Their study focuses on solving the multi-objective master planning problem for supply chains. Golroudbary and Zahraee (2015) investigated a system dynamics model for optimizing the recycling and collection of waste material in a closed-loop supply chain. Their research aims at evaluating the system behavior of an electrical manufacturing company as a case of study, by using System Dynamics (SD) simulating of closed-loop supply chain.

Xu *et al.* (2014) examined a closed-loop supply chain problem with retailing and recycling competition. Shi *et al.* (2010) investigated the coordination of production and recycling decisions with stochastic demand and return. Shi and Min (2013) studied the product weight and collection rate in closed-loop supply chains with recycling. Throughout their research, numerous managerial insights and policy implications are provided.

Georgiadis and Besiou (2008) explored sustainability in electrical and electronic equipment closed-loop supply chains. In their research they examine the impact of ecological motivation and technological innovations on the long-term behavior of a closed-loop supply chain with recycling activities. Georgiadis and Besiou (2010) examined environmental and economical sustainability of WEEE closed-loop supply chains with recycling. In their manuscript, they investigate the significance of the factors that comprise the environmental sustainability strategies (environmental legislation and green image) and the operational features of the closed-loop supply chain. Huang *et al.* (2013) discussed the analysis for strategy of closed-loop supply chain with dual recycling channel. Their research investigates optimal strategies of a closed-loop supply chain with dual recycling channel.

Yuan and Zhang (2015) discussed recycler reaction for the government behavior in closed-loop supply chain distribution network, based on the system dynamics. Besiou *et al.* (2012) examined the official recycling by using data from a real world closed-loop supply chain that operates in Greece. Hong *et al.* (2014) performed a socially optimal and fund-balanced advanced recycling fees and subsidies in a competitive forward and reverse supply chain. They demonstrate that their results outperform the current practice by using numerical examples for the laptop computer market in Taiwan.

Kannan *et al.* (2010) performed a genetic algorithm approach for solving a closed loop supply chain model with a case of battery recycling. The objective of their research is to develop a multi echelon, multi period, multi product closed loop supply chain network model for product returns and the decisions are made regarding material procurement, production, distribution, recycling and disposal. Schultmann *et al.* (2003) examined closed-loop supply chains for spent batteries. Their results show that almost complete recycling of spent batteries can be achieved by transforming the current structure into a modified recovery network. Sasikumar and Haq (2011) examined the integration of closed loop distribution supply chain network and 3PRLP selection for the case of battery recycling.

RFID related researches of CLSC have been made by many researchers. For instance, Ondemir *et al.* (2012) examined an optimal end-of-life management in closed-loop supply chains using RFID and sensors. Kim and Glock (2014) assessed the use of RFID in the management of reusable containers in closed-loop supply chains under stochastic container return quantities. Their research studies a closed-loop supply chain that uses containers for transporting products from a supplier to a retailer. Lee and Chan (2009) developed an RFID-based reverse logistics system. The significance of their research is the proposal of RFID-based reverse logistics framework and optimization of locations of collection points which allow economically and ecologically reasonable recycling. Zhou and Piramuthu (2013) observed the remanufacturing with RFID item-level information: Optimization, waste reduction and quality improvement. With RFID in a reverse supply chain, they observe the power shift from waste-driven to market-driven system.

Kumar and Chan (2011) performed a superiority search and optimization algorithm to solve RFID and an environmental factor embedded closed loop logistics model. Kumar and Rahman (2014) explored RFID-enabled process reengineering of closed-loop supply chains in the healthcare industry of Singapore. Their research demonstrates an application of the RFID-enabled process reengineering in sustainable healthcare system design.

Researches of CLSC with an uncertainty component have been made by many researchers. For example, Amaro and Barbosa-Povoa (2009) researched the effect of uncertainty on the optimal closed-loop supply chain planning under different partnerships structure. Hu *et al.* (2015) investigated a dynamic closed-loop vehicle routing problem with uncertainty and incompatible goods. The experimental results provide insights for managers who need to solve the dynamic closed-loop VRP with uncertain pickup and incompatible goods.

Ma and Chen (2014) researched the complexity uncertain analysis about three differences old and new product pricing oligarch retailers' closed-loop supply chain. For closed-loop supply chain research, the conclusions of the numerical simulation in their research not only have realistic guiding significance, but also have theoretical reference value. Amin and Zhang (2013) proposed a three-stage model for closed-loop supply chain configuration under uncertainty.

Zeballos *et al.* (2012) researched the uncertain quality and quantity of returns in closed-loop supply chains. Lieckens and Vandaele (2012) examined multi-level reverse logistics network design under uncertainty. The differential evolution algorithm with an enhanced constraint handling method is proposed as an appropriate heuristic to solve the model close to optimality. Chen *et al.* (2015) performed a supply chain design for unlocking the value of remanufacturing under uncertainty. Efendigil (2014) modeled product returns in a closed-loop supply chain under uncertainties by utilizing a neuro fuzzy approach. Shi *et al.* (2011) developed an optimal production planning for a multi-product closed loop system with uncertain demand and return. Computational results show that the proposed approach is highly promising for solving the problems.

Hasani *et al.* (2012) proposed a robust closed-loop supply chain network design for perishable goods in agile manufacturing under uncertainty. In their study, a general comprehensive model is proposed for strategic closed-loop supply chain network design under interval data uncertainty. Dubey *et al.* (2015) designed a responsive sustainable supply chain network under uncertainty. Chiu and Teng (2013) examined sustainable product and supply chain design decisions under uncertainties.

Qiang *et al.* (2013) investigated the closed-loop supply chain network with competition, distribution channel investment, and uncertainties. Amin and Zhang (2013) investigated a multi-objective facility location model for closed-loop supply chain network under uncertain demand and return. Their model is extended to consider environmental factors. In addition, they investigated the impact of demand and return uncertainties on the network configuration by stochastic programming.

Khatami *et al.* (2015) used Benders' decomposition for concurrent redesign of forward and closed-loop supply chain network with demand and return uncertainties. Fallah-Tafti *et al.* (2014) presented an interactive possibilistic programming approach for a multi-objective closed-loop supply chain network under uncertainty. Dai and Zheng (2015) designed a close-loop supply chain network under uncertainty using hybrid genetic algorithm by developing a fuzzy and chance-constrained programming model. The impacts of uncertainty in disposed rates, demands, and capacities on the overall profit of CLSC network are studied through sensitivity analysis. Wang and Hsu (2012) performed a possibilistic approach to the modeling and resolution of uncertain closed-loop logistics. Wang and Hsu (2010) researched the resolution of an uncertain closed-loop logistics model: An application to fuzzy linear programs with risk analysis. The resulting solution provides useful information on the expected solutions under a confidence level containing a degree of risk.

Vahdani *et al.* (2012) examined a reliable design of a forward/reverse logistics network under uncertainty by using a queuing model. Computational experiments are provided for a number of test problems using a realistic network instance. Vahdani *et al.* (2013) examined a reliable design of a closed loop supply chain network under uncertainty by using an interval fuzzy possibilistic chance-constrained model. Ramezani *et al.* (2013) proposed a robust design for a closed-loop supply chain network under an uncertain environment. Their research presents a robust design for a multi-product, multi-echelon, closed-loop logistic network model in an uncertain environment.

Researches of CLSC with a fuzzy component have been made by many researchers. For example, Demirel *et al.* (2014) proposed a genetic algorithm approach for optimizing a closed-loop supply chain network with crisp and fuzzy objectives. The research proposes a mixed integer programming model for a closed-loop supply chain network with multi-periods and multi-parts under two main policies as secondary market pricing and incremental incentive policies. Mehrbod *et al.* (2012) developed an interactive fuzzy goal programming for a multi-objective closed-loop logistics network. Jindal and Sangwan (2014) proposed a closed loop supply chain network design and optimization using fuzzy mixed integer linear programming model. In their paper, a multi-product, multi-facility capacitated closed-loop supply chain framework is proposed in an uncertain environment including reuse, refurbish, recycle and disposal of parts.

Vahdani *et al.* (2012) proposed a fuzzy possibilistic modeling for closed loop recycling collection networks. Olugu and Wong (2012) examined an expert fuzzy rule-based system for closed-loop supply chain performance assessment in the automotive industry. Mirakhorli (2014) presented a fuzzy multi-objective optimization for closed loop logistics network design in bread-producing industries. Ozceylan and Paksoy (2013) developed a fuzzy multi-objective linear programming approach for optimizing a closed-loop supply chain network.

Vahdani (2015) developed an optimization model for multi-objective closed-loop supply chain network under uncertainty by using a hybrid fuzzy-stochastic programming method. Ozceylan and Paksoy (2014) developed interactive fuzzy programming approaches to the strategic and tactical planning of a closed-loop supply chain under uncertainty. In their paper, a closed-loop supply chain network model consisting of various conflicting decisions of forward and reverse facilities is considered. Zhang and Zhao (2015) constructed a fuzzy robust control for an uncertain switched dual-channel closed-loop supply chain model. Zhang *et al.* (2014) studied a dynamic model and fuzzy robust control of uncertain closed-loop supply chain with time-varying delay in remanufacturing. A dynamic model and a fuzzy robust control strategy of an uncertain closed-loop supply chain system with a time-varying delay in remanufacturing are studied in their article.

Subulan *et al.* (2012) examined a fuzzy mixed integer programming model for medium-term planning in a closed-loop supply chain with remanufacturing option. Subulan *et al.* (2015) designed an environmentally conscious tire closed-loop supply chain network with multiple recovery options using interactive fuzzy goal programming. Phuc *et al.* (2013) presented a study on optimizing the fuzzy closed-loop supply chain for electrical and electronic equipments.

Ramezani *et al.* (2014) developed a closed-loop supply chain network design under a fuzzy environment. In their research, they address the application of fuzzy sets to design a multi-product, multi-period, closed-loop supply chain network. Tavakkoli-Moghaddam *et al.* (2015) developed a hybrid fuzzy approach for the closed-loop supply chain network design under uncertainty. Their research presents a bi-objective model in order to design a network of bi-directional facilities in logistics network under uncertainties.

Pricing or price related researches of CLSC have been made by many researchers. For example, Xie *et al.* (2015) examined dynamic acquisition pricing policy under uncertain remanufactured-product demand. He (2015) modeled acquisition pricing and remanufacturing decisions in a closed-loop supply chain. Their research models a closed-loop supply chain with a manufacturer and its supply channels - recycle channel and reliable supply channel. Hong *et al.* (2015) examined joint advertising, pricing and collection decisions in a closed-loop supply chain. They analytically show that local advertising strongly influences channel members' pricing strategies, used-product collection decisions and profits. Chen and Chang (2013) studied the dynamic pricing for new and remanufactured products in a closed-loop supply chain. Their study deals with managing two differentiated versions of the same product by developing analytical models.

Li *et al.* (2012) examined reverse channel design by considering the impacts of differential pricing and extended producer responsibility. They find that the most effective way to collect used products is through the manufacturer. Atamer *et al.* (2013) investigated the optimal pricing and production decisions in utilizing reusable containers. They investigated the optimal pricing and production decisions in order to maximize the manufacturer's profit. Yoo *et al.* (2015) explored pricing and return policy under various supply contracts in a closed-loop supply chain. The main purpose of the study is to explore how each supply contract affects the retailer's decision on pricing and return policies, which in turn influence the profits of the entire supply chain and of its members. Mahmoudzadeh *et al.* (2013) investigated robust optimal dynamic production/pricing policies in a closed-loop system. A numerical example is defined and sensitivity analysis is performed on both basic parameters and parameters associated with uncertainty to create managerial views.

Jena and Sarmah (2014) studied price competition and co-operation in a duopoly closed-loop supply chain. Their research studies co-operation and competition issues in a closed-loop supply chain. Shi *et al.* (2011) analyzed optimal production and pricing policy for a closed loop system. Through a numerical example, the impacts of the uncertainties of both demand and return on the production plan, selling price, and the acquisition price of used products are analyzed. Wei *et al.* (2015) analyzed pricing and collecting decisions in a closed-loop supply chain with symmetric and asymmetric information. The results obtained under symmetric and asymmetric information conditions and some key model parameters used in the research are analyzed using a numerical approach by which some managerial analysis are given.

Yang *et al.* (2013) studied collaboration for a closed-loop deteriorating inventory supply chain with multi-retailer and price-sensitive demand. Rezapour *et al.* (2015) presented a competitive closed-loop supply chain network design with price-dependent demands. Zhang *et al.* (2015) analyzed retail services and pricing decisions in a closed-loop supply chain with remanufacturing. Their results show that the retail services have a great impact on the manufacturer and the retailer's pricing strategies.

Zhao *et al.* (2013) explored pricing and remanufacturing decisions of a decentralized fuzzy supply chain. The purpose of the research is to explore how the manufacturer and the two retailers make their own decisions about wholesale price, retail prices, and the remanufacturing rates in the expected value model. Wei *et al.* (2012) explored pricing decisions for a closed-loop supply chain in a fuzzy environment. Wei and Zhao (2011) investigated pricing decisions with retail competition in a fuzzy closed-loop supply chain. Some insights into the economic behavior of firms are given, which can serve as the basis for further study in the future. Zou and Ye (2015) examined pricing-decision and coordination contract considering product design and quality of recovery product in a closed-loop supply chain. Fahimnia *et al.* (2013) studied the impact of carbon pricing on a closed-loop supply chain: an Australian case study. Their study is one of the first to evaluate the forward and reverse supply chain influences on the carbon footprint.

Product life cycle related researches of CLSC have been made by many researchers. For example, Chuang *et al.* (2014) studied closed-loop supply chain models for a high-tech product under alternative reverse channel and collection cost structures. In their research they study closed-loop supply chain models for a high-tech product which is featured with a short life-cycle and volatile demand. Morana and Seuring (2007) presented end-of-life returns of long-lived products from end customer - insights from an ideally set up closed-loop supply chain. Das and Posinasetti (2015) addressed environmental concerns in closed loop supply chain design and planning. The research proposes to collect end-of-life and other customer returned products through retailers by motivating retailers and customers with an incentive scheme. Amin and Zhang (2012) proposed a mathematical model for closed-loop network configuration based on product life cycle.

Georgiadis et al. (2006) analyzed the impact of product lifecycle on capacity planning of closed-loop supply chains with remanufacturing. Georgiadis and Athanasiou (2010) studied the impact of two-product joint lifecycles on capacity planning of remanufacturing networks.

Hong and Yeh (2012) proposed a modeling of closed-loop supply chains in the electronics industry with a retailer collection application. Li et al. (2015) researched product whole life-cycle and omni-channels data convergence oriented enterprise networks integration in a sensing environment.

Hu and Bidanda (2009) modeled sustainable product lifecycle decision support systems. They formulate a product lifecycle evolution system based on stochastic dynamic programming. Kumar and Putnam (2008) examined reverse logistics strategies and opportunities across three industry sectors. Cruz-Rivera and Ertel (2009) examined a reverse logistics network design for the collection of End-of-Life vehicles in Mexico. Low et al. (2014) demonstrated Product Structure-Based Integrated Life Cycle Analysis (PSILA), which is a technique for cost modelling and analysis of closed-loop production systems.

Chung et al. (2014) performed a modular design approach to improve product life cycle performance based on the optimization of a closed-loop supply chain. Accorsi et al. (2015) reported the design of closed-loop networks for product life cycle management by considering economic, environmental and geography.

Fashion related researches of CLSC have been made by many researchers. For example, Wang et al. (2014) addressed the channel choice for a remanufacturing fashion supply chain with government subsidy. In their paper, they address the problem of choosing an appropriate channel for the marketing channel structure of remanufactured fashion products. Nie et al. (2013) developed collective recycling responsibility in closed-loop fashion supply chains with a third party. They develop three closed-loop supply chain models where manufacturers can utilize financial or physical support to push a third party to collect the used fashion product for remanufacturing. Oh and Jeong (2014) proposed profit analysis and supply chain planning model for closed-loop supply chain in fashion industry. Hu et al. (2014) researched sustainable rent-based closed-loop supply chain for fashion products.

Many other researches of CLSC have been made by many researchers. For instance, Raj et al. (2013) presented divide and conquer optimization for closed loop supply chains. Ozceylan and Paksoy (2013) proposed a mixed integer programming model for a closed-loop supply-chain network. In their research, a new mixed integer mathematical model for a closed-loop supply-chain network that includes both forward and reverse flows with multi-periods and multi-parts is proposed. Pishvaee *et al.* (2011) proposed a robust optimization approach to closed-loop supply chain network design under uncertainty. Their research proposes a robust optimization model for handling the inherent uncertainty of input data in a closed-loop supply chain network design problem. Pishvaee *et al.* (2010) proposed a memetic algorithm for bi-objective integrated forward/reverse logistics network design. Subramanian *et al.* (2013) presented a methodology: PRIority based SiMulated annealing (PRISM) for a closed loop supply chain network design problem. Sim *et al.* (2004) presented a generic network design for a closed-loop supply chain using genetic algorithm. Soleimani and Kannan (2015) presented a hybrid particle swarm optimization and genetic algorithm for closed-loop supply chain network design in large-scale networks. Soleimani *et al.* (2013) designed and planned a multi-echelon, multi-period, multi-product closed-loop supply chain utilizing genetic algorithm. Pishvaee and Torabi (2010) presented a possibilistic programming approach for closed-loop supply chain network design under uncertainty. Zarandi *et al.* (2011) designed a closed-loop supply chain (CLSC) model using an interactive fuzzy goal programming. Zhang *et al.* (2012) used a lagrangian relaxation based approach for the capacitated lot sizing problem in closed-loop supply chain. Their research investigates the capacitated lot sizing problem in closed-loop supply chain considering setup costs, product returns, and remanufacturing. Wang and Huang (2013) used a two-stage robust programming approach to demand-driven disassembly planning for a closed-loop supply chain system. Wang and Hsu (2010) demonstrated a closed-loop logistic model with a spanning-tree based genetic algorithm. Zhang *et al.* (2015) presented swarm intelligence applied in green logistics as a literature review. Vieira *et al.* (2015) designed closed-loop supply chains with nonlinear dimensioning factors using ant colony optimization. Uster *et al.* (2007) used Benders decomposition with alternative multiple cuts for a multi-product closed-loop supply chain network design model. Zhou *et al.* (2015) demonstrated intelligent optimization algorithms with a stochastic closed-loop supply chain network problem involving oligopolistic competition for multiproducts and their product flow routings. Tokhmehchi *et al.* (2015) used a hybrid approach to solve a model of closed-loop supply chain. Subulan and Tasan (2013) used Taguchi method for analyzing the tactical planning model in a closed-loop supply chain considering remanufacturing option. Their proposed model is applied to an illustrative case and solved by an optimization solver.

Seitz (2007) presented a critical assessment of motives for product recovery with the case of engine remanufacturing. Their research is based on in-depth case studies within the remanufacturing facilities of a major European vehicle manufacturer. Qiu *et al.* (2012) presented the dynamic model of closed-loop supply chain with product recovering and its robust control. Savaskan *et al.* (2004) examined closed-loop supply chain models with product remanufacturing. In their research, they address the problem of choosing the appropriate reverse channel structure for the collection of used products from customers. Qiang (2015) presented a study on the closed-loop supply chain network with competition and design for remanufactureability. Through a series of case studies, they answer several important research questions, such as the impact of remanufactureability design and the consumers' perception of the remanufactured product on profitability and market share. Shi *et al.* (2015) investigated choosing reverse channels under collection responsibility sharing in a closed-loop supply chain with re-manufacturing. Hong *et al.* (2013) presented decision models of closed-loop supply chain with remanufacturing under hybrid dual-channel collection. Galbreth and Blackburn (2010) considered offshore remanufacturing with variable used product condition. Xiong *et al.* (2013) investigated the supplier and remanufacturing issues.

Paksoy *et al.* (2011) investigated operational and environmental performance measures in a multi-product closed-loop supply chain. Their research investigates a number of operational and environmental performance measures, in particular those related to transportation operations, within a closed-loop supply chain. Schmidt and Schwegler (2008) proposed a recursive ecological indicator system for the supply chain of a company. Their research proposes the concept of cumulative eco-intensity with which environmental or sustainability indicators are related to the added value of economic activities.

Sharma and Yu (2013) developed multi-stage data envelopment analysis congestion model. Their research develops a multi-stage data envelopment analysis congestion model to measure the efficiency and congestion of supply chain. Sheriff *et al.* (2014) considered combined location and routing problems for designing the quality-dependent and multi-product reverse logistics network. Schultmann *et al.* (2006) modeled reverse logistic tasks within closed-loop supply chains with an example from the automotive industry. In their contribution, the peculiarities of establishing a closed-loop supply chain are presented, based on an example considering the end-of-life vehicle treatment in Germany.

Turrisi *et al.* (2013) examined the impact of reverse logistics on supply chain performance. The authors propose a novel replenishment rule that accurately coordinates the upstream and downstream flows in a supply chain. Zuidwijk and Krikke (2008) studied strategic response to EEE returns. In their research they study how industry should strategically respond to imposed producer responsibility by regulation such as the WEEE-directive. Yan and Sun (2012) analyzed optimal stackelberg strategies for closed-loop supply chain with third-party reverse logistics. Wei and Zhao (2013) investigated reverse channel decisions for a fuzzy closed-loop supply chain. Subulan *et al.* (2014) studied an improved decoding procedure and seeker optimization algorithm for reverse logistics network design problem. Govindan *et al.* (2015) reviewed reverse logistics and closed-loop supply chain to explore the future. Kassem and Chen (2013) solved reverse logistics vehicle routing problems with time windows. A mixed integer programming model is proposed to formulate the considered problem. Jayant *et al.* (2012) examined reverse logistics with perspectives, empirical studies and research directions. Hosseini *et al.* (2015) studied reverse logistics in the construction industry. Das and Chowdhury (2012) designed a reverse logistics network for optimal collection, recovery and quality-based product-mix planning. Eskandarpour *et al.* (2014) modeled a reverse logistics network for recovery systems and a robust metaheuristic solution approach. Kim *et al.* (2013) presented lot-streaming policy for forward-reverse logistics with recovery capacity investment. In their research, they consider the problem where after usage the finished goods are collected at a constant rate and returned to the manufacturer who uses them to produce new products. Lee and Lam (2012) researched on managing reverse logistics to enhance sustainability of industrial marketing. In their paper, a sustainable industrial marketing framework of latest requirement of green and sustainable operation is proposed.Huang and Su (2013) examined the impact of product proliferation on the reverse supply chain. In their study, they develop a mathematical model for analyzing a capacitated reverse supply chain consisting of a single manufacturer and multiple retailers. Lee and Lee (2012) examined integrated forward and reverse logistics model as a case study in distilling and sale company in Korea. Das and Dutta (2013) developed a system dynamics framework for integrated reverse supply chain with three way recovery and product exchange policy.

Wu (2015) analyzed strategic and operational decisions under sales competition and collection competition for end-of-use products in remanufacturing. Toktay and Wei (2011) researched cost allocation in manufacturing-remanufacturing operations. Zanoni *et al.* (2006) presented cost performance and bullwhip effect in a hybrid manufacturing and remanufacturing system with different control policies. Wang *et al.* (2015) examined remanufacturer-manufacturer collaboration in a supply chain with the manufacturer plays the leader role. Zhou *et al.* (2013) explored the bright side of manufacturing-remanufacturing conflict in a decentralised closed-loop supply chain. Zhou *et al.* (2012) demonstrated remanufacturing closed-loop supply chain network design based on genetic particle swarm optimization algorithm. Zhang *et al.* (2011) proposed a capacitated production planning problem for closed-loop supply chain with remanufacturing. Subramanian and Subramanyam (2012) identified key factors in the market for remanufactured products. Their work contributes to the closed-loop supply chain research stream in operations management by empirically examining market factors that have not been studied before. Chen and Chang (2012) dealt with the economics of a closed-loop supply chain with remanufacturing. Their study deals with joint decisions on pricing and production lot-sizing in a closed-loop supply chain consisting of manufacturing and remanufacturing operations. Atasu *et al.* (2013) examined the impact of collection cost structure on the optimal reverse channel choice of manufacturers who remanufacture their own products.

Diabat *et al.* (2013) introduced strategic closed-loop facility location problem with carbon market trading. In their research, they introduce a multiechelon multicommodity facility location problem with a trading price of carbon emissions and a cost of procurement. Zhang *et al.* (2014) demonstrated the closed-loop supply chain network equilibrium with products lifetime and carbon emission constraints in multiperiod planning horizon. Li *et al.* (2014) presented the carbon subsidy analysis in remanufacturing closed-loop supply chain.Li *et al.* (2014) developed an incentive model for closed-loop supply chain under the EPR law. Jain *et al.* (2013) used a hierarchical approach for evaluating energy trade-offs in supply chains. Their research presents a hierarchical simulation based approach for estimating the energy consumption to keep the products flowing through a supply chain. Garg *et al.* (2015) researched a multi-criteria optimization approach to manage environmental issues in closed loop supply chain network design. Devika *et al.* (2014) designed a sustainable closed-loop supply chain network based on triple bottom line approach with a comparison of metaheuristics hybridization techniques. Defee *et al.* (2009) researched leveraging closed-loop orientation and leadership for environmental sustainability. Bell *et al.* (2013) discussed the natural resource scarcity and brought it into focus as a relevant supply chain topic related to closed-loop supply chain capabilities and the internal firm level resources needed to ensure performance in a changing world.

Toyasaki *et al.* (2013) identified the value of information systems for product recovery management. Their article sheds light on the role of information systems in product recovery management. Ketzenberg (2009) explored the value of information in a capacitated closed loop supply chain. Jayaraman *et al.* (2008) presented the role of information technology and collaboration in reverse logistics supply chains.   In their paper, they identify the reverse logistics supply-chain channels, identify problems that companies face when they handle product returns along these channels and present the critical role that information technology and collaboration can play to mitigate many of the problems and deficiencies.

Li *et al.* (2014) performed evolutionary game analysis of remanufacturing closed-loop supply chain with asymmetric information. Kumar *et al.* (2014) studied forecasting return products in an integrated forward/reverse supply chain utilizing an ANFIS. Krapp *et al.* (2013) used forecasts and managerial accounting information to enhance closed-loop supply chain management. Kannan *et al.* (2009) analysed closed loop supply chain using genetic algorithm and particle swarm optimization. Mehrbod *et al.* (2015) used a straight priority-based genetic algorithm for a logistics network. Huang *et al.* (2009) considered dynamic models of closed-loop supply chain and robust H control strategies. Guo and Ma (2013) researched on game model and complexity of retailer collecting and selling in closed-loop supply chain. The results have significant theoretical and practical application value. Akcali *et al.* (2009) proposed an annotated bibliography of models and solution approaches for a network design for reverse and closed-loop supply chains. They provided a critical review of the published literature and they identified potential areas for future research. Amin and Zhang (2012) proposed a multi-objective approach as an integrated model for closed-loop supply chain configuration and supplier selection. The objective functions maximize profit and weights of suppliers, and one of them minimizes defect rates. De Giovanni and Zaccour (2014) examined a two-period game of a closed-loop supply chain.

Beamon and Fernandes (2004) performed the supply-chain network configuration for product recovery. Ashayeri *et al.* (2015) provided a redesign of a warranty distribution network with recovery processes. Atasu *et al.* (2008) presented product reuse economics in closed-loop supply chain research. They provided a critical review of analytic research on the business economics of product reuse inspired by industrial practice. Alfonso-Lizarazo *et al.* (2013) provided a case of the palm oil supply chain by modeling reverse logistics process in the agro-industrial sector.Acar *et al.* (2015) evaluated the location of regional return centers in reverse logistics through integration of GIS, AHP and integer programming. In their study, the locations of regional return centers are determined. Abdallah *et al.* (2012) formulated a closed-loop supply chain design and performed sensitivity analysis.

Das and Dutta (2015) designed and analysed a closed-loop supply chain in presence of promotional offer. The findings of the study provide several insights to the decision-makers that lead to better performance of the entire closed-loop system. Bottani *et al.* (2015) investigated the issue of optimizing the asset management process in a real closed-loop supply chain, consisting of a pallet provider, a manufacturer and seven retailers., the results are expected to be useful to logistics and supply chain managers, to support the evaluation of the performance of CLSCs as the present research is grounded on a real CLSC. Dobos *et al.* (2013) designed contract parameters in a closed-loop supply chain. They study an extended joint economic lot size problem which incorporates the return flow of remanufacturable used products. Chen and Chang (2012) presented the co-operative strategy of a closed-loop supply chain with remanufacturing. Bhattacharjee and Cruz (2015) examined economic sustainability of closed loop supply chains and provided a holistic model for decision and policy analysis. Their model can be used at a company or industry level to conduct cost-benefit scenarios for all participants of the CLSC, and make decisions based on systemic value. De Giovanni (2014) researched environmental collaboration in a closed-loop supply chain with a reverse revenue sharing contract. Their findings suggest that green advertising should aim to increase customers' knowledge and awareness about the return policy because collaboration is successful only when the returns' residual value is large while the sharing parameter is not too high.

Eskandarpour *et al.* (2013) proposed a parallel variable neighborhood search for the multi-objective sustainable post-sales network design problem. In their study, a multi-objective post-sales network design model considering strategic and tactical decisions is proposed to minimize total fixed and variable costs, total tardiness, and environmental pollution. Easwaran and Uster (2009) used tabu search and benders decomposition approaches for a capacitated closed-loop supply chain network design problem.

Easwaran and Uster (2010) developed a closed-loop supply chain network design problem with integrated forward and reverse channel decisions. Feng *et al.* (2014) examined the equilibrium of closed-loop supply chain supernetwork with time-dependent parameters.

Ghayebloo *et al.* (2015) developed a bi-objective model of the closed-loop supply chain network with green supplier selection and disassembly of products by taking into account of the impact of parts reliability and product greenness on the recovery network. Flores-Cadena *et al.* (2012) presented the emergence of after-sales spare parts supply chain variability in a telecom firm as a complex system approach. In their article, a complex system approach to analyze the variability behavior is presented. Kim *et al.* (2014) studied a closed-loop supply chain for deteriorating products under stochastic container return times. Their research studies a two-stage supply chain where returnable transport items (RTIs) are used to ship finished products from the supplier to the buyer.

Ma *et al.* (2013) researched dual-channel closed-loop supply chain with government consumption-subsidy. Morana and Seuring (2011) analyzed a three level framework for closed-loop supply chain management-linking society, chain and actor level. Mota *et al.* (2015) presented supply chain sustainability, with views from economic, environmental and social design and planning. The relevance of their model as a decision support system is highlighted with its application to a real case study of a Portuguese battery producer and distributor. Metta and Badurdeen (2013) researched integrating sustainable product and supply chain design from the view of modeling issues and challenges.

Minner and Kiesmuller (2012) presented dynamic product acquisition in closed loop supply chains. Oraiopoulos *et al.* (2012) presented relicensing as a secondary market strategy. They analyze the OEM's strategy in both the monopoly and the duopoly cases, characterize the optimal relicensing fee set by the OEM, and draw conclusions on the conditions that favor stimulating or deterring the secondary market. Zhang *et al.* (2014) examined designing contracts for a closed-loop supply chain under information asymmetry. Their research studies the problem of designing contracts in a closed-loop supply chain when the cost of collection effort is the retailer's private information. Wang *et al.* (2013) presented a study on understanding the purchase intention towards remanufactured product in closed-loop supply chains with an empirical study in China. Zhao *et al.* (2008) investigated disruption coordination of closed-loop supply chain network (i) - models and theorems. Zhao *et al.* (2009) investigated disruption coordination of closed-loop supply chain network.

Litvinchev *et al.* (2014) used multiperiod and stochastic formulations for a closed loop supply chain with incentives. Min *et al.* (2006) proposed the spatial and temporal consolidation of returned products in a closed-loop supply chain network. Ozkir and Basligil (2013) proposed a multi-objective optimization of closed-loop supply chains in uncertain environment. Their research reveals the main features of establishing a closed loop supply chain including recovery processes. Nakashima and Gupta (2012) presented a study on the risk management of multi Kanban system in a closed loop supply chain. Moghaddam (2015) proposed supplier selection and order allocation in closed-loop supply chain systems using hybrid Monte Carlo simulation and goal programming. Yang *et al.* (2009) formulated and optimized the equilibrium state of the closed-loop supply chain network. The objective of their research is to formulate and optimize the equilibrium state of the network by using the theory of variational inequalities. Yoo *et al.* (2009) proposed service level management of nonstationary supply chain using direct neural network controller. Their research proposes a closed loop supply chain control based on a direct neural network controller.

Mondragon *et al.* (2011) defined measures for auditing performance and integration in closed-loop supply chains. The contribution of the research on closed-loop supply chains is a methodology that defines performance measures for auditing purposes of the forward and reverse components of supply chains and assists in assessing the importance of integration between different tiers of supply chains.

Ramezani *et al.* (2014) presented a financial approach in closed-loop supply chain network design. Rajamani *et al.* (2006) provided a framework to analyze cash supply chains. Their study provides the framework to study the cash supply chain structure and analyzes it as a closed-loop supply chain. Ramezani *et al.* (2015) presented a study on interrelating physical and financial flows in a bi-objective closed-loop supply chain network problem with uncertainty. Ostlin *et al.* (2008) presented importance of closed-loop supply chain relationships for product remanufacturing. Ozceylan *et al.* (2014) modeled and optimized the integrated problem of closed-loop supply chain network design and disassembly line balancing. Ozkir and Basligil (2012) modeled product-recovery processes in closed-loop supply-chain network design. Stindt and Sahamie (2014) performed review of research on closed loop supply chain management in the process industry. Su *et al.* (2012) presented component commonality in closed-loop manufacturing systems. Wikner and Tang (2008) presented a structural framework for closed-loop supply chains. Tanimizu and Shimizu (2014) demonstrated a study on closed-loop supply chain model for parts reuse with economic efficiency.

Sahamie *et al.* (2013) performed transdisciplinary research in sustainable operations and application to closed-loop supply chains. Sahyouni and Savaskan (2007) researched a facility location model for bidirectional flows. Pan *et al.* (2009) presented capacitated dynamic lot sizing problems in closed-loop supply chain. Zhang *et al.* (2015) developed a capacitated facility location model with bidirectional flows. van Nunen and Zuidwijk (2004) presented e-enabled closed-loop supply chains. Zaarour *et al.* (2014) developed the optimal determination of the collection period for returned products in the sustainable supply chain. Zhou *et al.* (2012) presented competition and integration in closed-loop supply chain network with variational inequality. Zhou *et al.* (2014) developed closed-loop supply chain network under oligopolistic competition with multiproducts, uncertain demands, and returns.

Lundin (2012) examined redesigning a closed-loop supply chain exposed to risks. Soleimani *et al.* (2014) researched on incorporating risk measures in closed-loop supply chain network design. Paksoy *et al.* (2012) presented fuzzy multi-objective optimization of a green supply chain network with risk management that includes environmental hazards. Xiao *et al.* (2012) proposed an optimization approach to risk decision-making of closed-loop logistics based on SCOR model.

Ghosh *et al.* (2010) researched heavy traffic analysis of a simple closed-loop supply chain. French and LaForge (2006) explored closed-loop supply chains in process industries with an empirical study of producer re-use issues. Guide *et al.* (2003) built contingency planning for closed-loop supply chains with product recovery.

Giri and Sharma (2015) considered optimizing a closed-loop supply chain with manufacturing defects and quality dependent return rate. Their research considers a closed-loop serial supply chain consisting of a raw material supplier, a manufacturer, a retailer and a collector who collects the used product from consumers. Guide and Van Wassenhove (2009) introduced the evolution of closed-loop supply chain research. Hammond and Beullens (2007) dealt with closed-loop supply chain network equilibrium under legislation. Their research expands previous work dealing with oligopolistic supply chains to the field of closed-loop supply chains.

Krikke *et al.* (2003) presented concurrent product and closed-loop supply chain design with an application to refrigerators. Hashemi *et al.* (2014) studied process planning for closed-loop aerospace manufacturing supply chain and environmental impact reduction. Their research studies an integrated system of manufacturing and remanufacturing using a capacitated facility in the aerospace industry, where products are returned after certain flight hours or cycles for overhaul. Kenne *et al.* (2012) proposed production planning of a hybrid manufacturing-remanufacturing system under uncertainty within a closed-loop supply chain. Ma and Wang (2014) performed complexity analysis of dynamic noncooperative game models for closed-loop supply chain with product recovery. In their paper, they consider a closed-loop supply chain with product recovery, which is composed of one manufacturer and one retailer. Hellstrom and Johansson (2010) explored the impact of control strategies on the management of returnable transport items.

Hosoda *et al.* (2015) demonstrated the impact of information sharing, random yield, correlation, and lead times in closed loop supply chains. Ma and Guo (2014) researched on third-party collecting game model with competition in closed-loop supply chain based on complex systems theory. Their research studied system dynamics characteristics of closed-loop supply chain using repeated game theory and complex system theory. Huang *et al.* (2011) researched on supply chain coordination for false failure returns with a quantity discount contract approach. Ilgin and Gupta (2011) presented performance improvement potential of sensor embedded products in environmental supply chains. Lehr *et al.* (2013) examined a system dynamics model for strategic decision-making in closed-loop supply chains. The research contributes to a better understanding of supply chain dynamics, the impact of product backflow, and reverse logistics.

Kalaitzidou *et al.* (2015) introduced optimal design of closed-loop supply chain networks with multifunctional nodes. Their research introduces a general mathematical programming framework that employs an innovative generalized supply chain network composition coupled with forward and reverse logistics activities. Kumar and Malegeant (2006) researched strategic alliance in a closed-loop supply chain, a case of manufacturer and eco-non-profit organization. Krikke (2011) examined impact of closed-loop network configurations on carbon footprints with a case study in copiers. Krikke *et al.* (2004) presented product modularity and the design of closed-loop supply chains. Li (2013) worked on an integrated approach to evaluating the production system in closed-loop supply chains.

le Blanc *et al.* (2006) discussed vehicle routing concepts in the closed-loop container network of ARN as a case study. In their research they discuss a real-life case study to optimize the logistics network for the collection of containers from end-of-life vehicle dismantlers in the Netherlands. Li *et al.* (2015) investigated decision-making based on consumers' perceived value in different remanufacturing modes.

Based on the foregoing, the present study adds another dimension to the existing literature. This study adopts a CLSC model and develops a GAMS model of a more realistic supply chain system.

3. The model description

In this section, the generic CLSC mathematical model formulation is adopted from Soleimani and Kannan (2015). The mathematical model is converted to a functioning GAMS source code by the author (See the appendix section).

The complete mathematical model is a mixed integer linear programming (MILP) model (See the model constraints). The complete mathematical model is presented as follows.

Sets

S	potential number of suppliers, indexed by "s"
F	potential number of manufacturers, indexed by "f"
W	potential number of warehouses, indexed by "w"
D	potential number of distributors, indexed by "d"
C	potential number of first customers (retailers), indexed by "c"
A	potential number of disassembly centers, indexed by "a"
R	potential number of redistributors, indexed by "r"
P	potential number of disposal locations, indexed by "p"
K	potential number of second customers, indexed by "k"
U	number of product, indexed by "u"
T	number of period, indexed by "t"

Parameters

D_{cut}	demand of product "u" by the first customer "c" in period "t"
D_{kut}	demand of product "u" of the second customer "k" in period "t"
P_{cut}	unit price of product "u" at the first customer "c" in period "t"
PH_{cut}	purchasing cost of product "u" at the first customer "c" in period "t"
P_{kut}	unit price of product "u" at the second customer "k" in period "t"
F_i	Fixed cost of the opening location "i"
DS_{ij}	distance between any two locations "i" and "j"
SC_{sut}	capacity of supplier "s" of product "u" in period "t"
SRC_{sut}	recycling capacity of supplier "s" of product "u" in period "t"
FC_{fut}	manufacturing capacity of manufacturer "f" of product "u" in period "t"
RFC_{fut}	remanufacturing capacity of manufacturer "f" of product "u" in period "t"
WC_{wut}	warehouse capacity in hours of warehouse "w" of product "u" in period "t"
DC_{dut}	capacity of distributor "d" of product "u" in period "t"
AC_{aut}	capacity of disassembly "a" of product "u" in period "t"
RC_{rut}	capacity of redistributor "r" of product "u" in period "t"
PC_{put}	capacity of disposal center "p" of product "u" in period "t"
Mc_{sut}	material cost of product "u" per unit supplied by supplier "s" in period "t"
Rc_{sut}	recycling cost of product "u" per unit recycled by supplier "s" in period "t"
Fc_{fut}	manufacturing cost of product "u" per unit manufactured by manufacturer
RFc_{fut}	remanufacturing cost of product "u" per unit by manufacturer "f" in period
DAc_{aut}	disassembly cost of product "u" per unit by disassembly center "a" in period
RPc_{aut}	repairing cost of product "u" per unit repaired by disassembly location "a"
Pc_{aut}	disposal cost of product "u" per unit disposed by disposal location "p" in
Nc_{fut}	non-utilized manufacturing capacity cost of product "u" of manufacturer "f"
RNc_{fut}	non-utilized remanufacturing cost of product "u" of manufacturer "f" in
Sc_{ut}	shortage cost of product "u" per unit in period "t"
Fh_{fu}	manufacturing time of product "u" per unit at manufacturer "f"
RFh_{fu}	remanufacturing time of product "u" per unit at manufacturer "f"
Rc_{sut}	recycling cost of supplier "s" of product "u" in period "t"
WH_{wut}	holding cost of product "u" per unit at the warehouse "w" in period "t"
DH_{dut}	holding cost of product "u" per unit at distributor store "d" store in period
B_{*u}	batch size of product "u" from supplier "s", manufacturer "f", distributor
Tc_{ut}	transportation cost of product "u" per unit per kilometer in period "t"
RR_{ut}	return ratio of product "u" at the first customers in period "t"
Rc	recycling ratio
Rm	remanufacturing ratio
Rr	repairing ratio
Rp	disposal ratio
M	is a large number

Decision variables

L_i	binary variable equals "1" if location "i" is open and "0" otherwise
Li_{ij}	binary variable equals "1" if a transportation link is established between any
Q_{ijut}	flow of batches of product "u" from location "i" to location "j" in period "t"
R_{wut}	the residual inventory of product "u" at warehouse "w" in period "t"
R_{dut}	the residual inventory of product "u" at distributor "d" in period "t"

Objective function

In this CLSC problem, the profit is considered as the objective function, therefore all sales and costs are included in the model.

Total sales (Sales of all products)

Equation 1: First products sale (flows from distributors, manufacturers, and warehouses)

Math model	$$\sum_{d\in D}\sum_{c\in C}\sum_{u\in U}\sum_{t\in T}Q_{dcut}B_{du}P_{cut} + \sum_{f\in F}\sum_{c\in C}\sum_{u\in U}\sum_{t\in T}Q_{fcut}B_{fu}P_{cut} + \sum_{w\in W}\sum_{c\in C}\sum_{u\in U}\sum_{t\in T}Q_{wcut}B_{wu}P_{cut}$$

GAMS code	`sum(d, sum(c, sum(u, sum(t, Q(d,c,u,t)*B(d,u)*P_(c,u,t))))) +`
	`sum(f, sum(c, sum(u, sum(t, Q(f,c,u,t)*B(f,u)*P_(c,u,t))))) +`
	`sum(w, sum(c, sum(u, sum(t, Q(w,c,u,t)*B(w,u)*P_(c,u,t)))))`

Equation 2: Second products sale (flows from redistributors, manufacturers, and warehouses)

Math model	$$\sum_{r\in R}\sum_{k\in K}\sum_{u\in U}\sum_{t\in T}Q_{rkut}B_{ru}P_{kut} + \sum_{f\in F}\sum_{k\in K}\sum_{u\in U}\sum_{t\in T}Q_{fkut}B_{fu}P_{kut} + \sum_{w\in W}\sum_{k\in K}\sum_{u\in U}\sum_{t\in T}Q_{wkut}B_{wu}P_{kut}$$

GAMS code	`sum(r, sum(k, sum(u, sum(t, Q(r,k,u,t)*B(r,u)*P_(k,u,t))))) +`
	`sum(f, sum(k, sum(u, sum(t, Q(f,k,u,t)*B(f,u)*P_(k,u,t))))) +`
	`sum(w, sum(k, sum(u, sum(t, Q(w,k,u,t)*B(w,u)*P_(k,u,t)))))`

The total cost includes the following costs in a generic CLSC system.

Total costs = [(fixed costs) + (material costs) + (manufacturing costs) + (non-utilized capacity costs) + (shortage costs) + (purchasing costs) + (disassembly costs) + (recycling costs) + (remanufacturing costs) + (repairing costs) + (disposal costs) + (transportation costs) + (inventory holding costs)]

Equation 3: Fixed costs (location costs)

Math model	$$\sum_{s\in S}F_sL_s + \sum_{f\in F}F_fL_f + \sum_{d\in D}F_dL_d + \sum_{a\in A}F_aL_a + \sum_{r\in R}F_rL_r + \sum_{p\in P}F_pL_p + \sum_{w\in W}F_wL_w$$

GAMS code	`sum(s, F_(s)*L(s)) + sum(f, F_(f)*L(f)) + sum(d, F_(d)*L(d)) + sum(a,`
	`F_(a)*L(a)) +`
	`sum(r, F_(r)*L(r)) + sum(p, F_(p)*L(p)) + sum(w, F_(w)*L(w))`

Equation 4: Material costs

Math model	$$\sum_{s \in S}\sum_{f \in F}\sum_{u \in U}\sum_{t \in T} Q_{sfut} B_{su} Mc_{sut} - \sum_{a \in A}\sum_{s \in S}\sum_{u \in U}\sum_{t \in T} Q_{asut} B_{au}\left(Mc_{sut} - Rc_{sut}\right)$$
GAMS code	```sum(s, sum(f, sum(u, sum(t, Q(s,f,u,t)*B(s,u)*Mc(s,u,t))))) - sum(a, sum(s, sum(u, sum(t, Q(a,s,u,t)*B(a,u)*(Mc(s,u,t)-Rc_(s,u,t)))))))```

Equation 5: Manufacturing costs

Math model	$$\sum_{f \in F}\sum_{d \in D}\sum_{u \in U}\sum_{t \in T} Q_{fdut} B_{fu} Fc_{fut} + \sum_{f \in F}\sum_{w \in W}\sum_{u \in U}\sum_{t \in T} Q_{fwut} B_{fu} Fc_{fut} + \sum_{f \in F}\sum_{c \in C}\sum_{u \in U}\sum_{t \in T} Q_{fcut} B_{fu} Fc_{fut} + \sum_{f \in F}\sum_{k \in K}\sum_{u \in U}\sum_{t \in T} Q_{fkut} B_{fu} Fc_{fut}$$
GAMS code	```sum(f, sum(d, sum(u, sum(t, Q(f,d,u,t)*B(f,u)*Fc(f,u,t))))) + sum(f, sum(w, sum(u, sum(t, Q(f,w,u,t)*B(f,u)*Fc(f,u,t))))) + sum(f, sum(c, sum(u, sum(t, Q(f,c,u,t)*B(f,u)*Fc(f,u,t))))) + sum(f, sum(k, sum(u, sum(t, Q(f,k,u,t)*B(f,u)*Fc(f,u,t)))))```

Equation 6: Non-utilized capacity costs (for manufacturers)

Math model	$$\sum_{f \in F}\left(\sum_{u \in U}\left(\sum_{t \in T}\left(\left(\frac{FC_{fut}}{Fh_{fu}}\right)L_f - \sum_{d \in D}\left(Q_{fdut}B_{fu}\right) - \sum_{w \in W}\left(Q_{fwut}B_{fu}\right) - \sum_{c \in C}\left(Q_{fcut}B_{fu}\right) + \sum_{w \in W}\sum_{r \in R}Q_{wrut}B_{wu} + \sum_{w \in W}\sum_{k \in K}Q_{wkut}B_{wu}\right)Nc_{fut}\right)\right)$$ $$+ \sum_{f \in F}\left(\sum_{u \in U}\left(\sum_{t \in T}\left(\left(\frac{RFC_{fut}}{RFh_{fu}}\right)L_f - \sum_{r \in R}\left(Q_{frut}B_{fu}\right) - \sum_{k \in K}\left(Q_{fkut}B_{fu}\right) - \sum_{w \in W}\sum_{r \in R}Q_{wrut}B_{wu} + \sum_{w \in W}\sum_{k \in K}Q_{wkut}B_{wu}\right)RNc_{fut}\right)\right)$$
GAMS code	```sum(f, sum(u, sum(t,((FC(f,u,t)/Fh(f,u))*L(f)
    - sum(d, Q(f,d,u,t)*B(f,u))
    - sum(w, Q(f,w,u,t)*B(f,u))
    - sum(c, Q(f,c,u,t)*B(f,u))
    + sum(w, sum(r, Q(w,r,u,t)*B(w,u)))
    + sum(w, sum(k, Q(w,k,u,t)*B(w,u))))*Nc(f,u,t)))) +

sum(f, sum(u, sum(t,((RFC(f,u,t)/RFh(f,u))*L(f)
    - sum(r, Q(f,r,u,t)*B(f,u))
    - sum(k, Q(f,k,u,t)*B(f,u))
    + sum(w, sum(r, Q(w,r,u,t)*B(w,u)))
    + sum(w, sum(k, Q(w,k,u,t)*B(w,u))))*RNc(f,u,t))))``` |

Equation 7: Shortage costs (for distributor)

Math model	$$\left(\sum_{c \in C} \left(\sum_{u \in U} \left(\sum_{t \in T} \left(\sum_{t-1}^{t} D_{cut} - \sum_{t-1}^{t} \sum_{d \in D} Q_{dcut} B_{du} - \sum_{t-1}^{t} \sum_{f \in F} Q_{fcut} B_{fu} - \sum_{t-1}^{t} \sum_{w \in W} Q_{wcut} B_{wu} \right) Sc_{ut} \right) \right) \right)$$

GAMS code	```sum(c, sum(u, sum(t, sum(t-1, D_(c,u,t)```
	``` - sum(t-1, sum(d, Q(d,c,u,t)*B(d,u)))```
	``` - sum(t-1, sum(f, Q(f,c,u,t)*B(f,u)))```
	``` - sum(t-1, sum(w, Q(w,c,u,k)*B(w,u))))*Sc_(u,t))))```

Equation 8: Purchasing costs

Math model	$$\sum_{c \in C} \sum_{a \in A} \sum_{u \in U} \sum_{t \in T} Q_{caut} PH_{cut} B_{cu}$$
GAMS code	```sum(c, sum(a, sum(u, sum(t, Q(c,a,u,t)*PH(c,u,t)*B(c,u)))))```

Equation 9: Disassembly costs

Math model	$$\sum_{c \in C} \sum_{a \in A} \sum_{u \in U} \sum_{t \in T} Q_{caut} B_{cu} DAc_{aut}$$
GAMS code	```sum(c, sum(a, sum(u, sum(t, Q(c,a,u,t)*B(c,u)*DAc(a,u,t)))))```

Equation 10: Recycling costs

Math model	$$\sum_{a \in A} \sum_{s \in S} \sum_{u \in U} \sum_{t \in T} Q_{asut} B_{au} Rc_{sut}$$
GAMS code	```sum(a, sum(s, sum(u, sum(t, Q(a,s,u,t)*B(a,u)*Rc_(s,u,t)))))```

Equation 11: Remanufacturing costs

Math model	$$\sum_{a \in A}\sum_{f \in F}\sum_{u \in U}\sum_{t \in T} Q_{afut} B_{au} RFc_{fut}$$
GAMS code	`sum(a, sum(f, sum(u, sum(t, Q(a,f,u,t)*B(a,u)*RFc(f,u,t)))))`

Equation 12: Repairing costs

Math model	$$\sum_{a \in A}\sum_{r \in R}\sum_{u \in U}\sum_{t \in T} Q_{arut} B_{au} RPc_{aut}$$
GAMS code	`sum(a, sum(r, sum(u, sum(t, Q(a,r,u,t)*B(a,u)*RPc(a,u,t)))))`

Equation 13: Disposal costs

Math model	$$\sum_{a \in A}\sum_{p \in P}\sum_{u \in U}\sum_{t \in T} Q_{aput} B_{au} Pc_{put}$$
GAMS code	`sum(a, sum(p, sum(u, sum(t, Q(a,p,u,t)*B(a,u)*Pc_(p,u,t)))))`

Equation 14: Transportation costs

Math model	$$\sum_{t \in T}\sum_{u \in U}\sum_{s \in S}\sum_{f \in F} Q_{sfut} B_{su} Tc_{ut} DS_{sf} + \sum_{t \in T}\sum_{u \in U}\sum_{f \in F}\sum_{d \in D} Q_{fdut} B_{fu} Tc_{ut} DS_{fd} + \sum_{t \in T}\sum_{u \in U}\sum_{f \in F}\sum_{w \in W} Q_{fwut} B_{fu} Tc_{ut} DS_{fw}$$ $$\sum_{t \in T}\sum_{u \in U}\sum_{f \in F}\sum_{c \in C} Q_{fcut} B_{fu} Tc_{ut} DS_{fc} + \sum_{t \in T}\sum_{u \in U}\sum_{f \in F}\sum_{k \in K} Q_{fkut} B_{fu} Tc_{ut} DS_{fk} + \sum_{t \in T}\sum_{u \in U}\sum_{w \in W}\sum_{c \in C} Q_{wcut} B_{wu} Tc_{ut} DS_{wc}$$ $$\sum_{t \in T}\sum_{u \in U}\sum_{w \in W}\sum_{k \in K} Q_{wkut} B_{wu} Tc_{ut} DS_{wk} + \sum_{t \in T}\sum_{u \in U}\sum_{d \in D}\sum_{c \in C} Q_{dcut} B_{du} Tc_{ut} DS_{dc} + \sum_{t \in T}\sum_{u \in U}\sum_{a \in A}\sum_{s \in S} Q_{asut} B_{au} Tc_{ut} DS_{as}$$ $$\sum_{t \in T}\sum_{a \in A}\sum_{u \in U}\sum_{f \in F} Q_{afut} B_{au} Tc_{ut} DS_{af} + \sum_{t \in T}\sum_{u \in U}\sum_{a \in A}\sum_{p \in P} Q_{aput} B_{au} Tc_{ut} DS_{ap} + \sum_{t \in T}\sum_{u \in U}\sum_{a \in A}\sum_{r \in R} Q_{arut} B_{au} Tc_{ut} DS_{ar}$$ $$\sum_{t \in T}\sum_{a \in A}\sum_{f \in F}\sum_{r \in R} Q_{frut} B_{fu} Tc_{ut} DS_{fr} + \sum_{t \in T}\sum_{u \in U}\sum_{w \in W}\sum_{r \in R} Q_{wrut} B_{wu} Tc_{ut} DS_{wr} + \sum_{t \in T}\sum_{u \in U}\sum_{r \in R}\sum_{k \in K} Q_{rkut} B_{ru} Tc_{ut} DS_{ruk}$$ $$\sum_{t \in T}\sum_{u \in U}\sum_{c \in C}\sum_{a \in A} Q_{caut} B_{cu} Tc_{ut} DS_{ca} + \sum_{t \in T}\sum_{u \in U}\sum_{w \in W}\sum_{d \in D} Q_{wdut} B_{wu} Tc_{ut} DS_{wd} + \sum_{t \in T}\sum_{u \in U}\sum_{a \in A}\sum_{k \in K} Q_{akut} B_{au} Tc_{ut} DS_{ak}$$

GAMS code

```
*Transportation costs 1
sum(t, sum(u, sum(s, sum(f, Q(s,f,u,t)*B(s,u)*Tc(u,t)*DS(s,f))))) +
sum(t, sum(u, sum(f, sum(d, Q(f,d,u,t)*B(f,u)*Tc(u,t)*DS(f,d))))) +
sum(t, sum(u, sum(f, sum(w, Q(f,w,u,t)*B(f,u)*Tc(u,t)*DS(f,w))))) +
*Transportation costs 2
sum(t, sum(u, sum(f, sum(c, Q(f,c,u,t)*B(f,u)*Tc(u,t)*DS(f,c))))) +
sum(t, sum(u, sum(f, sum(k, Q(f,k,u,t)*B(f,u)*Tc(u,t)*DS(f,k))))) +
sum(t, sum(u, sum(w, sum(c, Q(w,c,u,t)*B(w,u)*Tc(u,t)*DS(w,c))))) +
*Transportation costs 3
sum(t, sum(u, sum(w, sum(k, Q(w,k,u,t)*B(w,u)*Tc(u,t)*DS(w,k))))) +
sum(t, sum(u, sum(d, sum(c, Q(d,c,u,t)*B(d,u)*Tc(u,t)*DS(d,c))))) +
sum(t, sum(u, sum(a, sum(s, Q(a,s,u,t)*B(a,u)*Tc(u,t)*DS(a,s))))) +
*Transportation costs 4
sum(t, sum(a, sum(u, sum(f, Q(a,f,u,t)*B(a,u)*Tc(u,t)*DS(a,f))))) +
sum(t, sum(u, sum(a, sum(p, Q(a,p,u,t)*B(a,u)*Tc(u,t)*DS(a,p))))) +
sum(t, sum(u, sum(a, sum(r, Q(a,r,u,t)*B(a,u)*Tc(u,t)*DS(a,r))))) +
*Transportation costs 5
sum(t, sum(u, sum(f, sum(r, Q(f,r,u,t)*B(f,u)*Tc(u,t)*DS(f,r))))) +
sum(t, sum(u, sum(w, sum(r, Q(w,r,u,t)*B(w,u)*Tc(u,t)*DS(w,r))))) +
sum(t, sum(u, sum(r, sum(k, Q(r,k,u,t)*B(r,u)*Tc(u,t)*DS(r,k))))) +
*Transportation costs 6
sum(t, sum(u, sum(c, sum(a, Q(c,a,u,t)*B(c,u)*Tc(u,t)*DS(c,a))))) +
sum(t, sum(u, sum(w, sum(d, Q(w,d,u,t)*B(w,u)*Tc(u,t)*DS(w,d))))) +
sum(t, sum(u, sum(a, sum(k, Q(a,k,u,t)*B(a,u)*Tc(u,t)*DS(a,k)))))
```

Equation 15: Inventory holding costs

Math model	$$\sum_{w \in W} \sum_{u \in U} \sum_{t \in T} R_{wut} WH_{wut} + \sum_{d \in D} \sum_{u \in U} \sum_{t \in T} R_{dut} DH_{dut}$$
| GAMS code | ```
sum(w, sum(u, sum(t, R_(w,u,t)*WH(w,u,t)))) +
sum(d, sum(u, sum(t, R_(d,u,t)*DH(d,u,t))))
``` |

Subject to (Constraints)

All of the constraints of the generic CLSC model are presented as follows.

The constraint equations 16 through 25, 62, and 26 through 28 are balance constraints. There is a necessity of balancing in each entity. At each node, all entering flows of every product per period should be equal to all issuing flows from that node. Hence, the constraint equation 16 is the balance constraints of manufacturers.

Equation 16: Balance of manufacturers

Math model	$$\sum_{s \in S} Q_{sfut} B_{su} = \sum_{d \in D} Q_{fdut} B_{fu} + \sum_{w \in W} Q_{fwut} B_{fu} + \sum_{c \in C} Q_{fcut} B_{fu}, \quad \forall t \in T, \ \forall u \in U, \ \forall f \in F$$
| GAMS code | ```
balance_manufacturers(t,u,f)..
sum(s, Q(s,f,u,t)*B(s,u)) =e= sum(d, Q(f,d,u,t)*B(f,u)) + sum(w,
Q(f,w,u,t)*B(f,u)) + sum(c, Q(f,c,u,t)*B(f,u));
``` |

The constraint equation 17 is the balance constraints of warehouses.

Equation 17: Balance of warehouses

Math model	$$\sum_{f \in F} Q_{fwut} B_{fu} + R_{wu(t-1)} = R_{wut} + \sum_{d \in D} Q_{wdut} B_{wu} + \sum_{c \in C} Q_{wcut} B_{wu} + \sum_{k \in K} Q_{wkut} B_{wu}, \quad \forall t \in T, \ \forall u \in U, \ \forall w \in W$$
| GAMS code | ```
balance_warehouses(t,u,w)..
sum(f, Q(f,w,u,t)*B(f,u)) + R_(w,u,t-1) =e= R_(w,u,t) + sum(d,
Q(w,d,u,t)*B(w,u)) + sum(c, Q(w,c,u,t)*B(w,u)) + sum(k,
Q(w,k,u,t)*B(w,u));
``` |

The constraint equation 18 is the balance constraints of distributors.
```

Equation 18: Balance of distributors

Math model	$$\sum_{f \in F} Q_{fdut} B_{fu} + \sum_{w \in W} Q_{wdut} B_{wu} + R_{du(t-1)} = R_{dut} + \sum_{c \in C} Q_{dcut} B_{du}, \quad \forall t \in T, \ \forall u \in U, \ \forall d \in D$$
| GAMS code | ```
balance_distributors(t,u,d)..
sum(f, Q(f,d,u,t)*B(f,u)) + sum(w, Q(w,d,u,t)*B(w,u)) + R_(d,u,t-1)
=e= R_(d,u,t) + sum(c, Q(d,c,u,t)*B(d,u));
``` |

The constraint equation 19 is the balance constraint for customers who consider 70% service level requirement.

Equation 19: Balance of customer service level

Math model	$$\sum_{d \in D} Q_{dcut} B_{du} + \sum_{f \in F} Q_{fcut} B_{fu} + \sum_{w \in W} Q_{wcut} B_{wu} \geq 0.7 \ x \ D_{cut}, \quad \forall t \in T, \ \forall u \in U, \ \forall c \in C$$
| GAMS code | ```
balance_customer_service_level(t,u,c)..
sum(d, Q(d,c,u,t)*B(d,u)) + sum(f, Q(f,c,u,t)*B(f,u)) + sum(w,
Q(w,c,u,t)*B(w,u)) =g= 0.7*D_(c,u,t);
``` |

The constraint equation 20 is the balance constraints of disassembly centers (inputs).

Equation 20: Balance of disassembly centers (inputs)

Math model	$$\sum_{a \in A} Q_{caut} B_{cu} \leq \left(\sum_{d \in D} Q_{dcut} B_{du} + \sum_{f \in F} Q_{fcut} B_{fu} + \sum_{w \in W} Q_{wcut} B_{wu} \right) RR_{ut}, \quad \forall t \in T, \ \forall u \in U, \ \forall c \in C$$
| GAMS code | ```
balance_disassembly_centers_inputs(t,u,c)..
sum(a, Q(c,a,u,t)*B(c,u)) =l= (sum(d, Q(d,c,u,t)*B(d,u)) + sum(f,
Q(f,c,u,t)*B(f,u)) + sum(w, Q(w,c,u,t)*B(w,u)))*RR(u,t);
``` |

The constraint equation 21 is the balance constraints of disassembly centers (outputs).
```

Equation 21: Balance of disassembly centers (outputs)

Math model	$\sum_{c \in C} Q_{caut} B_{cu} = \sum_{s \in S} Q_{asut} B_{au} + \sum_{f \in F} Q_{afut} B_{au} + \sum_{r \in R} Q_{arut} B_{au} + \sum_{p \in P} Q_{aput} B_{au} + \sum_{k \in K} Q_{akut} B_{au}, \quad \forall t \in T, \ \forall u \in U, \ \forall a \in A$
GAMS code	```balance_disassembly_centers_outputs(t,u,a)..``` ```sum(c, Q(c,a,u,t)*B(c,u)) =e= (sum(s, Q(a,s,u,t)*B(a,u)) + sum(f, Q(a,f,u,t)*B(a,u)) + sum(r, Q(a,r,u,t)*B(a,u)) + sum(p, Q(a,p,u,t)*B(a,u)) + sum(k, Q(a,k,u,t)*B(a,u)));```

The constraint equations 22 through 25, 62, and 26 through 28 are recycling rate constraint equation 22, remanufacturing rate constraint equation 23, repairing rate constraint equation 24, disposal rate constraint equation 25, manufacturers reverse flows equation 26, redistributors equation 27, and, ultimately, second customers balance constraint equation 28.

Equation 22: Balance of recycling rate

Math model	$\sum_{c \in C} (Q_{caut} B_{cu}) Rc = \sum_{s \in S} (Q_{asut} B_{au}), \quad \forall t \in T, \ \forall u \in U, \ \forall a \in A$
GAMS code	```balance_recycling_rate(t,u,a)..``` ```sum(c, Q(c,a,u,t)*B(c,u))*R_c =e= sum(s, Q(a,s,u,t)*B(a,u));```

Equation 23: Balance of remanufacture rate

Math model	$\sum_{c \in C} (Q_{caut} B_{cu}) Rm = \sum_{f \in F} (Q_{afut} B_{au}), \quad \forall t \in T, \ \forall u \in U, \ \forall a \in A$
GAMS code	```balance_remanufacture_rate(t,u,a)..``` ```sum(c, Q(c,a,u,t)*B(c,u))*R_m =e= sum(f, Q(a,f,u,t)*B(a,u));```

Equation 24: Balance of repair rate

Math model	$\sum_{c \in C} (Q_{caut} B_{cu}) Rr = \sum_{r \in R} (Q_{arut} B_{au}), \quad \forall t \in T, \ \forall u \in U, \ \forall a \in A$
GAMS code	```balance_repair_rate(t,u,a)..``` ```sum(c, Q(c,a,u,t)*B(c,u))*R_r =e= sum(r, Q(a,r,u,t)*B(a,u));```

Equation 25: Balance of disposal rate

Math model	$$\sum_{c\in C}\left(Q_{caut}B_{cu}\right)Rp = \sum_{p\in P}\left(Q_{aput}B_{au}\right), \; \forall t \in T, \; \forall u \in U, \; \forall a \in A$$

GAMS code	```balance_disposal_rate(t,u,a)..```
	
	```sum(c, Q(c,a,u,t)*B(c,u))*R_p =e= sum(p, Q(a,p,u,t)*B(a,u));```

The sum of all assigning rates via disassembly centers should be equal to 1 (the constraint equation 62). It should be mentioned about constraint equation 62 that return products are collected from customers to disassembly centers, and then there are four options available based on the quality of return products. First, some are sent to suppliers for recycling (Rc), second, some are sent to manufacturers for remanufacturing (Rm), third, some are appropriate to be sent to second markets and so they are transferred to redistributors after repairs (Rr), and finally, the rest are disposed through the disposal centers to ensure environmentally friendly disposal (Rp).

Equation 62: Balance of recycling (Rc), remanufacturing (Rm), repairs (Rr), disposal (Rp)

Math model	$$Rc + Rm + Rr + Rp = 1$$

GAMS code	```balance_Rc_Rm_Rr_Rp..```
	
	```(R_c + R_m + R_r + R_p) =e= 1;```

Equation 26: Balance of manufacturers reverse flows

Math model	$$\sum_{a\in A}\left(Q_{afut}B_{au}\right) = \sum_{r\in R}\left(Q_{frut}B_{fu}\right) + \sum_{k\in K}\left(Q_{fkut}B_{fu}\right) + \sum_{w\in W}\sum_{k\in K}\left(Q_{wkut}B_{wu}\right) + \sum_{w\in W}\sum_{r\in R}\left(Q_{wrut}B_{wu}\right), \; \forall t \in T, \; \forall u \in U, \; \forall f \in F$$

GAMS code	```balance_manufacturers_reverse_flows(t,u,f)..```
	
	```sum(a, Q(a,f,u,t)*B(a,u)) =e= sum(r, Q(f,r,u,t)*B(f,u)) + sum(k, Q(f,k,u,t)*B(f,u)) + sum(w, sum(k, Q(w,k,u,t)*B(w,u))) + sum(w, sum(r, Q(w,r,u,t)*B(w,u)));```

Equation 27: Balance of redistributors

Math model	$\sum_{a\in A}\left(Q_{arut}B_{au}\right)+\sum_{f\in F}\left(Q_{frut}B_{fu}\right)+\sum_{w\in W}\left(Q_{wrut}B_{wu}\right)=\sum_{k\in K}\left(Q_{rkut}B_{ru}\right),\ \forall t\in T,\ \forall u\in U,\ \forall r\in R$

GAMS code	`balance_redistributors(t,u,r)..`
	
	`sum(a, Q(a,r,u,t)*B(a,u)) + sum(f, Q(f,r,u,t)*B(f,u)) + sum(w, Q(w,r,u,t)*B(w,u)) =e= sum(k, Q(r,k,u,t)*B(r,u));`

Equation 28: Balance of second customers

Math model	$\sum_{r\in R}\left(Q_{rkut}B_{ru}\right)\le D_{kut},\ \forall t\in T,\ \forall u\in U,\ \forall k\in K$

GAMS code	`balance_second_customers(t,u,k)..`
	
	`sum(r, Q(r,k,u,t)*B(r,u)) =l= D_(k,u,t);`

The constraint equations 29 through 37 are capacity constraints controlling maximum flows that can enter into or issue from each entity (node). The constraint equation 29 controls suppliers' output capacities for each product per period. The constraint equations 30 through 37 are for capacities of manufacturers, warehouses, distributors, redistributors, suppliers, disposal centers, and warehouses inputs.

Equation 29: Capacity of suppliers output

Math model	$\sum_{f\in F}Q_{sfut}B_{su}\le SC_{sut}L_s,\ \forall t\in T,\ \forall u\in U,\ \forall s\in S$

GAMS code	`capacity_suppliers_output(t,u,s)..`
	
	`sum(f, Q(s,f,u,t)*B(s,u)) =l= SC(s,u,t)*L(s);`

Equation 30: Capacity of manufacturers

Math model	$\left(\sum_{d\in D}Q_{fdut}B_{fu}+\sum_{w\in W}Q_{fwut}B_{fu}+\sum_{c\in C}Q_{fcut}B_{fu}+\sum_{k\in K}Q_{fkut}B_{fu}\right)Fh_{fu}\le FC_{fut}L_f,\ \forall t\in T,\ \forall u\in U,\ \forall f\in F$

GAMS code	`capacity_manufacturers(t,u,f)..`
	
	`(sum(d, Q(f,d,u,t)*B(f,u)) + sum(w, Q(f,w,u,t)*B(f,u)) + sum(c, Q(f,c,u,t)*B(f,u)) + sum(k, Q(f,k,u,t)*B(f,u)))*Fh(f,u) =l= FC(f,u,t)*L(f);`

Equation 31: Capacity of suppliers

Math model	$R_{wut} \leq SC_{wut}L_w, \quad \forall t \in T, \quad \forall u \in U, \quad \forall w \in W$

GAMS code	`capacity_suppliers(t,u,w)..`
	
	`R_(w,u,t) =l= SC(w,u,t)*L(w);`

Equation 32: Capacity of distributors

Math model	$\sum_{f \in F} Q_{fdut}B_{fu} + \sum_{w \in W} Q_{wdut}B_{wu} + R_{du(t-1)} \leq DC_{dut}L_d, \quad \forall t \in T, \quad \forall u \in U, \quad \forall d \in D$

GAMS code	`capacity_distributors(t,u,d)..`
	
	`sum(f, Q(f,d,u,t)*B(f,u)) + sum(w, Q(w,d,u,t)*B(w,u)) + R_(d,u,t-1)`
	`=l= DC(d,u,t)*L(d);`

Equation 33: Capacity of disassembly centers

Math model	$\sum_{s \in S} Q_{asut}B_{au} + \sum_{f \in F} Q_{afut}B_{au} + \sum_{r \in R} Q_{arut}B_{au} + \sum_{p \in P} Q_{aput}B_{au} \leq AC_{aut} \; x \; L_a, \quad \forall t \in T, \quad \forall u \in U, \quad \forall a \in A$

GAMS code	`capacity_disassembly_centers(t,u,a)..`
	
	`(sum(s, Q(a,s,u,t)*B(a,u)) + sum(f, Q(a,f,u,t)*B(a,u)) + sum(r,`
	`Q(a,r,u,t)*B(a,u)) + sum(p, Q(a,p,u,t)*B(a,u))) =l= AC(a,u,t)*L(a);`

Equation 34: Capacity of redistributors

Math model	$\sum_{k \in K} Q_{rkut}B_{ru} \leq RC_{rut} \; x \; L_r, \quad \forall t \in T, \quad \forall u \in U, \quad \forall r \in R$
GAMS code	`capacity_redistributors(t,u,r)..` `sum(k, Q(r,k,u,t)*B(r,u)) =l=` `RC(r,u,t)*L(r);`

Equation 35: Capacity of recycling suppliers

Math model	$\sum_{a \in A} Q_{asut}B_{au} \leq SRC_{sut} \; x \; L_s, \quad \forall t \in T, \quad \forall u \in U, \quad \forall s \in S$
GAMS code	`capacity_recycling_suppliers(t,u,s).. sum(a, Q(a,s,u,t)*B(a,u)) =l=` `SRC(s,u,t)*L(s);`

Equation 36: Capacity of disposal centers

Math model	$\sum_{a \in A} Q_{aput} B_{au} \leq PC_{put} \; x \; L_p, \quad \forall t \in T, \; \forall u \in U, \; \forall p \in P$
GAMS code	`capacity_disposal_centers(t,u,p)..    sum(a, Q(a,p,u,t)*B(a,u)) =l= PC(p,u,t)*L(p);`

Equation 37: Capacity of warehouses

Math model	$\sum_{f \in F} Q_{fwut} B_{fu} \leq WC_{wut} \; x \; L_w, \quad \forall t \in T, \; \forall u \in U, \; \forall w \in W$
GAMS code	`capacity_warehouses(t,u,w)..    sum(f, Q(f,w,u,t)*B(f,u)) =l= WC(w,u,t)*L(w);`

The constraint equations 38 through 54 manage the links between entities. For instance, when the left of constraint equation 38 is considered and when there are no flows between a supplier and a manufacturer (all products in all periods), there should be no link between both entities. When, based on the right-hand-side of the same constraint, there is no real link/shipping way between the supplier and the manufacturer, flows cannot be expected there. Hence, such constraints guarantee no links between nodes without actual real flows and no flows between two nodes sans an actual link.

Equation 38: Link management for every s and f

Math model	$Li_{sf} \leq \sum_{u \in U} \sum_{t \in T} Q_{sfut} \leq M \; Li_{sf}, \quad \forall s \in S, \; \forall f \in F$
GAMS code	`link_management_sf(s,f)..    Li(s,f) =l= sum(u, sum(t, Q(s,f,u,t)));` `link_management_sf2(s,f)..    sum(u, sum(t, Q(s,f,u,t))) =l= M*Li(s,f);`

Equation 39: Link management for every f and d

Math model	$Li_{fd} \leq \sum_{u \in U} \sum_{t \in T} Q_{fdut} \leq M \; Li_{fd}, \; \forall f \in F, \; \forall d \in D$
GAMS code	`link_management_fd(f,d)..    Li(f,d) =l= sum(u, sum(t, Q(f,d,u,t)));` `link_management_fd2(f,d)..    sum(u, sum(t, Q(f,d,u,t))) =l= M*Li(f,d);`

Equation 40: Link management for every f and w

Math model	$Li_{fw} \leq \sum_{u \in U} \sum_{t \in T} Q_{fwut} \leq M\ Li_{fw}, \quad \forall f \in F, \ \forall w \in W$
| GAMS code | ```
link_management_fw(f,w)..      Li(f,w) =l= sum(u, sum(t, Q(f,w,u,t)));
link_management_fw2(f,w)..     sum(u, sum(t, Q(f,w,u,t))) =l=
M*Li(f,w);
``` |

Equation 41: Link management for every f and c

Math model	$Li_{fc} \leq \sum_{u \in U} \sum_{t \in T} Q_{fcut} \leq M\ Li_{fc}, \quad \forall f \in F, \ \forall c \in C$
| GAMS code | ```
link_management_fc(f,c)..      Li(f,c) =l= sum(u, sum(t, Q(f,c,u,t)));
link_management_fc2(f,c)..     sum(u, sum(t, Q(f,c,u,t))) =l=
M*Li(f,c);
``` |

Equation 42: Link management for every f and k

Math model	$Li_{fk} \leq \sum_{u \in U} \sum_{t \in T} Q_{fkut} \leq M\ Li_{fk}, \quad \forall f \in F, \ \forall k \in K$
| GAMS code | ```
link_management_fk(f,k)..      Li(f,k) =l= sum(u, sum(t,
Q(f,k,u,t)));
link_management_fk2(f,k)..     sum(u, sum(t, Q(f,k,u,t))) =l=
M*Li(f,k);
``` |

Equation 43: Link management for every f and r

Math model	$Li_{fr} \leq \sum_{u \in U} \sum_{t \in T} Q_{frut} \leq M\ Li_{fr}, \quad \forall r \in R, \ \forall f \in F$
| GAMS code | ```
link_management_fr(f,r)..      Li(f,r) =l= sum(u, sum(t,
Q(f,r,u,t)));
link_management_fr2(f,r)..     sum(u, sum(t, Q(f,r,u,t))) =l=
M*Li(f,r);
``` |

Equation 44: Link management for every w and d

Math model	$Li_{wd} \leq \sum_{u \in U} \sum_{t \in T} Q_{wdut} \leq M\, Li_{wd}, \ \forall w \in W, \ \forall d \in D$
GAMS code	link_management_wd(w,d).. Li(w,d) =l= sum(u, sum(t, Q(w,d,u,t))); link_management_wd2(w,d).. sum(u, sum(t, Q(w,d,u,t))) =l= M*Li(w,d);

Equation 45: Link management for every w and c

Math model	$Li_{wc} \leq \sum_{u \in U} \sum_{t \in T} Q_{wcut} \leq M\, Li_{wc}, \ \forall w \in W, \ \forall c \in C$
GAMS code	link_management_wc(w,c).. Li(w,c) =l= sum(u, sum(t, Q(w,c,u,t))); link_management_wc2(w,c).. sum(u, sum(t, Q(w,c,u,t))) =l= M*Li(w,c);

Equation 46: Link management for every w and k

Math model	$Li_{wk} \leq \sum_{u \in U} \sum_{t \in T} Q_{wkut} \leq M\, Li_{wk}, \ \forall w \in W, \ \forall k \in K$
GAMS code	link_management_wk(w,k).. Li(w,k) =l= sum(u, sum(t, Q(w,k,u,t))); link_management_wk2(w,k).. sum(u, sum(t, Q(w,k,u,t))) =l= M*Li(w,k);

Equation 47: Link management for every w and r

Math model	$Li_{wr} \leq \sum_{u \in U} \sum_{t \in T} Q_{wrut} \leq M\, Li_{wr}, \ \forall w \in W, \ \forall r \in R$
GAMS code	link_management_wr(w,r).. Li(w,r) =l= sum(u, sum(t, Q(w,r,u,t))); link_management_wr2(w,r).. sum(u, sum(t, Q(w,r,u,t))) =l= M*Li(w,r);

Equation 48: Link management for every d and c

Math model	$Li_{dc} \leq \sum\limits_{u \in U} \sum\limits_{t \in T} Q_{dcut} \leq M\, Li_{dc}, \ \forall d \in D, \ \forall c \in C$
| GAMS code | ```
link_management_dc(d,c)..          Li(d,c) =l= sum(u, sum(t,
Q(d,c,u,t)));
link_management_dc2(d,c)..          sum(u, sum(t, Q(d,c,u,t))) =l=
M*Li(d,c);
``` |

Equation 49: Link management for every c and a

Math model	$Li_{ca} \leq \sum\limits_{u \in U} \sum\limits_{t \in T} Q_{caut} \leq M\, Li_{ca}, \ \forall a \in A, \ \forall c \in C$
| GAMS code | ```
link_management_ca(c,a)..          Li(c,a) =l= sum(u, sum(t,
Q(c,a,u,t)));
link_management_ca2(c,a)..          sum(u, sum(t, Q(c,a,u,t))) =l=
M*Li(c,a);
``` |

Equation 50: Link management for every a and s

Math model	$Li_{as} \leq \sum\limits_{u \in U} \sum\limits_{t \in T} Q_{asut} \leq M\, Li_{as}, \ \forall s \in S, \ \forall a \in A$
| GAMS code | ```
link_management_as(a,s)..          Li(a,s) =l= sum(u, sum(t,
Q(a,s,u,t)));
link_management_as2(a,s)..          sum(u, sum(t, Q(a,s,u,t))) =l=
M*Li(a,s);
``` |

Equation 51: Link management for every a and f

Math model	$Li_{af} \leq \sum\limits_{u \in U} \sum\limits_{t \in T} Q_{afut} \leq M\, Li_{af}, \ \forall f \in F, \ \forall a \in A$
| GAMS code | ```
link_management_af(a,f)..          Li(a,f) =l= sum(u, sum(t,
Q(a,f,u,t)));
link_management_af2(a,f)..          sum(u, sum(t, Q(a,f,u,t))) =l=
M*Li(a,f);
``` |

Equation 52: Link management for every a and r

Math model	$Li_{ar} \leq \sum_{u \in U} \sum_{t \in T} Q_{arut} \leq M \, Li_{ar}, \ \forall r \in R, \ \forall a \in A$

GAMS code	`link_management_ar(a,r)..` `Li(a,r) =l= sum(u, sum(t,`
	`Q(a,r,u,t)));`
	`link_management_ar2(a,r)..` `sum(u, sum(t, Q(a,r,u,t))) =l=`
	`M*Li(a,r);`

Equation 53: Link management for every a and p

Math model	$Li_{ap} \leq \sum_{u \in U} \sum_{t \in T} Q_{aput} \leq M \, Li_{ap}, \ \forall p \in P, \ \forall a \in A$

GAMS code	`link_management_ap(a,p)..` `Li(a,p) =l= sum(u, sum(t,`
	`Q(a,p,u,t)));`
	`link_management_ap2(a,p)..` `sum(u, sum(t, Q(a,p,u,t))) =l=`
	`M*Li(a,p);`

Equation 54: Link management for every r and k

Math model	$Li_{rk} \leq \sum_{u \in U} \sum_{t \in T} Q_{rkut} \leq M \, Li_{rk}, \ \forall k \in K, \ \forall r \in R$

GAMS code	`link_management_rk(r,k)..` `Li(r,k) =l= sum(u, sum(t,`
	`Q(r,k,u,t)));`
	`link_management_rk2(r,k)..` `sum(u, sum(t, Q(r,k,u,t))) =l=`
	`M*Li(r,k);`

The constraint equations 55 through 61 manage the maximum number of allowable locations. Though there are limitations on the number of activated locations, they cope with them and do not allow a supply chain to establish more nodes than relative possible limitations.

Equation 55: Maximum number of allowable locations for suppliers (s)

Math model	$\sum_{s \in S} L_s \leq S$
GAMS code	`max_number_allowable_locations_suppliers_s..` `sum(s, L(s)) =l=` `S__;`

Equation 56: Maximum number of allowable locations for manufacturers (f)

Math model	$$\sum_{f \in F} L_f \leq F$$
GAMS code	`max_number_allowable_locations_manufacturers_f..` **`sum(f, L(f))`** `=l= F__;`

Equation 57: Maximum number of allowable locations for distributors (d)

Math model	$$\sum_{d \in D} L_d \leq D$$
GAMS code	`max_number_allowable_locations_distributors_d..` **`sum(d, L(d))`** `=l= D__;`

Equation 58: Maximum number of allowable locations for warehouses (w)

Math model	$$\sum_{w \in W} L_w \leq W$$
GAMS code	`max_number_allowable_locations_warehouses_w..` **`sum(w, L(w))`** `=l= W__;`

Equation 59: Maximum number of allowable locations for disassembly centers (a)

Math model	$$\sum_{a \in A} L_a \leq A$$
GAMS code	`max_number_allowable_locations_disassembly_centers_a..` **`sum(a, L(a))`** `=l= A__;`

Equation 60: Maximum number of allowable locations for redistributors (r)

Math model	$\sum_{r \in R} L_r \leq R$
GAMS code	max_number_allowable_locations_redistributors_r.. **sum**(r, L(r)) =l= R__;

Equation 61: Maximum number of allowable locations for disposal locations (p)

Math model	$\sum_{p \in P} L_p \leq P$
GAMS code	max_number_allowable_locations_disposal_locations_p.. **sum**(p, L(p)) =l= P__;

4. Initialization with sample data and running the simulation

In this section the CLSC model described in the previous section is initialized with sample data to visualize the optimization problem. Sample data initialization requires some steps as illustrated in figure 2. For solving the CLSC optimization problem with the initialized data, Interior Point Optimization (IpOpt) algorithm is used. The IpOpt algorithm with the MUltifrontal Massively Parallel sparse (MUMPS) direct solver (for more information visit graal.ens-lyon.fr/MUMPS website) implements an interior point line search filter method that aims to find a solution of the objective function, which is subject to some constraints.

The mathematical details of the IpOpt algorithm are in various publications (Waechter 2002; Waechter and Biegler 2005, 2006; Nocedal et al. 2005). It should be noted that the IpOpt algorithm is intended to solve non-linear problems; however, it is also capable of providing solutions for various other types of problems as well as linear problems. In this case, the CLSC model is declared and solved as relaxed mixed integer programming (RMIP).

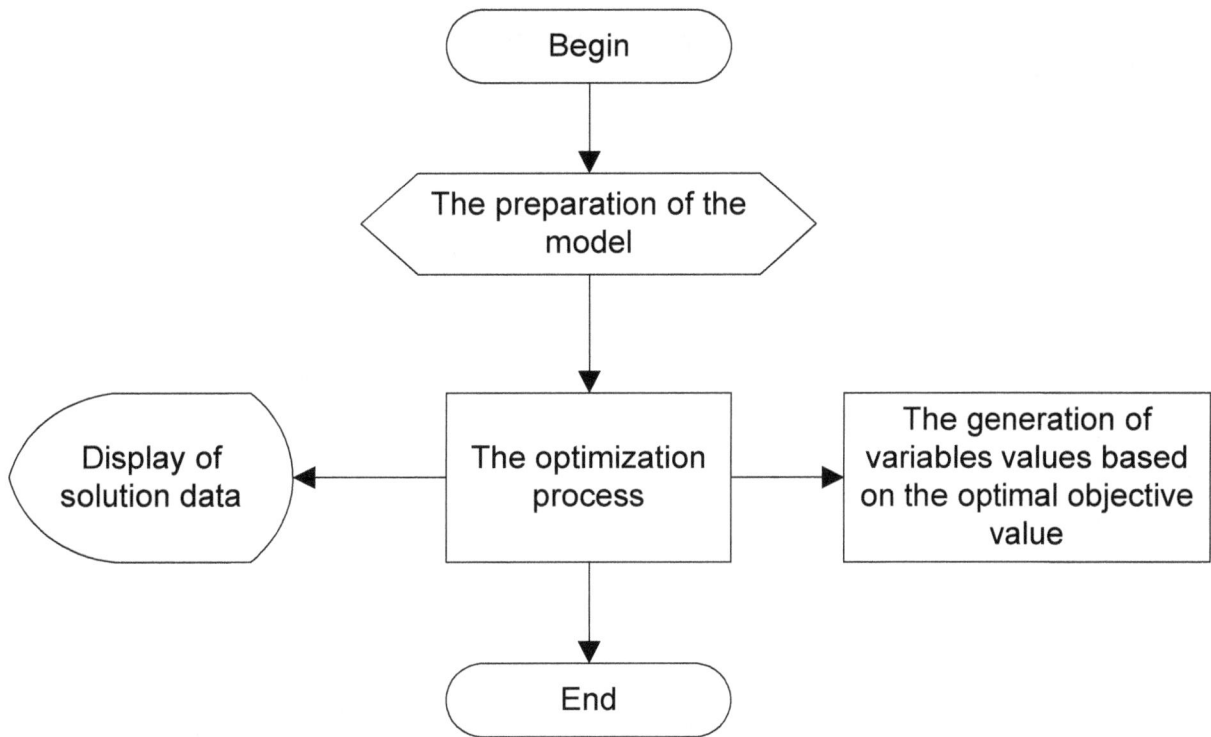

Figure 2. The flowchart of the optimization process

The next part shows a solution to the CLSC optimization problem initialized with a sample data. In this case, the total number of variables is 64793 (See table 3). Table 2 illustrates the optimization process.

Table 2. The optimization process (IpOpt algorithm)

Iter. No.	Objective	Inf_pr	Inf_du	Lg(mu)	||d||	Lg(rg)	Alpha_du	Alpha_pr	Ls
0	1.6705906e+003	1.41e+000	2.05e+000	0.0	0.00e+000	–	0.00e+000	0.00e+000	0
1	9.2402669e+002	1.41e-002	7.13e+000	-7.1	6.80e-001	–	1.10e-001	9.90e-001h	1
2	9.0978385e+002	1.41e-004	5.71e-002	-7.2	7.74e-003	–	9.74e-001	9.90e-001h	1
3	7.0700206e+002	1.40e-006	5.07e-004	-8.7	2.29e-001,	–	9.88e-001	9.90e-001f	1
4	4.2444326e+001	3.68e-015	7.81e-005	-10.3	7.35e-001	–	9.83e-001	1.00e+000f	1
5	3.3932618e+000	2.78e-016	7.34e-003	-11.0	7.68e-001	–	9.92e-001	1.00e+000f	1
6	3.3638466e-002	9.16e-016	1.99e-004	-11.0	3.22e-001	–	9.73e-001	9.93e-001f	1
7	1.3521673e-005	5.55e-015	2.33e-003	-11.0	6.17e-001	–	5.87e-001	1.00e+000f	1
8	1.0994275e-005	1.07e-014	3.25e-004	-11.0	1.47e+000	–	9.96e-001	1.00e+000h	1
9	1.0994251e-005	7.11e-015	3.13e-004	-11.0	5.92e+001	–	4.82e-002	1.01e-001h	1
10	1.0994144e-005	1.07e-014	1.34e-004	-11.0	4.00e-001	–	1.00e+000	5.00e-001h	2
11	1.0994038e-005	1.24e-014	5.70e-014	-11.0	1.52e-001	–	1.00e+000	1.00e+000h	1

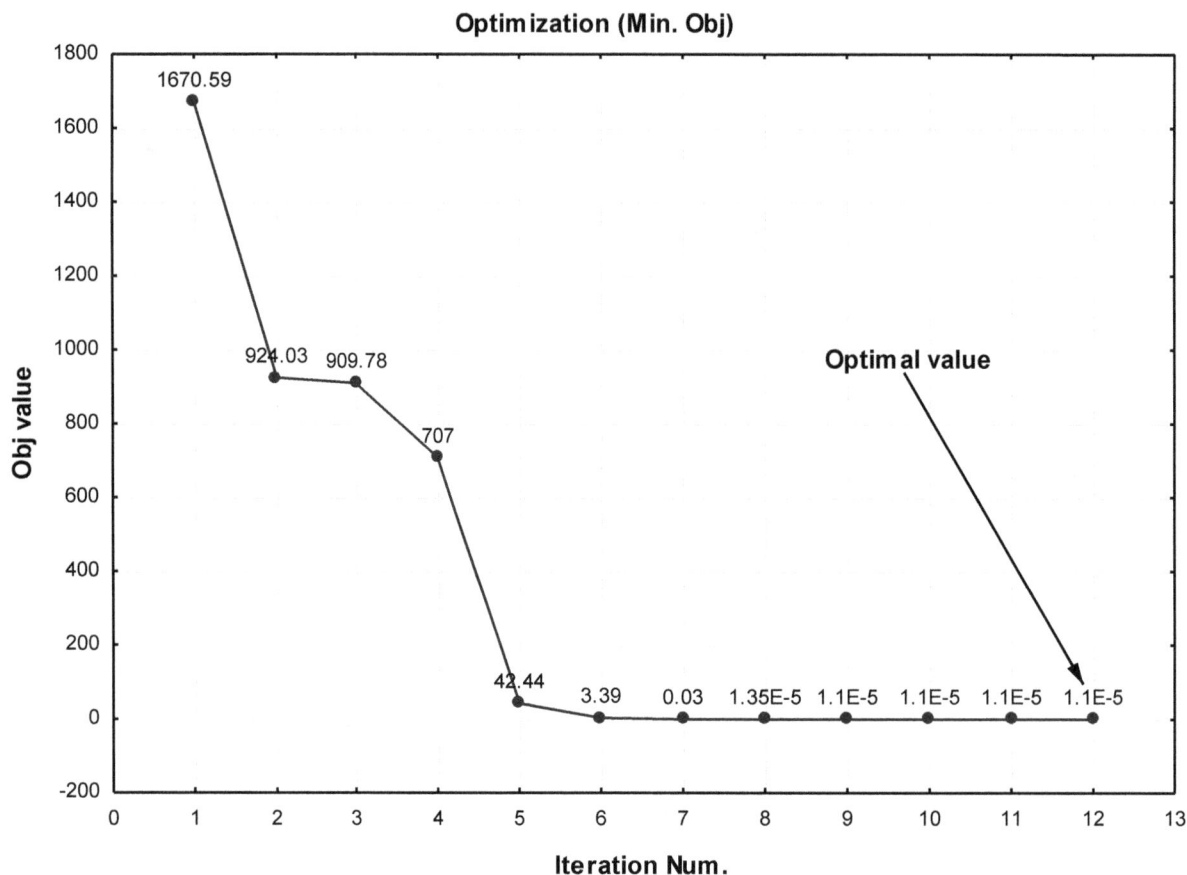

Figure 3. The optimization of the objective function

Table 2 and figure 3 show the CLSC optimization process. Table 3 and Table 4 illustrate the summary of the optimization problem. The column "*iter*" is the iteration counter and the column "*objective*" is the current value of the objective function. Other values of the columns are primal infeasibility (*inf_pr*), dual infeasibility (*inf_du*), logarithm of current barrier parameter (*lg(mu)*), max-norm of the primal search direction ($||d||$), logarithm of Hessian perturbation (*lg(rg)*), dual step size (*alpha_du*), primal step size (*alpha_pr*), and number of backtracking steps (*ls*).

Table 3. About the optimized model

Number of nonzeros in equality constraint Jacobian	1350
Number of nonzeros in inequality constraint Jacobian	132327
Number of nonzeros in Lagrangian Hessian	0
Total number of variables	64793
variables with only lower bounds	64350
variables with lower and upper bounds	443
variables with only upper bounds	0
Total number of equality constraints	750
Total number of inequality constraints	4455
inequality constraints with only lower bounds	0
inequality constraints with lower and upper bounds	0
inequality constraints with only upper bounds	4455
Number of objective function evaluations	17
Number of objective gradient evaluations	12
Number of equality constraint evaluations	17
Number of inequality constraint evaluations	17
Number of equality constraint Jacobian evaluations	12
Number of inequality constraint Jacobian evaluations	12
Number of Lagrangian Hessian evaluations	11
Total CPU secs in IPOPT (w/o function evaluations)	9.828
Total CPU secs in NLP function evaluations	0.172

Ipopt shows information (see Tables 2, 3, 4) of the optimization procedure and returns with some information about its computational effort.

Table 4. The final output of the optimization process

	(scaled)	(unscaled)
Objective	3.5861640258387167e-008	1.0994037975512871e-005
Dual infeasibility	5.6969147212507381e-014	1.7464928078419619e-011
Constraint violation	1.3875751694286426e-048	1.3875751694286426e-048
Complementarity	1.0032221801876969e-011	3.0755600322215693e-009
Overall NLP error	1.0032221801876969e-011	3.0755600322215693e-009

Technically, in the table 2, the first column *"iter"* is the iteration counter. It is not reset after an update of the barrier parameter or when the algorithm switched between the restoration phase and the regular algorithm. The next two columns *"objective"* and *"inf_pr"* indicate the value of the objective function. The fourth column *"inf_du"* is a measure of optimality; as the Ipopt aims to find a point satisfying the optimality conditions. The last column *"ls"* is an indication about how many trial points needed to be evaluated. As the value of the barrier parameter is going to zero, there is a decrease in the number in column *"lg(mu)"*. Additionally, the larger the step sizes of columns *"alpha_du"* and *"alpha_pr"*, the better is the usually the progress. (For more details about the algorithm, please refer to https://projects.coin-or.org/Ipopt).

The process provides an optimal solution for the sample model having over thousands of variables and constraints (See Table 2).

5. Conclusion

This study presented a model of a more realistic supply chain system with an optimization source code. This study considered the main elements of a closed-loop supply chain system and stated an optimization model for reducing the total costs of the whole system. In this model, strategic and tactical management levels are considered.

The IpOpt optimization method is applied on the CLSC model. A sample CLSC system model having more than thousands of variables/constraints is solved. An optimal solution for a sample CLSC logistics problem is obtained. The optimal solutions are found at the last iteration of the CLSC optimization process. The optimal solution found at the last iteration provides critical information, which addresses the best CLSC system configuration and optimal conditions.

6. References

Abdallah, T., Diabat, A., & Simchi-Levi, D. (2012). Sustainable supply chain design: a closed-loop formulation and sensitivity analysis. *Production Planning & Control, 23*(2-3), 120-133. doi: 10.1080/09537287.2011.591622

Acar, A. Z., Onden, I., & Kara, K. (2015). Evaluating the Location of Regional Return Centers in Reverse Logistics through Integration of GIS, AHP and Integer Programming. *International Journal of Industrial Engineering-Theory Applications and Practice, 22*(4), 399-411.

Accorsi, R., Manzini, R., Pini, C., & Penazzi, S. (2015). On the design of closed-loop networks for product life cycle management: Economic, environmental and geography considerations. *Journal of Transport Geography, 48*, 121-134. doi: 10.1016/j.jtrangeo.2015.09.005

Akcali, E., & Cetinkaya, S. (2011). Quantitative models for inventory and production planning in closed-loop supply chains. *International Journal of Production Research, 49*(8), 2373-2407. doi: 10.1080/00207541003692021

Akcali, E., Cetinkaya, S., & Uster, H. (2009). Network Design for Reverse and Closed-Loop Supply Chains: An Annotated Bibliography of Models and Solution Approaches. *Networks, 53*(3), 231-248. doi: 10.1002/net.20267

Alfonso-Lizarazo, E. H., Montoya-Torres, J. R., & Gutierrez-Franco, E. (2013). Modeling reverse logistics process in the agro-industrial sector: The case of the palm oil supply chain. *Applied Mathematical Modelling, 37*(23), 9652-9664. doi: 10.1016/j.apm.2013.05.015

Amaro, A. C. S., & Barbosa-Povoa, A. (2009). The effect of uncertainty on the optimal closed-loop supply chain planning under different partnerships structure. *Computers & Chemical Engineering, 33*(12), 2144-2158. doi: 10.1016/j.compchemeng.2009.06.003

Amin, S. H., & Zhang, G. Q. (2012). An integrated model for closed-loop supply chain configuration and supplier selection: Multi-objective approach. *Expert Systems with Applications, 39*(8), 6782-6791. doi: 10.1016/j.eswa.2011.12.056

Amin, S. H., & Zhang, G. Q. (2012). A proposed mathematical model for closed-loop network configuration based on product life cycle. *International Journal of Advanced Manufacturing Technology, 58*(5-8), 791-801. doi: 10.1007/s00170-011-3407-2

Amin, S. H., & Zhang, G. Q. (2013). A multi-objective facility location model for closed-loop supply chain network under uncertain demand and return. *Applied Mathematical Modelling, 37*(6), 4165-4176. doi: 10.1016/j.apm.2012.09.039

Amin, S. H., & Zhang, G. Q. (2013). A three-stage model for closed-loop supply chain configuration under uncertainty. *International Journal of Production Research, 51*(5), 1405-1425. doi: 10.1080/00207543.2012.693643

Ashayeri, J., Ma, N., & Sotirov, R. (2015). The redesign of a warranty distribution network with recovery processes. *Transportation Research Part E-Logistics and Transportation Review, 77*, 184-197. doi: 10.1016/j.tre.2015.02.017

Atamer, B., Bakal, I. S., & Bayindir, Z. P. (2013). Optimal pricing and production decisions in utilizing reusable containers. *International Journal of Production Economics, 143*(2), 222-232. doi: 10.1016/j.ijpe.2011.08.007

Atasu, A., Guide, V. D. R., & Van Wassenhove, L. N. (2008). Product reuse economics in closed-loop supply chain research. *Production and Operations Management, 17*(5), 483-496. doi: 10.3401/poms.1080.0051

Atasu, A., Toktay, L. B., & Van Wassenhove, L. N. (2013). How Collection Cost Structure Drives a Manufacturer's Reverse Channel Choice. *Production and Operations Management, 22*(5), 1089-1102. doi: 10.1111/j.1937-5956.2012.01426.x

Beamon, B. M., & Fernandes, C. (2004). Supply-chain network configuration for product recovery. *Production Planning & Control, 15*(3), 270-281. doi: 10.1080/09537280410001697701

Bell, J. E., Mollenkopf, D. A., & Stolze, H. J. (2013). Natural resource scarcity and the closed-loop supply chain: a resource-advantage view. *International Journal of*

Physical Distribution & Logistics Management, 43(5-6), 351-379. doi: 10.1108/ijpdlm-03-2012-0092

Besiou, M., Georgiadis, P., & Van Wassenhove, L. N. (2012). Official recycling and scavengers: Symbiotic or conflicting? *European Journal of Operational Research, 218*(2), 563-576. doi: 10.1016/j.ejor.2011.11.030

Bhattacharjee, S., & Cruz, J. (2015). Economic sustainability of closed loop supply chains: A holistic model for decision and policy analysis. *Decision Support Systems, 77*, 67-86. doi: 10.1016/j.dss.2015.05.011

Bottani, E., Montanan, R., Rinaldi, M., & Vignali, G. (2015). Modeling and multi-objective optimization of closed loop supply chains: A case study. *Computers & Industrial Engineering, 87*, 328-342. doi: 10.1016/j.cie.2015.05.009

Chen, J. M., & Chang, C. I. (2012). The co-opetitive strategy of a closed-loop supply chain with remanufacturing. *Transportation Research Part E-Logistics and Transportation Review, 48*(2), 387-400. doi: 10.1016/j.tre.2011.10.001

Chen, J. M., & Chang, C. I. (2012). The economics of a closed-loop supply chain with remanufacturing. *Journal of the Operational Research Society, 63*(10), 1323-1335. doi: 10.1057/jors.2011.142

Chen, J. M., & Chang, C. I. (2013). Dynamic pricing for new and remanufactured products in a closed-loop supply chain. *International Journal of Production Economics, 146*(1), 153-160. doi: 10.1016/j.ijpe.2013.06.017

Chen, W. Y., Kucukyazici, B., Verter, V., & Saenz, M. J. (2015). Supply chain design for unlocking the value of remanufacturing under uncertainty. *European Journal of Operational Research, 247*(3), 804-819. doi: 10.1016/j.ejor.2015.06.062

Chen, Y. T., Chan, F. T. S., & Chung, S. H. (2015). An integrated closed-loop supply chain model with location allocation problem and product recycling decisions. *International Journal of Production Research, 53*(10), 3120-3140. doi: 10.1080/00207543.2014.975849

Chern, C. C., Lei, S. T., & Huang, K. L. (2014). Solving a multi-objective master planning problem with substitution and a recycling process for a capacitated multi-commodity supply chain network. *Journal of Intelligent Manufacturing, 25*(1), 1-25. doi: 10.1007/s10845-012-0667-0

Chiu, M. C., & Teng, L. W. (2013). Sustainable Product and Supply Chain Design Decisions under Uncertainties. *International Journal of Precision Engineering and Manufacturing, 14*(11), 1953-1960. doi: 10.1007/s12541-013-0265-x

Chuang, C. H., Wang, C. X., & Zhao, Y. B. (2014). Closed-loop supply chain models for a high-tech product under alternative reverse channel and collection cost structures. *International Journal of Production Economics, 156*, 108-123. doi: 10.1016/j.ijpe.2014.05.008

Chung, S. L., Wee, H. M., & Yang, P. C. (2008). Optimal policy for a closed-loop supply chain inventory system with remanufacturing. *Mathematical and Computer Modelling, 48*(5-6), 867-881. doi: 10.1016/j.mcm.2007.11.014

Chung, W. H., Kremer, G. E. O., & Wysk, R. A. (2014). A Modular Design Approach to Improve Product Life Cycle Performance Based on the Optimization of a Closed-Loop Supply Chain. *Journal of Mechanical Design, 136*(2), 20. doi: 10.1115/1.4025022

Cruz-Rivera, R., & Ertel, J. (2009). Reverse logistics network design for the collection of End-of-Life Vehicles in Mexico. *European Journal of Operational Research, 196*(3), 930-939. doi: 10.1016/j.ejor.2008.04.041

Dai, Z., & Zheng, X. T. (2015). Design of close-loop supply chain network under uncertainty using hybrid genetic algorithm: A fuzzy and chance-constrained programming model. *Computers & Industrial Engineering, 88*, 444-457. doi: 10.1016/j.cie.2015.08.004

Das, D., & Dutta, P. (2013). A system dynamics framework for integrated reverse supply chain with three way recovery and product exchange policy. *Computers & Industrial Engineering, 66*(4), 720-733. doi: 10.1016/j.cie.2013.09.016

Das, D., & Dutta, P. (2015). Design and analysis of a closed-loop supply chain in presence of promotional offer. *International Journal of Production Research, 53*(1), 141-165. doi: 10.1080/00207543.2014.942007

Das, K., & Chowdhury, A. H. (2012). Designing a reverse logistics network for optimal collection, recovery and quality-based product-mix planning. *International Journal of Production Economics, 135*(1), 209-221. doi: 10.1016/j.ijpe.2011.07.010

Das, K., & Posinasetti, N. R. (2015). Addressing environmental concerns in closed loop supply chain design and planning. *International Journal of Production Economics, 163*, 34-47. doi: 10.1016/j.ijpe.2015.02.012

De Giovanni, P. (2014). Environmental collaboration in a closed-loop supply chain with a reverse revenue sharing contract. *Annals of Operations Research, 220*(1), 135-157. doi: 10.1007/s10479-011-0912-5

De Giovanni, P., & Zaccour, G. (2014). A two-period game of a closed-loop supply chain. *European Journal of Operational Research, 232*(1), 22-40. doi:

10.1016/j.ejor.2013.06.032

Defee, C. C., Esper, T., & Mollenkopf, D. (2009). Leveraging closed-loop orientation and leadership for environmental sustainability. *Supply Chain Management-an International Journal, 14*(2), 87-98. doi: 10.1108/13598540910941957

Demirel, N., Ozceylan, E., Paksoy, T., & Gokcen, H. (2014). A genetic algorithm approach for optimising a closed-loop supply chain network with crisp and fuzzy objectives. *International Journal of Production Research, 52*(12), 3637-3664. doi: 10.1080/00207543.2013.879616

Devika, K., Jafarian, A., & Nourbakhsh, V. (2014). Designing a sustainable closed-loop supply chain network based on triple bottom line approach: A comparison of metaheuristics hybridization techniques. *European Journal of Operational Research, 235*(3), 594-615. doi: 10.1016/j.ejor.2013.12.032

Diabat, A., Abdallah, T., Al-Refaie, A., Svetinovic, D., & Govindan, K. (2013). Strategic Closed-Loop Facility Location Problem With Carbon Market Trading. *Ieee Transactions on Engineering Management, 60*(2), 398-408. doi: 10.1109/tem.2012.2211105

Diabat, A., Abdallah, T., & Henschel, A. (2015). A closed-loop location-inventory problem with spare parts consideration. *Computers & Operations Research, 54*, 245-256. doi: 10.1016/j.cor.2013.08.023

Dobos, I., Gobsch, B., Pakhomova, N., Pishchulov, G., & Richter, K. (2013). Design of contract parameters in a closed-loop supply chain. *Central European Journal of Operations Research, 21*(4), 713-727. doi: 10.1007/s10100-013-0308-5

Dubey, R., Gunasekaran, A., & Childe, S. J. (2015). The design of a responsive sustainable supply chain network under uncertainty. *International Journal of Advanced Manufacturing Technology, 80*(1-4), 427-445. doi: 10.1007/s00170-015-6967-8

Easwaran, G., & Uster, H. (2009). Tabu Search and Benders Decomposition Approaches for a Capacitated Closed-Loop Supply Chain Network Design Problem. *Transportation Science, 43*(3), 301-320. doi: 10.1287/trsc.1090.0267

Easwaran, G., & Uster, H. (2010). A closed-loop supply chain network design problem with integrated forward and reverse channel decisions. *Iie Transactions, 42*(11), 779-792. doi: 10.1080/0740817x.2010.504689

Efendigil, T. (2014). Modelling Product Returns in a Closed-Loop Supply Chain Under Uncertainties: A Neuro Fuzzy Approach. *Journal of Multiple-Valued Logic and Soft Computing, 23*(3-4), 407-426.

Eskandarpour, M., Masehian, E., Soltani, R., & Khosrojerdi, A. (2014). A reverse logistics network for recovery systems and a robust metaheuristic solution approach. *International Journal of Advanced Manufacturing Technology, 74*(9-12), 1393-1406. doi: 10.1007/s00170-014-6045-7

Eskandarpour, M., Zegordi, S. H., & Nikbakhsh, E. (2013). A parallel variable neighborhood search for the multi-objective sustainable post-sales network design problem. *International Journal of Production Economics, 145*(1), 117-131. doi: 10.1016/j.ijpe.2012.10.013

Fahimnia, B., Sarkis, J., Dehghanian, F., Banihashemi, N., & Rahman, S. (2013). The impact of carbon pricing on a closed-loop supply chain: an Australian case study. *Journal of Cleaner Production, 59*, 210-225. doi: 10.1016/j.jclepro.2013.06.056

Fallah-Tafti, A., Sahraeian, R., Tavakkoli-Moghaddam, R., & Moeinipour, M. (2014). An interactive possibilistic programming approach for a multi-objective closed-loop supply chain network under uncertainty. *International Journal of Systems Science, 45*(3), 283-299. doi: 10.1080/00207721.2012.720296

Feng, Z. F., Wang, Z. P., & Chen, Y. (2014). The equilibrium of closed-loop supply chain supernetwork with time-dependent parameters. *Transportation Research Part E-Logistics and Transportation Review, 64*, 1-11. doi: 10.1016/j.tre.2014.01.009

Flores-Cadena, M., Morales-Matamoros, O., Tejeida-Padilla, R., Badillo-Pina, I., & Mejia-Tellez, J. D. (2012). The Emergence Of After-Sales Spare Parts Supply Chain Variability In A Telecom Firm - A Complex System Approach. *Fractals-Complex Geometry Patterns and Scaling in Nature and Society, 20*(1), 1-16. doi: 10.1142/s0218348x11005488

French, M. L., & LaForge, R. L. (2006). Closed-loop supply chains in process industries: An empirical study of producer re-use issues. *Journal of Operations Management, 24*(3), 271-286. doi: 10.1016/j.jom.204.07.012

Galbreth, M. R., & Blackburn, J. D. (2010). Offshore Remanufacturing with Variable Used Product Condition. *Decision Sciences, 41*(1), 5-20.

Garg, K., Kannan, D., Diabat, A., & Jha, P. C. (2015). A multi-criteria optimization approach to manage environmental issues in closed loop supply chain network design. *Journal of Cleaner Production, 100*, 297-314. doi: 10.1016/j.jclepro.2015.02.075

Georgiadis, P., & Athanasiou, E. (2010). The impact of two-product joint lifecycles on capacity planning of remanufacturing networks. *European Journal of Operational*

Research, 202(2), 420-433. doi: 10.1016/j.ejor.2009.05.022

Georgiadis, P., & Besiou, M. (2008). Sustainability in electrical and electronic equipment closed-loop supply chains: A System Dynamics approach. *Journal of Cleaner Production, 16*(15), 1665-1678. doi: 10.1016/j.jclepro.2008.04.019

Georgiadis, P., & Besiou, M. (2010). Environmental and economical sustainability of WEEE closed-loop supply chains with recycling: a system dynamics analysis. *International Journal of Advanced Manufacturing Technology, 47*(5-8), 475-493. doi: 10.1007/s00170-009-2362-7

Georgiadis, P., Vlachos, D., & Tagaras, G. (2006). The impact of product lifecycle on capacity planning of closed-loop supply chains with remanufacturing. *Production and Operations Management, 15*(4), 514-527.

Ghayebloo, S., Tarokh, M. J., Venkatadri, U., & Diallo, C. (2015). Developing a bi-objective model of the closed-loop supply chain network with green supplier selection and disassembly of products: The impact of parts reliability and product greenness on the recovery network. *Journal of Manufacturing Systems, 36*, 76-86. doi: 10.1016/j.jmsy.2015.02.011

Ghosh, A., Ryan, S. M., Wang, L. Z., & Weerasinghe, A. (2010). Heavy Traffic Analysis of a Simple Closed-Loop Supply Chain. *Stochastic Models, 26*(4), 549-593. doi: 10.1080/15326349.2010.519665

Giri, B. C., & Sharma, S. (2015). Optimizing a closed-loop supply chain with manufacturing defects and quality dependent return rate. *Journal of Manufacturing Systems, 35*, 92-111. doi: 10.1016/j.jmsy.2014.11.014

Golroudbary, S. R., & Zahraee, S. M. (2015). System dynamics model for optimizing the recycling and collection of waste material in a closed-loop supply chain. *Simulation Modelling Practice and Theory, 53*, 88-102. doi: 10.1016/j.simpat.2015.02.001

Govindan, K., Soleimani, H., & Kannan, D. (2015). Reverse logistics and closed-loop supply chain: A comprehensive review to explore the future. *European Journal of Operational Research, 240*(3), 603-626. doi: 10.1016/j.ejor.2014.07.012

Guide, V. D. R., Jayaraman, V., & Linton, J. D. (2003). Building contingency planning for closed-loop supply chains with product recovery. *Journal of Operations Management, 21*(3), 259-279. doi: 10.1016/s0272-6963(02)00110-9

Guide, V. D. R., & Van Wassenhove, L. N. (2009). The Evolution of Closed-Loop Supply Chain Research. *Operations Research, 57*(1), 10-18. doi: 10.1287/opre.1080.0628

Guo, Y. H., & Ma, J. H. (2013). Research on game model and complexity of retailer collecting and selling in closed-loop supply chain. *Applied Mathematical Modelling, 37*(7), 5047-5058. doi: 10.1016/j.apm.2012.09.034

Hammond, D., & Beullens, P. (2007). Closed-loop supply chain network equilibrium under legislation. *European Journal of Operational Research, 183*(2), 895-908. doi: 10.1016/j.ejor.2006.10.033

Hasani, A., Zegordi, S. H., & Nikbakhsh, E. (2012). Robust closed-loop supply chain network design for perishable goods in agile manufacturing under uncertainty. *International Journal of Production Research, 50*(16), 4649-4669. doi: 10.1080/00207543.2011.625051

Hashemi, V., Chen, M. Y., & Fang, L. P. (2014). Process planning for closed-loop aerospace manufacturing supply chain and environmental impact reduction. *Computers & Industrial Engineering, 75*, 87-95. doi: 10.1016/j.cie.2014.06.005

He, Y. J. (2015). Acquisition pricing and remanufacturing decisions in a closed-loop supply chain. *International Journal of Production Economics, 163*, 48-60. doi: 10.1016/j.ijpe.2015.02.002

Hellstrom, D., & Johansson, O. (2010). The impact of control strategies on the management of returnable transport items. *Transportation Research Part E-Logistics and Transportation Review, 46*(6), 1128-1139. doi: 10.1016/j.tre.2010.05.006

Hong, I. H., Lee, Y. T., & Chang, P. Y. (2014). Socially optimal and fund-balanced advanced recycling fees and subsidies in a competitive forward and reverse supply chain. *Resources Conservation and Recycling, 82*, 75-85. doi: 10.1016/j.resconrec.2013.10.018

Hong, I. H., & Yeh, J. S. (2012). Modeling closed-loop supply chains in the electronics industry: A retailer collection application. *Transportation Research Part E-Logistics and Transportation Review, 48*(4), 817-829. doi: 10.1016/j.tre.2012.01.006

Hong, X. P., Wang, Z. J., Wang, D. Z., & Zhang, H. G. (2013). Decision models of closed-loop supply chain with remanufacturing under hybrid dual-channel collection. *International Journal of Advanced Manufacturing Technology, 68*(5-8), 1851-1865. doi: 10.1007/s00170-013-4982-1

Hong, X. P., Xu, L., Du, P., & Wang, W. J. (2015). Joint advertising, pricing and collection decisions in a closed-loop supply chain. *International Journal of Production Economics, 167*, 12-22. doi: 10.1016/j.ijpe.2015.05.001

Hosoda, T., Disney, S. M., & Gavirneni, S. (2015). The impact of information sharing, random yield, correlation, and lead times in closed loop supply chains. *European Journal of Operational Research, 246*(3), 827-836. doi: 10.1016/j.ejor.2015.05.036

Hosseini, M. R., Rameezdeen, R., Chileshe, N., & Lehmann, S. (2015). Reverse logistics in the construction industry. *Waste Management & Research, 33*(6), 499-514. doi: 10.1177/0734242x15584842

Hu, G. P., & Bidanda, B. (2009). Modeling sustainable product lifecycle decision support systems. *International Journal of Production Economics, 122*(1), 366-375. doi: 10.1016/j.ijpe.2009.06.011

Hu, Z. H., Li, Q., Chen, X. J., & Wang, Y. F. (2014). Sustainable Rent-Based Closed-Loop Supply Chain for Fashion Products. *Sustainability, 6*(10), 7063-7088. doi: 10.3390/su6107063

Hu, Z. H., Sheu, J. B., Zhao, L., & Lu, C. C. (2015). A dynamic closed-loop vehicle routing problem with uncertainty and incompatible goods. *Transportation Research Part C-Emerging Technologies, 55,* 273-297. doi: 10.1016/j.trc.2015.01.010

Huang, M., Song, M., Lee, L. H., & Ching, W. K. (2013). Analysis for strategy of closed-loop supply chain with dual recycling channel. *International Journal of Production Economics, 144*(2), 510-520. doi: 10.1016/j.ijpe.2013.04.002

Huang, S. M., & Su, J. C. P. (2013). Impact of product proliferation on the reverse supply chain. *Omega-International Journal of Management Science, 41*(3), 626-639. doi: 10.1016/j.omega.2012.08.003

Huang, X. M., Choi, S. M., Ching, W. K., Siu, T. K., & Huang, M. (2011). On supply chain coordination for false failure returns: A quantity discount contract approach. *International Journal of Production Economics, 133*(2), 634-644. doi: 10.1016/j.ijpe.2011.04.031

Huang, X. Y., Yan, N. N., & Qiu, R. Z. (2009). Dynamic models of closed-loop supply chain and robust H control strategies. *International Journal of Production Research, 47*(9), 2279-2300. doi: 10.1080/00207540701636355

Ilgin, M. A., & Gupta, S. M. (2011). Performance improvement potential of sensor embedded products in environmental supply chains. *Resources Conservation and Recycling, 55*(6), 580-592. doi: 10.1016/j.resconrec.2010.05.001

Jain, S., Lindskog, E., Andersson, J., & Johansson, B. (2013). A hierarchical approach for evaluating energy trade-offs in supply chains. *International Journal of*

Production Economics, 146(2), 411-422. doi: 10.1016/j.ijpe.2013.03.015

Jayant, A., Gupta, P., & Garg, S. K. (2012). Reverse Logistics: Perspectives, Empirical Studies And Research Directions. *International Journal of Industrial Engineering-Theory Applications and Practice, 19*(10), 369-388.

Jayaraman, V., Ross, A. D., & Agarwal, A. (2008). Role of information technology and collaboration in reverse logistics supply chains. *International Journal of Logistics-Research and Applications, 11*(6), 409-425. doi: 10.1080/13675560701694499

Jena, S. K., & Sarmah, S. P. (2014). Price competition and co-operation in a duopoly closed-loop supply chain. *International Journal of Production Economics, 156*, 346-360. doi: 10.1016/j.ijpe.2014.06.018

Jindal, A., & Sangwan, K. S. (2014). Closed loop supply chain network design and optimisation using fuzzy mixed integer linear programming model. *International Journal of Production Research, 52*(14), 4156-4173. doi: 10.1080/00207543.2013.861948

Kalaitzidou, M. A., Longinidis, P., & Georgiadis, M. C. (2015). Optimal design of closed-loop supply chain networks with multifunctional nodes. *Computers & Chemical Engineering, 80*, 73-91. doi: 10.1016/j.compchemeng.2015.05.009

Kannan, G., Haq, A. N., & Devika, M. (2009). Analysis of closed loop supply chain using genetic algorithm and particle swarm optimisation. *International Journal of Production Research, 47*(5), 1175-1200. doi: 10.1080/00207540701543585

Kannan, G., Sasikumar, P., & Devika, K. (2010). A genetic algorithm approach for solving a closed loop supply chain model: A case of battery recycling. *Applied Mathematical Modelling, 34*(3), 655-670. doi: 10.1016/j.apm.2009.06.021

Kassem, S., & Chen, M. Y. (2013). Solving reverse logistics vehicle routing problems with time windows. *International Journal of Advanced Manufacturing Technology, 68*(1-4), 57-68. doi: 10.1007/s00170-012-4708-9

Kenne, J. P., Dejax, P., & Gharbi, A. (2012). Production planning of a hybrid manufacturing-remanufacturing system under uncertainty within a closed-loop supply chain. *International Journal of Production Economics, 135*(1), 81-93. doi: 10.1016/j.ijpe.2010.10.026

Ketzenberg, M. (2009). The value of information in a capacitated closed loop supply chain. *European Journal of Operational Research, 198*(2), 491-503. doi: 10.1016/j.ejor.2008.09.028

Khatami, M., Mahootchi, M., & Farahani, R. Z. (2015). Benders' decomposition for concurrent redesign of forward and closed-loop supply chain network with demand and return uncertainties. *Transportation Research Part E-Logistics and Transportation Review, 79,* 1-21. doi: 10.1016/j.tre.2015.03.003

Kim, T., & Glock, C. H. (2014). On the use of RFID in the management of reusable containers in closed-loop supply chains under stochastic container return quantities. *Transportation Research Part E-Logistics and Transportation Review, 64,* 12-27. doi: 10.1016/j.tre.2014.01.011

Kim, T., Glock, C. H., & Kwon, Y. (2014). A closed-loop supply chain for deteriorating products under stochastic container return times. *Omega-International Journal of Management Science, 43,* 30-40. doi: 10.1016/j.omega.2013.06.002

Kim, T., Goyal, S. K., & Kim, C. H. (2013). Lot-streaming policy for forward-reverse logistics with recovery capacity investment. *International Journal of Advanced Manufacturing Technology, 68*(1-4), 509-522. doi: 10.1007/s00170-013-4748-9

Krapp, M., Nebel, J., & Sahamie, R. (2013). Using forecasts and managerial accounting information to enhance closed-loop supply chain management. *Or Spectrum, 35*(4), 975-1007. doi: 10.1007/s00291-013-0345-4

Krikke, H. (2011). Impact of closed-loop network configurations on carbon footprints: A case study in copiers. *Resources Conservation and Recycling, 55*(12), 1196-1205. doi: 10.1016/j.resconrec.2011.07.001

Krikke, H., Bloemhof-Ruwaard, J., & Van Wassenhove, L. N. (2003). Concurrent product and closed-loop supply chain design with an application to refrigerators. *International Journal of Production Research, 41*(16), 3689-3719. doi: 10.1080/0020754031000120087

Krikke, H., le Blanc, I., & van de Velde, S. (2004). Product modularity and the design of closed-loop supply chains. *California Management Review, 46*(2), 23-+. doi: 10.2307/41166208

Kumar, A., & Rahman, S. (2014). RFID-enabled process reengineering of closed-loop supply chains in the healthcare industry of Singapore. *Journal of Cleaner Production, 85,* 382-394. doi: 10.1016/j.jclepro.2014.04.037

Kumar, D. T., Soleimani, H., & Kannan, G. (2014). Forecasting Return Products In An Integrated Forward/Reverse Supply Chain Utilizing An Anfis. *International Journal of Applied Mathematics and Computer Science, 24*(3), 669-682. doi: 10.2478/amcs-2014-0049

Kumar, S., & Malegeant, P. (2006). Strategic alliance in a closed-loop supply chain, a case of manufacturer and eco-non-profit organization. *Technovation, 26*(10), 1127-

1135. doi: 10.1016/j.technovation.2005.08.002

Kumar, S., & Putnam, V. (2008). Cradle to cradle: Reverse logistics strategies and opportunities across three industry sectors. *International Journal of Production Economics, 115*(2), 305-315. doi: 10.1016/j.ijpe.2007.11.015

Kumar, V. V., & Chan, F. T. S. (2011). A superiority search and optimisation algorithm to solve RFID and an environmental factor embedded closed loop logistics model. *International Journal of Production Research, 49*(16), 4807-4831. doi: 10.1080/00207543.2010.503201

le Blanc, I., van Krieken, M., Krikke, H., & Fleuren, H. (2006). Vehicle routing concepts in the closed-loop container network of ARN - a case study. *OR Spectrum, 28*(1), 53-71. doi: 10.1007/s00291-005-0003-6

Lee, C. K. M., & Chan, T. M. (2009). Development of RFID-based Reverse Logistics System. *Expert Systems with Applications, 36*(5), 9299-9307. doi: 10.1016/j.eswa.2008.12.002

Lee, C. K. M., & Lam, J. S. L. (2012). Managing reverse logistics to enhance sustainability of industrial marketing. *Industrial Marketing Management, 41*(4), 589-598. doi: 10.1016/j.indmarman.2012.04.006

Lee, J. E., & Lee, K. D. (2012). Integrated Forward and Reverse Logistics Model: A Case Study in Distilling and Sale Company in Korea. *International Journal of Innovative Computing Information and Control, 8*(7A), 4483-4495.

Lehr, C. B., Thun, J. H., & Milling, P. M. (2013). From waste to value - a system dynamics model for strategic decision-making in closed-loop supply chains. *International Journal of Production Research, 51*(13), 4105-4116. doi: 10.1080/00207543.2013.774488

Li, C. (2013). An integrated approach to evaluating the production system in closed-loop supply chains. *International Journal of Production Research, 51*(13), 4045-4069. doi: 10.1080/00207543.2013.774467

Li, J., Du, W. H., Yang, F. M., & Hua, G. W. (2014). The Carbon Subsidy Analysis in Remanufacturing Closed-Loop Supply Chain. *Sustainability, 6*(6), 3861-3877. doi: 10.3390/su6063861

Li, J., Du, W. H., Yang, F. M., & Hua, G. W. (2014). Evolutionary Game Analysis of Remanufacturing Closed-Loop Supply Chain with Asymmetric Information. *Sustainability, 6*(9), 6312-6324. doi: 10.3390/su6096312

Li, Q., Luo, H., Xie, P. X., Feng, X. Q., & Du, R. Y. (2015). Product whole life-cycle and omni-channels data convergence oriented enterprise networks integration in a sensing environment. *Computers in Industry, 70*, 23-45. doi: 10.1016/j.compind.2015.01.011

Li, S. R., Shi, L., Feng, X. J., & Li, K. P. (2012). Reverse channel design: the impacts of differential pricing and extended producer responsibility. *International Journal of Shipping and Transport Logistics, 4*(4), 357-375. doi: 10.1504/ijstl.2012.049310

Li, W., Wu, H., & Deng, L. R. (2015). Decision-Making Based on Consumers' Perceived Value in Different Remanufacturing Modes. *Discrete Dynamics in Nature and Society*, 8. doi: 10.1155/2015/278210

Li, X., Li, Y., & Govindan, K. (2014). An incentive model for closed-loop supply chain under the EPR law. *Journal of the Operational Research Society, 65*(1), 88-96. doi: 10.1057/jors.2012.179

Lieckens, K., & Vandaele, N. (2012). Multi-level reverse logistics network design under uncertainty. *International Journal of Production Research, 50*(1), 23-40. doi: 10.1080/00207543.2011.571442

Litvinchev, I., Rios, Y. A., Ozdemir, D., & Hernandez-Landa, L. G. (2014). Multiperiod and stochastic formulations for a closed loop supply chain with incentives. *Journal of Computer and Systems Sciences International, 53*(2), 201-211. doi: 10.1134/s1064230714020129

Low, J. S. C., Lu, W. F., & Song, B. (2014). Product Structure-Based Integrated Life Cycle Analysis (PSILA): a technique for cost modelling and analysis of closed-loop production systems. *Journal of Cleaner Production, 70*, 105-117. doi: 10.1016/j.jclepro.2014.02.037

Lundin, J. F. (2012). Redesigning a closed-loop supply chain exposed to risks. *International Journal of Production Economics, 140*(2), 596-603. doi: 10.1016/j.ijpe.2011.01.010

Ma, J. H., & Chen, B. (2014). The Complexity Uncertain Analysis about Three Differences Old and New Product Pricing Oligarch Retailers Closed-Loop Supply Chain. *Abstract and Applied Analysis*, 11. doi: 10.1155/2014/891624

Ma, J. H., & Guo, Y. H. (2014). Research on Third-Party Collecting Game Model with Competition in Closed-Loop Supply Chain Based on Complex Systems Theory. *Abstract and Applied Analysis*, 22. doi: 10.1155/2014/750179

Ma, J. H., & Wang, H. W. (2014). Complexity analysis of dynamic noncooperative game models for closed-loop supply chain with product recovery. *Applied Mathematical Modelling, 38*(23), 5562-5572. doi: 10.1016/j.apm.2014.02.027

Ma, W. M., Zhao, Z., & Ke, H. (2013). Dual-channel closed-loop supply chain with government consumption-subsidy. *European Journal of Operational Research, 226*(2), 221-227. doi: 10.1016/j.ejor.2012.10.033

Mahmoudzadeh, M., Sadjadi, S. J., & Mansour, S. (2013). Robust optimal dynamic production/pricing policies in a closed-loop system. *Applied Mathematical Modelling, 37*(16-17), 8141-8161. doi: 10.1016/j.apm.2013.03.008

Mehrbod, M., Tu, N., Miao, L. X., & Dai, W. J. (2012). Interactive fuzzy goal programming for a multi-objective closed-loop logistics network. *Annals of Operations Research, 201*(1), 367-381. doi: 10.1007/s10479-012-1192-4

Mehrbod, M., Xue, Z. J., Miao, L. X., & Lin, W. H. (2015). A straight priority-based genetic algorithm for a logistics network. *Rairo-Operations Research, 49*(2), 243-264. doi: 10.1051/ro/2014032

Metta, H., & Badurdeen, F. (2013). Integrating Sustainable Product and Supply Chain Design: Modeling Issues and Challenges. *Ieee Transactions on Engineering Management, 60*(2), 438-446. doi: 10.1109/tem.2012.2206392

Min, H., Ko, C. S., & Ko, H. J. (2006). The spatial and temporal consolidation of returned products in a closed-loop supply chain network. *Computers & Industrial Engineering, 51*(2), 309-320. doi: 10.1016/j.cie.2006.02.010

Minner, S., & Kiesmuller, G. P. (2012). Dynamic product acquisition in closed loop supply chains. *International Journal of Production Research, 50*(11), 2836-2851. doi: 10.1080/00207543.2010.539280

Mirakhorli, A. (2014). Fuzzy multi-objective optimization for closed loop logistics network design in bread-producing industries. *International Journal of Advanced Manufacturing Technology, 70*(1-4), 349-362. doi: 10.1007/s00170-013-5264-7

Mitra, S. (2012). Inventory management in a two-echelon closed-loop supply chain with correlated demands and returns. *Computers & Industrial Engineering, 62*(4), 870-879. doi: 10.1016/j.cie.2011.12.008

Moghaddam, K. S. (2015). Supplier selection and order allocation in closed-loop supply chain systems using hybrid Monte Carlo simulation and goal programming. *International Journal of Production Research, 53*(20), 6320-6338. doi: 10.1080/00207543.2015.1054452

Mondragon, A. E. C., Lalwani, C., & Mondragon, C. E. C. (2011). Measures for auditing performance and integration in closed-loop supply chains. *Supply Chain Management-an International Journal, 16*(1), 43-56. doi: 10.1108/13598541111103494

Morana, R., & Seuring, S. (2007). End-of-life returns of long-lived products from end customer - insights from an ideally set up closed-loop supply chain. *International Journal of Production Research, 45*(18-19), 4423-4437. doi: 10.1080/00207540701472736

Morana, R., & Seuring, S. (2011). A Three Level Framework for Closed-Loop Supply Chain Management-Linking Society, Chain and Actor Level. *Sustainability, 3*(4), 678-691. doi: 10.3390/su3040678

Mota, B., Gomes, M. I., Carvalho, A., & Barbosa-Povoa, A. P. (2015). Towards supply chain sustainability: economic, environmental and social design and planning. *Journal of Cleaner Production, 105*, 14-27. doi: 10.1016/j.jclepro.2014.07.052

Nakashima, K., & Gupta, S. M. (2012). A study on the risk management of multi Kanban system in a closed loop supply chain. *International Journal of Production Economics, 139*(1), 65-68. doi: 10.1016/j.ijpe.2012.03.016

Nie, J. J., Huang, Z. S., Zhao, Y. X., & Shi, Y. (2013). Collective Recycling Responsibility in Closed-Loop Fashion Supply Chains with a Third Party: Financial Sharing or Physical Sharing? *Mathematical Problems in Engineering*, 11. doi: 10.1155/2013/176130

Nocedal, J., Waechter, A., Waltz, R.A. (2005) Adaptive barrier strategies for nonlinear interior methods. Technical Report RC 23563, IBM T.J. Watson Research Center, Yorktown Heights, USA, March 2005.

Oh, J., & Jeong, B. (2014). Profit Analysis and Supply Chain Planning Model for Closed-Loop Supply Chain in Fashion Industry. *Sustainability, 6*(12), 9027-9056. doi: 10.3390/su6129027

Olugu, E. U., & Wong, K. Y. (2012). An expert fuzzy rule-based system for closed-loop supply chain performance assessment in the automotive industry. *Expert Systems with Applications, 39*(1), 375-384. doi: 10.1016/j.eswa.2011.07.026

Ondemir, O., Ilgin, M. A., & Gupta, S. M. (2012). Optimal End-of-Life Management in Closed-Loop Supply Chains Using RFID and Sensors. *Ieee Transactions on Industrial Informatics, 8*(3), 719-728. doi: 10.1109/tii.2011.2166767

Oraiopoulos, N., Ferguson, M. E., & Toktay, L. B. (2012). Relicensing as a Secondary Market Strategy. *Management Science, 58*(5), 1022-1037. doi: 10.1287/mnsc.1110.1456

Ostlin, J., Sundin, E., & Bjorkman, M. (2008). Importance of closed-loop supply chain relationships for product remanufacturing. *International Journal of Production Economics, 115*(2), 336-348. doi: 10.1016/j.ijpe.2008.02.020

Ozceylan, E., & Paksoy, T. (2013). Fuzzy multi-objective linear programming approach for optimising a closed-loop supply chain network. *International Journal of Production Research, 51*(8), 2443-2461. doi: 10.1080/00207543.2012.740579

Ozceylan, E., & Paksoy, T. (2013). A mixed integer programming model for a closed-loop supply-chain network. *International Journal of Production Research, 51*(3), 718-734. doi: 10.1080/00207543.2012.661090

Ozceylan, E., & Paksoy, T. (2014). Interactive fuzzy programming approaches to the strategic and tactical planning of a closed-loop supply chain under uncertainty. *International Journal of Production Research, 52*(8), 2363-2387. doi: 10.1080/00207543.2013.865852

Ozceylan, E., Paksoy, T., & Bektas, T. (2014). Modeling and optimizing the integrated problem of closed-loop supply chain network design and disassembly line balancing. *Transportation Research Part E-Logistics and Transportation Review, 61*, 142-164. doi: 10.1016/j.tre.2013.11.001

Ozkir, V., & Basligil, H. (2012). Modelling product-recovery processes in closed-loop supply-chain network design. *International Journal of Production Research, 50*(8), 2218-2233. doi: 10.1080/00207543.2011.575092

Ozkir, V., & Basligil, H. (2013). Multi-objective optimization of closed-loop supply chains in uncertain environment. *Journal of Cleaner Production, 41*, 114-125. doi: 10.1016/j.jclepro.2012.10.013

Paksoy, T., Bektas, T., & Ozceylan, E. (2011). Operational and environmental performance measures in a multi-product closed-loop supply chain. *Transportation Research Part E-Logistics and Transportation Review, 47*(4), 532-546. doi: 10.1016/j.tre.2010.12.001

Paksoy, T., Pehlivan, N. Y., & Ozceylan, E. (2012). Fuzzy Multi-Objective Optimization of a Green Supply Chain Network with Risk Management that Includes Environmental Hazards. *Human and Ecological Risk Assessment, 18*(5), 1120-1151. doi: 10.1080/10807039.2012.707940

Pan, Z. D., Tang, J. F., & Liu, O. (2009). Capacitated dynamic lot sizing problems in closed-loop supply chain. *European Journal of Operational Research, 198*(3), 810-821. doi: 10.1016/j.ejor.2008.10.018

Phuc, P. N. K., Yu, V. F., & Chou, S. Y. (2013). Optimizing the Fuzzy Closed-Loop Supply Chain for Electrical and Electronic Equipments. *International Journal of Fuzzy Systems, 15*(1), 9-21.

233

Pishvaee, M. S., Farahani, R. Z., & Dullaert, W. (2010). A memetic algorithm for bi-objective integrated forward/reverse logistics network design. *Computers & Operations Research, 37*(6), 1100-1112. doi: 10.1016/j.cor.2009.09.018

Pishvaee, M. S., Rabbani, M., & Torabi, S. A. (2011). A robust optimization approach to closed-loop supply chain network design under uncertainty. *Applied Mathematical Modelling, 35*(2), 637-649. doi: 10.1016/j.apm.2010.07.013

Pishvaee, M. S., & Torabi, S. A. (2010). A possibilistic programming approach for closed-loop supply chain network design under uncertainty. *Fuzzy Sets and Systems, 161*(20), 2668-2683. doi: 10.1016/j.fss.2010.04.010

Qiang, Q. (2015). The closed-loop supply chain network with competition and design for remanufactureability. *Journal of Cleaner Production, 105*, 348-356. doi: 10.1016/j.jclepro.2014.07.005

Qiang, Q., Ke, K., Anderson, T., & Dung, J. (2013). The closed-loop supply chain network with competition, distribution channel investment, and uncertainties. *Omega-International Journal of Management Science, 41*(2), 186-194. doi: 10.1016/j.omega.2011.08.011

Qiu, R. Z., Huang, X. Y., & Lu, Z. (2012). The Dynamic Model of Closed-Loop Supply Chain with Product Recovering and its Robust Control. *Przeglad Elektrotechniczny, 88*(9B), 1-4.

Raj, T. S., Lakshrninarayanan, S., & Forbes, J. F. (2013). Divide and Conquer Optimization for Closed Loop Supply Chains. *Industrial & Engineering Chemistry Research, 52*(46), 16267-16283. doi: 10.1021/ie400742s

Rajamani, D., Geismar, H. N., & Sriskandarajah, C. (2006). A framework to analyze cash supply chains. *Production and Operations Management, 15*(4), 544-552.

Ramezani, M., Bashiri, M., & Tavakkoli-Moghaddam, R. (2013). A robust design for a closed-loop supply chain network under an uncertain environment. *International Journal of Advanced Manufacturing Technology, 66*(5-8), 825-843. doi: 10.1007/s00170-012-4369-8

Ramezani, M., Kimiagari, A. M., & Karimi, B. (2014). Closed-loop supply chain network design: A financial approach. *Applied Mathematical Modelling, 38*(15-16), 4099-4119. doi: 10.1016/j.apm.2014.02.004

Ramezani, M., Kimiagari, A. M., & Karimi, B. (2015). Interrelating physical and financial flows in a bi-objective closed-loop supply chain network problem with uncertainty. *Scientia Iranica, 22*(3), 1278-1293.

Ramezani, M., Kimiagari, A. M., Karimi, B., & Hejazi, T. H. (2014). Closed-loop supply chain network design under a fuzzy environment. *Knowledge-Based Systems, 59*, 108-120. doi: 10.1016/j.knosys.2014.01.016

Rezapour, S., Farahani, R. Z., Fahimnia, B., Govindan, K., & Mansouri, Y. (2015). Competitive closed-loop supply chain network design with price-dependent demands. *Journal of Cleaner Production, 93*, 251-272. doi: 10.1016/j.jclepro.2014.12.095

Sahamie, R., Stindt, D., & Nuss, C. (2013). Transdisciplinary Research in Sustainable Operations An Application to Closed-Loop Supply Chains. *Business Strategy and the Environment, 22*(4), 245-268. doi: 10.1002/bse.1771

Sahyouni, K., & Savaskan, R. C. (2007). A facility location model for bidirectional flows. *Transportation Science, 41*(4), 484-499. doi: 10.1287/trsc.1070.0215

Sasikumar, P., & Haq, A. N. (2011). Integration of closed loop distribution supply chain network and 3PRLP selection for the case of battery recycling. *International Journal of Production Research, 49*(11), 3363-3385. doi: 10.1080/00207541003794876

Savaskan, R. C., Bhattacharya, S., & Van Wassenhove, L. N. (2004). Closed-loop supply chain models with product remanufacturing. *Management Science, 50*(2), 239-252. doi: 10.1287/mnsc.1030.0186

Schmidt, M., & Schwegler, R. (2008). A recursive ecological indicator system for the supply chain of a company. *Journal of Cleaner Production, 16*(15), 1658-1664. doi: 10.1016/j.jclepro.2008.04.006

Schultmann, F., Engels, B., & Rentz, O. (2003). Closed-loop supply chains for spent batteries. *Interfaces, 33*(6), 57-71. doi: 10.1287/inte.33.6.57.25183

Schultmann, F., Zumkeller, M., & Rentz, O. (2006). Modeling reverse logistic tasks within closed-loop supply chains: An example from the automotive industry. *European Journal of Operational Research, 171*(3), 1033-1050. doi: 10.1016/j.ejor.2005.01.016

Seitz, M. A. (2007). A critical assessment of motives for product recovery: the case of engine remanufacturing. *Journal of Cleaner Production, 15*(11-12), 1147-1157. doi: 10.1016/j.jclepro.2006.05.029

Sharma, M. J., & Yu, S. J. (2013). Multi-Stage data envelopment analysis congestion model. *Operational Research, 13*(3), 399-413. doi: 10.1007/s12351-012-0128-8

Sheriff, K. M. M., Nachiappan, S., & Min, H. (2014). Combined location and routing problems for designing the quality-dependent and multi-product reverse logistics network. *Journal of the Operational Research Society, 65*(6), 873-887. doi: 10.1057/jors.2013.22

Shi, J. M., Zhang, G. Q., & Sha, J. C. (2011). Optimal production and pricing policy for a closed loop system. *Resources Conservation and Recycling, 55*(6), 639-647. doi: 10.1016/j.resconrec.2010.05.016

Shi, J. M., Zhang, G. Q., & Sha, J. C. (2011). Optimal production planning for a multi-product closed loop system with uncertain demand and return. *Computers & Operations Research, 38*(3), 641-650. doi: 10.1016/j.cor.2010.08.008

Shi, J. M., Zhang, G. Q., Sha, J. C., & Amin, S. H. (2010). Coordinating Production And Recycling Decisions With Stochastic Demand And Return. *Journal of Systems Science and Systems Engineering, 19*(4), 385-407. doi: 10.1007/s11518-010-5147-5

Shi, W. B., & Min, K. J. (2013). A Study of Product Weight and Collection Rate in Closed-Loop Supply Chains With Recycling. *Ieee Transactions on Engineering Management, 60*(2), 409-423. doi: 10.1109/tem.2012.2214222

Shi, Y., Nie, J. J., Qu, T., Chu, L. K., & Sculli, D. (2015). Choosing reverse channels under collection responsibility sharing in a closed-loop supply chain with re-manufacturing. *Journal of Intelligent Manufacturing, 26*(2), 387-402. doi: 10.1007/s10845-013-0797-z

Sim, E., Jung, S., Kim, H., & Park, J. (2004). A generic network design for a closed-loop supply chain using genetic algorithm. In K. Deb, R. Poli, W. Banzhaf, H. G. Beyer, E. Burke, P. Darwen, D. Dasgupta, D. Floreano, O. Foster, M. Harman, O. Holland, P. L. Lanzi, L. Spector, A. Tettamanzi, D. Thierens & A. Tyrrell (Eds.), *Genetic and Evolutionary Computation Gecco 2004 , Pt 2, Proceedings* (Vol. 3103, pp. 1214-1225). Berlin: Springer-Verlag Berlin.

Soleimani, H., & Kannan, G. (2015). A hybrid particle swarm optimization and genetic algorithm for closed-loop supply chain network design in large-scale networks. *Applied Mathematical Modelling, 39*(14), 3990-4012. doi: 10.1016/j.apm.2014.12.016

Soleimani, H., Seyyed-Esfahani, M., & Kannan, G. (2014). Incorporating risk measures in closed-loop supply chain network design. *International Journal of Production Research, 52*(6), 1843-1867. doi: 10.1080/00207543.2013.849823

Soleimani, H., Seyyed-Esfahani, M., & Shirazi, M. A. (2013). Designing and planning a multi-echelon multi-period multi-product closed-loop supply chain utilizing genetic algorithm. *International Journal of Advanced Manufacturing Technology, 68*(1-4), 917-931. doi: 10.1007/s00170-013-4953-6

Stindt, D., & Sahamie, R. (2014). Review of research on closed loop supply chain management in the process industry. *Flexible Services and Manufacturing Journal, 26*(1-2), 268-293. doi: 10.1007/s10696-012-9137-4

Su, J. C. P., Lin, Y. C., & Lee, V. (2012). Component commonality in closed-loop manufacturing systems. *Journal of Intelligent Manufacturing, 23*(6), 2383-2396. doi: 10.1007/s10845-010-0485-1

Subramanian, P., Ramkumar, N., Narendran, T. T., & Ganesh, K. (2013). PRISM: PRIority based SiMulated annealing for a closed loop supply chain network design problem. *Applied Soft Computing, 13*(2), 1121-1135. doi: 10.1016/j.asoc.2012.10.004

Subramanian, R., & Subramanyam, R. (2012). Key Factors in the Market for Remanufactured Products. *M&Som-Manufacturing & Service Operations Management, 14*(2), 315-326. doi: 10.1287/msom.1110.0368

Subulan, K., Baykasoglu, A., & Saltabas, A. (2014). An improved decoding procedure and seeker optimization algorithm for reverse logistics network design problem. *Journal of Intelligent & Fuzzy Systems, 27*(6), 2703-2714. doi: 10.3233/ifs-141335

Subulan, K., & Tasan, A. S. (2013). Taguchi method for analyzing the tactical planning model in a closed-loop supply chain considering remanufacturing option. *International Journal of Advanced Manufacturing Technology, 66*(1-4), 251-269. doi: 10.1007/s00170-012-4322-x

Subulan, K., Tasan, A. S., & Baykasoglu, A. (2012). Fuzzy mixed integer programming model for medium-term planning in a closed-loop supply chain with remanufacturing option. *Journal of Intelligent & Fuzzy Systems, 23*(6), 345-368. doi: 10.3233/ifs-2012-0525

Subulan, K., Tasan, A. S., & Baykasoglu, A. (2015). Designing an environmentally conscious tire closed-loop supply chain network with multiple recovery options using interactive fuzzy goal programming. *Applied Mathematical Modelling, 39*(9), 2661-2702. doi: 10.1016/j.apm.2014.11.004

Tanimizu, Y., & Shimizu, Y. (2014). A study on closed-loop supply chain model for parts reuse with economic efficiency. *Journal of Advanced Mechanical Design Systems and Manufacturing, 8*(5), 12. doi: 10.1299/jamdsm.2014jamdsm0068

Tavakkoli-Moghaddam, R., Sadri, S., Pourmohammad-Zia, N., & Mohammadi, M. (2015). A hybrid fuzzy approach for the closed-loop supply chain network design under uncertainty. *Journal of Intelligent & Fuzzy Systems, 28*(6), 2811-2826. doi: 10.3233/ifs-151561

Tokhmehchi, N., Makui, A., & Sadi-Nezhad, S. (2015). A Hybrid Approach to Solve a Model of Closed-Loop Supply Chain. *Mathematical Problems in Engineering*, 18. doi: 10.1155/2015/179102

Toktay, L. B., & Wei, D. (2011). Cost Allocation in Manufacturing-Remanufacturing Operations. *Production and Operations Management, 20*(6), 841-847. doi: 10.1111/J.1937-5956.2011.01236.x

Toyasaki, F., Wakolbinger, T., & Kettinger, W. J. (2013). The value of information systems for product recovery management. *International Journal of Production Research, 51*(4), 1214-1235. doi: 10.1080/00207543.2012.695090

Turrisi, M., Bruccoleri, M., & Cannella, S. (2013). Impact of reverse logistics on supply chain performance. *International Journal of Physical Distribution & Logistics Management, 43*(7), 564-585. doi: 10.1108/ijpdlm-04-2012-0132

Uster, H., Easwaran, G., Akcali, E., & Cetinkaya, S. (2007). Benders decomposition with alternative multiple cuts for a multi-product closed-loop supply chain network design model. *Naval Research Logistics, 54*(8), 890-907. doi: 10.1002/nav.20262

Vahdani, B. (2015). An Optimization Model for Multi-Objective Closed-Loop Supply Chain Network under Uncertainty: A Hybrid Fuzzy-Stochastic Programming Method. *Iranian Journal of Fuzzy Systems, 12*(4), 33-57.

Vahdani, B., Razmi, J., & Tavakkoli-Moghaddam, R. (2012). Fuzzy Possibilistic Modeling for Closed Loop Recycling Collection Networks. *Environmental Modeling & Assessment, 17*(6), 623-637. doi: 10.1007/s10666-012-9313-7

Vahdani, B., Tavakkoli-Moghaddam, R., Jolai, F., & Baboli, A. (2013). Reliable design of a closed loop supply chain network under uncertainty: An interval fuzzy possibilistic chance-constrained model. *Engineering Optimization, 45*(6), 745-765. doi: 10.1080/0305215x.2012.704029

Vahdani, B., Tavakkoli-Moghaddam, R., Modarres, M., & Baboli, A. (2012). Reliable design of a forward/reverse logistics network under uncertainty: A robust-M/M/c queuing model. *Transportation Research Part E-Logistics and Transportation Review, 48*(6), 1152-1168. doi: 10.1016/j.tre.2012.06.002

van Nunen, J., & Zuidwijk, R. A. (2004). E-enabled closed-loop supply chains. *California Management Review, 46*(2), 40-+.

Vieira, P. F., Vieira, S. M., Gomes, M. I., Barbosa-Povoa, A. P., & Sousa, J. M. C. (2015). Designing closed-loop supply chains with nonlinear dimensioning factors using ant colony optimization. *Soft Computing, 19*(8), 2245-2264. doi: 10.1007/s00500-014-1405-7

Waechter A., Biegler, L.T. (2005) Line search filter methods for nonlinear programming: Local convergence. *SIAM Journal on Optimization*, 16(1):32-48

Waechter A., Biegler, L.T. (2006) On the implementation of a primal-dual interior point filter line search algorithm for large-scale nonlinear programming. *Mathematical Programming*, 106(1):25-57

Waechter, A. (2002) An Interior Point Algorithm for Large-Scale Nonlinear Optimization with Applications in Process Engineering. Ph.D. thesis, Carnegie Mellon University, Pittsburgh, PA, USA, January 2002.

Waechter, A., Biegler, L.T. (2005) Line search filter methods for nonlinear programming: Motivation and global convergence. *SIAM Journal on Optimization*, 16(1):1-31

Wang, H. F., & Hsu, H. W. (2010). A closed-loop logistic model with a spanning-tree based genetic algorithm. *Computers & Operations Research, 37*(2), 376-389. doi: 10.1016/j.cor.2009.06.001

Wang, H. F., & Hsu, H. W. (2010). Resolution of an uncertain closed-loop logistics model: An application to fuzzy linear programs with risk analysis. *Journal of Environmental Management, 91*(11), 2148-2162. doi: 10.1016/j.jenvman.2010.05.009

Wang, H. F., & Hsu, H. W. (2012). A possibilistic approach to the modeling and resolution of uncertain closed-loop logistics. *Fuzzy Optimization and Decision Making, 11*(2), 177-208. doi: 10.1007/s10700-012-9120-2

Wang, H. F., & Huang, Y. S. (2013). A two-stage robust programming approach to demand-driven disassembly planning for a closed-loop supply chain system. *International Journal of Production Research, 51*(8), 2414-2432. doi: 10.1080/00207543.2012.737940

Wang, K., Xiong, Z. K., Xiong, Y., & Yan, W. (2015). Remanufacturer-Manufacturer Collaboration in a Supply Chain: The Manufacturer Plays the Leader Role. *Asia-Pacific Journal of Operational Research, 32*(5), 17. doi: 10.1142/s0217595915500402

Wang, K. Z., Zhao, Y. X., Cheng, Y. H., & Choi, T. M. (2014). Cooperation or Competition? Channel Choice for a Remanufacturing Fashion Supply Chain with

Government Subsidy. *Sustainability, 6*(10), 7292-7310. doi: 10.3390/su6107292

Wang, Y. C., Wiegerinck, V., Krikke, H., & Zhang, H. D. (2013). Understanding the purchase intention towards remanufactured product in closed-loop supply chains An empirical study in China. *International Journal of Physical Distribution & Logistics Management, 43*(10), 866-888. doi: 10.1108/ijpdlm-01-2013-0011

Wei, J., Govindan, K., Li, Y. J., & Zhao, J. (2015). Pricing and collecting decisions in a closed-loop supply chain with symmetric and asymmetric information. *Computers & Operations Research, 54*, 257-265. doi: 10.1016/j.cor.2013.11.021

Wei, J., & Zhao, J. (2011). Pricing decisions with retail competition in a fuzzy closed-loop supply chain. *Expert Systems with Applications, 38*(9), 11209-11216. doi: 10.1016/j.eswa.2011.02.168

Wei, J., & Zhao, J. (2013). Reverse channel decisions for a fuzzy closed-loop supply chain. *Applied Mathematical Modelling, 37*(3), 1502-1513. doi: 10.1016/j.apm.2012.04.003

Wei, J., Zhao, J., & Li, Y. J. (2012). Pricing Decisions For A Closed-Loop Supply Chain In A Fuzzy Environment. *Asia-Pacific Journal of Operational Research, 29*(1), 30. doi: 10.1142/s0217595912400039

Wikner, J., & Tang, O. (2008). A structural framework for closed-loop supply chains. *International Journal of Logistics Management, 19*(3), 344-366. doi: 10.1108/09574090810919198

Wu, C. H. (2015). Strategic and operational decisions under sales competition and collection competition for end-of-use products in remanufacturing. *International Journal of Production Economics, 169*, 11-20. doi: 10.1016/j.ijpe.2015.07.020

Wu, X., & Ryan, S. M. (2014). Joint Optimization of Asset and Inventory Management in a Product-Service System. *Engineering Economist, 59*(2), 91-115. doi: 10.1080/0013791x.2013.873844

Xiao, R. B., Cai, Z. Y., & Zhang, X. H. (2012). An optimization approach to risk decision-making of closed-loop logistics based on SCOR model. *Optimization, 61*(10), 1221-1251. doi: 10.1080/02331934.2012.688827

Xie, J. P., Li, Z. J., Yao, Y., & Liang, L. (2015). Dynamic acquisition pricing policy under uncertain remanufactured-product demand. *Industrial Management & Data Systems, 115*(3), 521-540. doi: 10.1108/imds-11-2014-0333

Xiong, Y., Zhou, Y., Li, G. D., Chan, H. K., & Xiong, Z. K. (2013). Don't forget your supplier when remanufacturing. *European Journal of Operational Research,*

230(1), 15-25. doi: 10.1016/j.ejor.2013.03.034

Xu, C. C., Li, B., Lan, Y. F., & Tang, Y. (2014). A Closed-Loop Supply Chain Problem with Retailing and Recycling Competition. *Abstract and Applied Analysis*, 14. doi: 10.1155/2014/509825

Yan, N. N., & Sun, B. W. (2012). Optimal Stackelberg Strategies For Closed-Loop Supply Chain With Third-Party Reverse Logistics. *Asia-Pacific Journal of Operational Research, 29*(5), 21. doi: 10.1142/s0217595912500261

Yang, G. F., Wang, Z. P., & Li, X. Q. (2009). The optimization of the closed-loop supply chain network. *Transportation Research Part E-Logistics and Transportation Review, 45*(1), 16-28. doi: 10.1016/j.tre.2008.02.007

Yang, P. C., Chung, S. L., Wee, H. M., Zahara, E., & Peng, C. Y. (2013). Collaboration for a closed-loop deteriorating inventory supply chain with multi-retailer and price-sensitive demand. *International Journal of Production Economics, 143*(2), 557-566. doi: 10.1016/j.ijpe.2012.07.020

Yang, P. C., Wee, H. M., Chung, S. L., & Ho, P. C. (2010). Sequential and global optimization for a closed-loop deteriorating inventory supply chain. *Mathematical and Computer Modelling, 52*(1-2), 161-176. doi: 10.1016/j.mcm.2010.02.005

Yoo, J. S., Hong, S. R., & Kim, C. O. (2009). Service level management of nonstationary supply chain using direct neural network controller. *Expert Systems with Applications, 36*(2), 3574-3586. doi: 10.1016/j.eswa.2008.02.005

Yoo, S. H., Kim, D., & Park, M. S. (2015). Pricing and return policy under various supply contracts in a closed-loop supply chain. *International Journal of Production Research, 53*(1), 106-126. doi: 10.1080/00207543.2014.932927

Yuan, K. F., & Gao, Y. (2010). Inventory decision-making models for a closed-loop supply chain system. *International Journal of Production Research, 48*(20), 6155-6187. doi: 10.1080/00207540903173637

Yuan, K. F., Ma, S. H., He, B., & Gao, Y. (2015). Inventory decision-making models for a closed-loop supply chain system with different decision-making structures. *International Journal of Production Research, 53*(1), 183-219. doi: 10.1080/00207543.2014.946160

Yuan, X. G., & Zhang, X. Q. (2015). Recycler Reaction for the Government Behavior in Closed-Loop Supply Chain Distribution Network: Based on the System Dynamics. *Discrete Dynamics in Nature and Society*, 11. doi: 10.1155/2015/206149

Zaarour, N., Melachrinoudis, E., Solomon, M. M., & Min, H. (2014). The optimal determination of the collection period for returned products in the sustainable supply chain. *International Journal of Logistics-Research and Applications, 17*(1), 35-45. doi: 10.1080/13675567.2013.836160

Zanoni, S., Ferretti, I., & Tang, O. (2006). Cost performance and bullwhip effect in a hybrid manufacturing and remanufacturing system with different control policies. *International Journal of Production Research, 44*(18-19), 3847-3862. doi: 10.1080/00207540600857375

Zarandi, M. H. F., Sisakht, A. H., & Davari, S. (2011). Design of a closed-loop supply chain (CLSC) model using an interactive fuzzy goal programming. *International Journal of Advanced Manufacturing Technology, 56*(5-8), 809-821. doi: 10.1007/s00170-011-3212-y

Zeballos, L. J., Gomes, M. I., Barbosa-Povoa, A. P., & Novais, A. Q. (2012). Addressing the uncertain quality and quantity of returns in closed-loop supply chains. *Computers & Chemical Engineering, 47*, 237-247. doi: 10.1016/j.compchemeng.2012.06.034

Zhang, G. T., Sun, H., Hu, J. S., & Dai, G. X. (2014). The Closed-Loop Supply Chain Network Equilibrium with Products Lifetime and Carbon Emission Constraints in Multiperiod Planning Horizon. *Discrete Dynamics in Nature and Society*, 16. doi: 10.1155/2014/784637

Zhang, J., Liu, X., & Tu, Y. (2011). A capacitated production planning problem for closed-loop supply chain with remanufacturing. *International Journal of Advanced Manufacturing Technology, 54*(5-8), 757-766. doi: 10.1007/s00170-010-2948-0

Zhang, P., Xiong, Y., Xiong, Z. K., & Yan, W. (2014). Designing contracts for a closed-loop supply chain under information asymmetry. *Operations Research Letters, 42*(2), 150-155. doi: 10.1016/j.orl.2014.01.004

Zhang, S. T., & Zhao, X. W. (2015). Fuzzy Robust Control for an Uncertain Switched Dual-Channel Closed-Loop Supply Chain Model. *Ieee Transactions on Fuzzy Systems, 23*(3), 485-500. doi: 10.1109/tfuzz.2014.2315659

Zhang, S. T., Zhao, X. W., & Zhang, J. T. (2014). Dynamic Model and Fuzzy Robust Control of Uncertain Closed-Loop Supply Chain with Time-Varying Delay in Remanufacturing. *Industrial & Engineering Chemistry Research, 53*(23), 9805-9811. doi: 10.1021/ie404104c

Zhang, S. Z., Lee, C. K. M., Chan, H. K., Choy, K. L., & Wu, Z. (2015). Swarm intelligence applied in green logistics: A literature review. *Engineering Applications of Artificial Intelligence, 37,* 154-169. doi: 10.1016/j.engappai.2014.09.007

Zhang, Z. H., Berenguer, G., & Shen, Z. J. (2015). A Capacitated Facility Location Model with Bidirectional Flows. *Transportation Science, 49*(1), 114-129. doi: 10.1287/trsc.2013.0496

Zhang, Z. H., Jiang, H., & Pan, X. Z. (2012). A Lagrangian relaxation based approach for the capacitated lot sizing problem in closed-loop supply chain. *International Journal of Production Economics, 140*(1), 249-255. doi: 10.1016/j.ijpe.2012.01.018

Zhang, Z. Z., Wang, Z. J., & Liu, L. W. (2015). Retail Services and Pricing Decisions in a Closed-Loop Supply Chain with Remanufacturing. *Sustainability, 7*(3), 2373-2396. doi: 10.3390/su7032373

Zhao, J., Liu, W. Y., & Wei, J. (2013). Pricing and Remanufacturing Decisions of a Decentralized Fuzzy Supply Chain. *Discrete Dynamics in Nature and Society*, 10. doi: 10.1155/2013/986704

Zhao, L. D., Liu, M., & Qu, L. B. (2009). Disruption Coordination Of Closed-Loop Supply Chain Network (Ii) - Analysis And Simulations. *International Journal of Innovative Computing Information and Control, 5*(2), 511-520.

Zhao, L. D., Qu, L. B., & Liu, M. (2008). Disruption Coordination Of Closed-Loop Supply Chain Network (I) - Models And Theorems. *International Journal of Innovative Computing Information and Control, 4*(11), 2955-2964.

Zhou, G. G., Yang, Y. X., & Cao, J. (2012). Competition and Integration in Closed-Loop Supply Chain Network with Variational Inequality. *Mathematical Problems in Engineering*, 21. doi: 10.1155/2012/524809

Zhou, L., & Disney, S. M. (2006). Bullwhip and inventory variance in a closed loop supply chain. *Or Spectrum, 28*(1), 127-149. doi: 10.1007/s00291-005-0009-0

Zhou, W., & Piramuthu, S. (2013). Remanufacturing with RFID item-level information: Optimization, waste reduction and quality improvement. *International Journal of Production Economics, 145*(2), 647-657. doi: 10.1016/j.ijpe.2013.05.019

Zhou, X. C., Zhao, Z. X., Zhou, K. J., & He, C. H. (2012). Remanufacturing closed-loop supply chain network design based on genetic particle swarm optimization algorithm. *Journal of Central South University of Technology, 19*(2), 482-487. doi: 10.1007/s11771-012-1029-y

Zhou, Y., Chan, C. K., HungWong, K., & Lee, Y. C. E. (2014). Closed-Loop Supply Chain Network under Oligopolistic Competition with Multiproducts, Uncertain Demands, and Returns. *Mathematical Problems in Engineering*, 15. doi: 10.1155/2014/912914

Zhou, Y., Chan, C. K., Kar, K. H., & Lee, Y. C. E. (2015). Intelligent Optimization Algorithms: A Stochastic Closed-Loop Supply Chain Network Problem Involving Oligopolistic Competition for Multiproducts and Their Product Flow Routings. *Mathematical Problems in Engineering*, 22. doi: 10.1155/2015/918705

Zhou, Y., Xiong, Y., Li, G. D., Xiong, Z. K., & Beck, M. (2013). The bright side of manufacturingremanufacturing conflict in a decentralised closed-loop supply chain. *International Journal of Production Research, 51*(9), 2639-2651. doi: 10.1080/00207543.2012.737956

Zou, Q. M., & Ye, G. Y. (2015). Pricing-Decision and Coordination Contract considering Product Design and Quality of Recovery Product in a Closed-Loop Supply Chain. *Mathematical Problems in Engineering*, 14. doi: 10.1155/2015/593123

Zuidwijk, R., & Krikke, H. (2008). Strategic response to EEE returns: Product eco-design or new recovery processes? *European Journal of Operational Research, 191*(3), 1206-1222. doi: 10.1016/j.ejor.2007.08.004

Appendix

The GAMS source code is developed based on the mathematical model formulation of Soleimani and Kannan (2015).

```
$Title Closed-Loop Supply Chain Optimization
$Ontext

Model 1 - Multi-period multi-product large scale closed-loop supply chain optimization - T.Y. 2016


GAMS code for Model 1 - T.Y. 2016



ALL parameters should be initialized appropriately and checked before running the optimization.


$Offtext


Sets
s suppliers /s1*s3/
f manufacturers /m1*m3/
w warehouses /w1*w2/
d distributors /d1*d3/
c first customers (retailers) /c1*c3/
a disassembly centers /a1*a3/
r redistributors /r1*r2/
p disposal locations /p1*p3/
k second customers /k1*k3/
u products /u1*u3/
t periods /t1*t5/
;

Parameters

D_(*,u,t)   demand of product u by the first customer c or second customer k in period t
P_(*,u,t)   unit price of product u at the first customer c or second customer k in period t
PH(c,u,t)   purchasing cost of product u at the first customer c in period t

F_(*)       fixed cost of the opening location i
DS(*,*)     distance between any two locations i and j
SC(*,u,t)   capacity of supplier s or warehouse w of product u in period t
SRC(s,u,t)  recycling capacity of supplier s of product u in period t
FC(f,u,t)   manufacturing capacity of manufacturer f of product u in period t
RFC(f,u,t)  remanufacturing capacity of manufacturer f of product u in period t
WC(w,u,t)   warehouse capacity in hours of warehouse w of product u in period t

DC(d,u,t)   capacity of distributor d of product u in period t
AC(a,u,t)   capacity of disassembly a of product u in period t
RC(r,u,t)   capacity of redistributor r of product u in period t
PC(p,u,t)   capacity of disposal center p of product u in period t

Mc(s,u,t)   material cost of product u per unit supplied by s in period t
Rc_(s,u,t)  recycling cost of product u per unit recycled by supplier s in period t
*Rc(s,u,t)  recycling cost of supplier s of product u in period t
```

```
Fc(f,u,t)   manufacturing cost of product u per unit manufactured by manufacturer f in period t
RFc(f,u,t)  remanufacturing cost of product u per unit by manufacturer f in period t
DAc(a,u,t)  disassembly cost of product u per unit by disassembly center a in period t
RPc(a,u,t)  repairing cost of product u per unit repaired by disassembly location a in period t
Pc_(*,u,t)  disposal cost of product u per unit disposed by disposal location p or disassembly center
a in period t

Nc(f,u,t)   non-utilized manufacturing capacity cost of product u of manufacturer f in period t
RNc(f,u,t)  non-utilized remanufacturing cost of product u of manufacturer f in period t
Sc_(u,t)    shortage cost of product u per unit in period t
Fh(f,u)     manufacturing time of product u per unit at manufacturer f
RFh(f,u)    remanufacturing time of product u per unit at manufacturer f

WH(w,u,t)   holding cost of product u per unit at the warehouse w in period t
DH(d,u,t)   holding cost of product u per unit at distributor store d in period t

B(*,*)      batch size of product u from supplier s - manufacturer f - distributor d - disassembly a -
redistributor r - warehouse w - customer c

Tc(u,t)     transportation cost of product u per unit per kilometer in period t
RR(u,t)     return ratio of product u at the first customers in period t

R_c         recycling ratio      Note: (R_c + R_m + R_r + R_p) =e= 1
R_m         remanufacturing ratio
R_r         repairing ratio
R_p         disposal ratio

M           large number

S__         size of s - max_number_allowable_locations_suppliers_S
F__         size of f - max_number_allowable_locations_manufacturers_F
D__         size of d - max_number_allowable_locations_distributors_D
W__         size of w - max_number_allowable_locations_warehouses_W
A__         size of a - max_number_allowable_locations_disassembly_centers_A
R__         size of r - max_number_allowable_locations_redistributors_R
P__         size of p - max_number_allowable_locations_disposal_locations_P

;

* Initialize parameters
*D(c,u,t)      = normal(500,10);


Binary variables

L(*)    binary variable equals 1 if location i s f d a r p w is open - 0 otherwise
Li(*,*) binary variable equals 1 if a transpor link is established between any two locations i and j

;

Positive variables

Q(*,*,u,t) flow of batches of product u from location x to location y in period t

R_(*,u,t)   the residual inventory of product u at warehouse w or distributor d in period t

;

Variable
```

```
UU Objective

;

Equations

balance_manufacturers                      The balance constraint of manufacturers
balance_warehouses                         The balance constraint of warehouses
balance_distributors                       The balance constraint of distributors
balance_customer_service_level             The balance constraint of customer service level
balance_disassembly_centers_inputs         The balance constraint of disassembly centers' inputs
balance_disassembly_centers_outputs        The balance constraint of disassembly centers' outputs
balance_recycling_rate                     The balance constraint of recycling rate
balance_remanufacture_rate                 The balance constraint of remanufacture rate
balance_repair_rate                        The balance constraint of repair rate
balance_disposal_rate                      The balance constraint of disposal rate
balance_Rc_Rm_Rr_Rp                        The balance constraint of Rc Rm Rr Rp
balance_manufacturers_reverse_flows        The balance constraint of manufacturers reverse flows
balance_redistributors                     The balance constraint of redistributors
balance_second_customers                   The balance constraint of second customers


capacity_suppliers_output                  The capacity constraint of suppliers output
capacity_manufacturers                     The capacity constraint of manufacturers
capacity_suppliers                         The capacity constraint of suppliers
capacity_distributors                      The capacity constraint of distributors
capacity_disassembly_centers               The capacity constraint of disassembly centers
capacity_redistributors                    The capacity constraint of redistributors
capacity_recycling_suppliers               The capacity constraint of recycling suppliers
capacity_disposal_centers                  The capacity constraint of disposal centers
capacity_warehouses                        The capacity constraint of warehouses

link_management_sf                         The constraint for link_management_sf
link_management_fd                         The constraint for link_management_fd
link_management_fw                         The constraint for link_management_fw
link_management_fc                         The constraint for link_management_fc
link_management_fk                         The constraint for link_management_fk
link_management_fr                         The constraint for link_management_fr
link_management_wd                         The constraint for link_management_wd
link_management_wc                         The constraint for link_management_wc
link_management_wk                         The constraint for link_management_wk
link_management_wr                         The constraint for link_management_wr
link_management_dc                         The constraint for link_management_dc
link_management_ca                         The constraint for link_management_ca
link_management_as                         The constraint for link_management_as
link_management_af                         The constraint for link_management_af
link_management_ar                         The constraint for link_management_ar
link_management_ap                         The constraint for link_management_ap
link_management_rk                         The constraint for link_management_rk

link_management_sf2                        The constraint for link_management_sf2
link_management_fd2                        The constraint for link_management_fd2
link_management_fw2                        The constraint for link_management_fw2
link_management_fc2                        The constraint for link_management_fc2
link_management_fk2                        The constraint for link_management_fk2
link_management_fr2                        The constraint for link_management_fr2
link_management_wd2                        The constraint for link_management_wd2
link_management_wc2                        The constraint for link_management_wc2
link_management_wk2                        The constraint for link_management_wk2
```

```
link_management_wr2                              The constraint for link_management_wr2
link_management_dc2                              The constraint for link_management_dc2
link_management_ca2                              The constraint for link_management_ca2
link_management_as2                              The constraint for link_management_as2
link_management_af2                              The constraint for link_management_af2
link_management_ar2                              The constraint for link_management_ar2
link_management_ap2                              The constraint for link_management_ap2
link_management_rk2                              The constraint for link_management_rk2


max_number_allowable_locations_suppliers_s            max_number_allowable_locations_suppliers_s
max_number_allowable_locations_manufacturers_f        max_number_allowable_locations_manufacturers_f
max_number_allowable_locations_distributors_d         max_number_allowable_locations_distributors_d
max_number_allowable_locations_warehouses_w           max_number_allowable_locations_warehouses_w
max_number_allowable_locations_disassembly_centers_a  max_allowable_locations_disassembly_centers_a
max_number_allowable_locations_redistributors_r       max_allowable_locations_redistributors_r
max_number_allowable_locations_disposal_locations_p   max_allowable_locations_disposal_locations_p

* Objective Function
 obj Objective function
;

balance_manufacturers(t,u,f)..

sum(s, Q(s,f,u,t)*B(s,u)) =e= sum(d, Q(f,d,u,t)*B(f,u)) + sum(w, Q(f,w,u,t)*B(f,u)) + sum(c,
Q(f,c,u,t)*B(f,u));


balance_warehouses(t,u,w)..

sum(f, Q(f,w,u,t)*B(f,u)) + R_(w,u,t-1) =e= R_(w,u,t) + sum(d, Q(w,d,u,t)*B(w,u)) + sum(c,
Q(w,c,u,t)*B(w,u)) + sum(k, Q(w,k,u,t)*B(w,u));


balance_distributors(t,u,d)..

sum(f, Q(f,d,u,t)*B(f,u)) + sum(w, Q(w,d,u,t)*B(w,u)) + R_(d,u,t-1) =e= R_(d,u,t) + sum(c,
Q(d,c,u,t)*B(d,u));


balance_customer_service_level(t,u,c)..

sum(d, Q(d,c,u,t)*B(d,u)) + sum(f, Q(f,c,u,t)*B(f,u)) + sum(w, Q(w,c,u,t)*B(w,u)) =g= 0.7*D_(c,u,t);


balance_disassembly_centers_inputs(t,u,c)..

sum(a, Q(c,a,u,t)*B(c,u)) =l= (sum(d, Q(d,c,u,t)*B(d,u)) + sum(f, Q(f,c,u,t)*B(f,u)) + sum(w,
Q(w,c,u,t)*B(w,u)))*RR(u,t);


balance_disassembly_centers_outputs(t,u,a)..

sum(c, Q(c,a,u,t)*B(c,u)) =e= (sum(s, Q(a,s,u,t)*B(a,u)) + sum(f, Q(a,f,u,t)*B(a,u)) + sum(r,
Q(a,r,u,t)*B(a,u)) + sum(p, Q(a,p,u,t)*B(a,u)) + sum(k, Q(a,k,u,t)*B(a,u)));


balance_recycling_rate(t,u,a)..

sum(c, Q(c,a,u,t)*B(c,u))*R_c =e= sum(s, Q(a,s,u,t)*B(a,u));
```

```
balance_remanufacture_rate(t,u,a)..

sum(c, Q(c,a,u,t)*B(c,u))*R_m =e= sum(f, Q(a,f,u,t)*B(a,u));


balance_repair_rate(t,u,a)..

sum(c, Q(c,a,u,t)*B(c,u))*R_r =e= sum(r, Q(a,r,u,t)*B(a,u));


balance_disposal_rate(t,u,a)..

sum(c, Q(c,a,u,t)*B(c,u))*R_p =e= sum(p, Q(a,p,u,t)*B(a,u));


balance_Rc_Rm_Rr_Rp..

(R_c + R_m + R_r + R_p) =e= 1;


balance_manufacturers_reverse_flows(t,u,f)..

sum(a, Q(a,f,u,t)*B(a,u)) =e= sum(r, Q(f,r,u,t)*B(f,u)) + sum(k, Q(f,k,u,t)*B(f,u)) + sum(w, sum(k,
Q(w,k,u,t)*B(w,u))) + sum(w, sum(r, Q(w,r,u,t)*B(w,u)));


balance_redistributors(t,u,r)..

sum(a, Q(a,r,u,t)*B(a,u)) + sum(f, Q(f,r,u,t)*B(f,u)) + sum(w, Q(w,r,u,t)*B(w,u)) =e= sum(k,
Q(r,k,u,t)*B(r,u));


balance_second_customers(t,u,k)..

sum(r, Q(r,k,u,t)*B(r,u)) =l= D_(k,u,t);

capacity_suppliers_output(t,u,s)..

sum(f, Q(s,f,u,t)*B(s,u)) =l= SC(s,u,t)*L(s);


capacity_manufacturers(t,u,f)..

(sum(d, Q(f,d,u,t)*B(f,u)) + sum(w, Q(f,w,u,t)*B(f,u)) + sum(c, Q(f,c,u,t)*B(f,u)) + sum(k,
Q(f,k,u,t)*B(f,u)))*Fh(f,u) =l= FC(f,u,t)*L(f);


capacity_suppliers(t,u,w)..

R_(w,u,t) =l= SC(w,u,t)*L(w);


capacity_distributors(t,u,d)..

sum(f, Q(f,d,u,t)*B(f,u)) + sum(w, Q(w,d,u,t)*B(w,u)) + R_(d,u,t-1) =l= DC(d,u,t)*L(d);


capacity_disassembly_centers(t,u,a)..

(sum(s, Q(a,s,u,t)*B(a,u)) + sum(f, Q(a,f,u,t)*B(a,u)) + sum(r, Q(a,r,u,t)*B(a,u)) +  sum(p,
```

```
Q(a,p,u,t)*B(a,u))) =l= AC(a,u,t)*L(a);


capacity_redistributors(t,u,r)..          sum(k, Q(r,k,u,t)*B(r,u)) =l= RC(r,u,t)*L(r);
capacity_recycling_suppliers(t,u,s)..     sum(a, Q(a,s,u,t)*B(a,u)) =l= SRC(s,u,t)*L(s);
capacity_disposal_centers(t,u,p)..        sum(a, Q(a,p,u,t)*B(a,u)) =l= PC(p,u,t)*L(p);
capacity_warehouses(t,u,w)..              sum(f, Q(f,w,u,t)*B(f,u)) =l= WC(w,u,t)*L(w);


link_management_sf(s,f)..      Li(s,f) =l= sum(u, sum(t, Q(s,f,u,t)));
link_management_sf2(s,f)..     sum(u, sum(t, Q(s,f,u,t))) =l= M*Li(s,f);
link_management_fd(f,d)..      Li(f,d) =l= sum(u, sum(t, Q(f,d,u,t)));
link_management_fd2(f,d)..     sum(u, sum(t, Q(f,d,u,t))) =l= M*Li(f,d);

link_management_fw(f,w)..      Li(f,w) =l= sum(u, sum(t, Q(f,w,u,t)));
link_management_fw2(f,w)..     sum(u, sum(t, Q(f,w,u,t))) =l= M*Li(f,w);

link_management_fc(f,c)..      Li(f,c) =l= sum(u, sum(t, Q(f,c,u,t)));
link_management_fc2(f,c)..     sum(u, sum(t, Q(f,c,u,t))) =l= M*Li(f,c);

link_management_fk(f,k)..      Li(f,k) =l= sum(u, sum(t, Q(f,k,u,t)));
link_management_fk2(f,k)..     sum(u, sum(t, Q(f,k,u,t))) =l= M*Li(f,k);

link_management_fr(f,r)..      Li(f,r) =l= sum(u, sum(t, Q(f,r,u,t)));
link_management_fr2(f,r)..     sum(u, sum(t, Q(f,r,u,t))) =l= M*Li(f,r);

link_management_wd(w,d)..      Li(w,d) =l= sum(u, sum(t, Q(w,d,u,t)));
link_management_wd2(w,d)..     sum(u, sum(t, Q(w,d,u,t))) =l= M*Li(w,d);

link_management_wc(w,c)..      Li(w,c) =l= sum(u, sum(t, Q(w,c,u,t)));
link_management_wc2(w,c)..     sum(u, sum(t, Q(w,c,u,t))) =l= M*Li(w,c);

link_management_wk(w,k)..      Li(w,k) =l= sum(u, sum(t, Q(w,k,u,t)));
link_management_wk2(w,k)..     sum(u, sum(t, Q(w,k,u,t))) =l= M*Li(w,k);

link_management_wr(w,r)..      Li(w,r) =l= sum(u, sum(t, Q(w,r,u,t)));
link_management_wr2(w,r)..     sum(u, sum(t, Q(w,r,u,t))) =l= M*Li(w,r);

link_management_dc(d,c)..      Li(d,c) =l= sum(u, sum(t, Q(d,c,u,t)));
link_management_dc2(d,c)..     sum(u, sum(t, Q(d,c,u,t))) =l= M*Li(d,c);

link_management_ca(c,a)..      Li(c,a) =l= sum(u, sum(t, Q(c,a,u,t)));
link_management_ca2(c,a)..     sum(u, sum(t, Q(c,a,u,t))) =l= M*Li(c,a);

link_management_as(a,s)..      Li(a,s) =l= sum(u, sum(t, Q(a,s,u,t)));
link_management_as2(a,s)..     sum(u, sum(t, Q(a,s,u,t))) =l= M*Li(a,s);

link_management_af(a,f)..      Li(a,f) =l= sum(u, sum(t, Q(a,f,u,t)));
link_management_af2(a,f)..     sum(u, sum(t, Q(a,f,u,t))) =l= M*Li(a,f);

link_management_ar(a,r)..      Li(a,r) =l= sum(u, sum(t, Q(a,r,u,t)));
link_management_ar2(a,r)..     sum(u, sum(t, Q(a,r,u,t))) =l= M*Li(a,r);

link_management_ap(a,p)..      Li(a,p) =l= sum(u, sum(t, Q(a,p,u,t)));
link_management_ap2(a,p)..     sum(u, sum(t, Q(a,p,u,t))) =l= M*Li(a,p);

link_management_rk(r,k)..      Li(r,k) =l= sum(u, sum(t, Q(r,k,u,t)));
link_management_rk2(r,k)..     sum(u, sum(t, Q(r,k,u,t))) =l= M*Li(r,k);

max_number_allowable_locations_suppliers_s..       sum(s, L(s)) =l= S__;
```

```
max_number_allowable_locations_manufacturers_f..        sum(f, L(f)) =l= F__;
max_number_allowable_locations_distributors_d..         sum(d, L(d)) =l= D__;
max_number_allowable_locations_warehouses_w..           sum(w, L(w)) =l= W__;
max_number_allowable_locations_disassembly_centers_a..  sum(a, L(a)) =l= A__;
max_number_allowable_locations_redistributors_r..       sum(r, L(r)) =l= R__;
max_number_allowable_locations_disposal_locations_p..   sum(p, L(p)) =l= P__;


* Objective Function
obj.. UU =e=
*Total sales
*Sales of all products
*First products sale (flows from distributors, manufacturers, and warehouses)
 sum(d, sum(c, sum(u, sum(t, Q(d,c,u,t)*B(d,u)*P_(c,u,t))))) +
 sum(f, sum(c, sum(u, sum(t, Q(f,c,u,t)*B(f,u)*P_(c,u,t))))) +
 sum(w, sum(c, sum(u, sum(t, Q(w,c,u,t)*B(w,u)*P_(c,u,t))))) +
*Second products sale (flows from redistributors, manufacturers, and warehouses)
 sum(r, sum(k, sum(u, sum(t, Q(r,k,u,t)*B(r,u)*P_(k,u,t))))) +
 sum(f, sum(k, sum(u, sum(t, Q(f,k,u,t)*B(f,u)*P_(k,u,t))))) +
 sum(w, sum(k, sum(u, sum(t, Q(w,k,u,t)*B(w,u)*P_(k,u,t))))) +


*Total costs
*Fixed costs (location costs)
 sum(s, F_(s)*L(s)) +
 sum(f, F_(f)*L(f)) +
 sum(d, F_(d)*L(d)) +
 sum(a, F_(a)*L(a)) +
 sum(r, F_(r)*L(r)) +
 sum(p, F_(p)*L(p)) +
 sum(w, F_(w)*L(w)) +
*Material costs
 sum(s, sum(f, sum(u, sum(t, Q(s,f,u,t)*B(s,u)*Mc(s,u,t))))) -
 sum(a, sum(s, sum(u, sum(t, Q(a,s,u,t)*B(a,u)*(Mc(s,u,t)-Rc_(s,u,t)))))) +
*Manufaturing costs
 sum(f, sum(d, sum(u, sum(t, Q(f,d,u,t)*B(f,u)*Fc(f,u,t))))) +
 sum(f, sum(w, sum(u, sum(t, Q(f,w,u,t)*B(f,u)*Fc(f,u,t))))) +
 sum(f, sum(c, sum(u, sum(t, Q(f,c,u,t)*B(f,u)*Fc(f,u,t))))) +
 sum(f, sum(k, sum(u, sum(t, Q(f,k,u,t)*B(f,u)*Fc(f,u,t))))) +
*Non-utilized capacity costs (for manufacturers)
 sum(f, sum(u, sum(t,((FC(f,u,t)/Fh(f,u))*L(f)
 - sum(d, Q(f,d,u,t)*B(f,u))
 - sum(w, Q(f,w,u,t)*B(f,u))
 - sum(c, Q(f,c,u,t)*B(f,u))
 + sum(w, sum(r, Q(w,r,u,t)*B(w,u)))
 + sum(w, sum(k, Q(w,k,u,t)*B(w,u))))*Nc(f,u,t)))) +
*Non-utilized capacity costs - Cont.'d - (for manufacturers)
 sum(f, sum(u, sum(t,((RFC(f,u,t)/RFh(f,u))*L(f)
 - sum(r, Q(f,r,u,t)*B(f,u))
 - sum(k, Q(f,k,u,t)*B(f,u))
 + sum(w, sum(r, Q(w,r,u,t)*B(w,u)))
 + sum(w, sum(k, Q(w,k,u,t)*B(w,u))))*RNc(f,u,t)))) +
* Shortage costs (for distributor)
*  sum(c, sum(u, sum(t, sum(t-1, D_(c,u,t)
*  - sum(t-1, sum(d, Q(d,c,u,t)*B(d,u)))
*  - sum(t-1, sum(f, Q(f,c,u,t)*B(f,u)))
*  - sum(t-1, sum(w, Q(w,c,u,k)*B(w,u))))*Sc_(u,t)))) +
*Purchasing costs (for distributors)
 sum(c, sum(a, sum(u, sum(t, Q(c,a,u,t)*PH(c,u,t)*B(c,u))))) +
*Disassembly costs
 sum(c, sum(a, sum(u, sum(t, Q(c,a,u,t)*B(c,u)*DAc(a,u,t))))) +
*Recycling costs
```

```
  sum(a, sum(s, sum(u, sum(t, Q(a,s,u,t)*B(a,u)*Rc_(s,u,t))))) +
*Remanufacturing costs
  sum(a, sum(f, sum(u, sum(t, Q(a,f,u,t)*B(a,u)*RFc(f,u,t))))) +
*Repairing costs
  sum(a, sum(r, sum(u, sum(t, Q(a,r,u,t)*B(a,u)*RPc(a,u,t))))) +
*Disposal costs
  sum(a, sum(p, sum(u, sum(t, Q(a,p,u,t)*B(a,u)*Pc_(p,u,t))))) +
*Transportation costs 1
  sum(t, sum(u, sum(s, sum(f, Q(s,f,u,t)*B(s,u)*Tc(u,t)*DS(s,f))))) +
  sum(t, sum(u, sum(f, sum(d, Q(f,d,u,t)*B(f,u)*Tc(u,t)*DS(f,d))))) +
  sum(t, sum(u, sum(f, sum(w, Q(f,w,u,t)*B(f,u)*Tc(u,t)*DS(f,w))))) +
*Transportation costs 2
  sum(t, sum(u, sum(f, sum(c, Q(f,c,u,t)*B(f,u)*Tc(u,t)*DS(f,c))))) +
  sum(t, sum(u, sum(f, sum(k, Q(f,k,u,t)*B(f,u)*Tc(u,t)*DS(f,k))))) +
  sum(t, sum(u, sum(w, sum(c, Q(w,c,u,t)*B(w,u)*Tc(u,t)*DS(w,c))))) +
*Transportation costs 3
  sum(t, sum(u, sum(w, sum(k, Q(w,k,u,t)*B(w,u)*Tc(u,t)*DS(w,k))))) +
  sum(t, sum(u, sum(d, sum(c, Q(d,c,u,t)*B(d,u)*Tc(u,t)*DS(d,c))))) +
  sum(t, sum(u, sum(a, sum(s, Q(a,s,u,t)*B(a,u)*Tc(u,t)*DS(a,s))))) +
*Transportation costs 4
  sum(t, sum(a, sum(u, sum(f, Q(a,f,u,t)*B(a,u)*Tc(u,t)*DS(a,f))))) +
  sum(t, sum(u, sum(a, sum(p, Q(a,p,u,t)*B(a,u)*Tc(u,t)*DS(a,p))))) +
  sum(t, sum(u, sum(a, sum(r, Q(a,r,u,t)*B(a,u)*Tc(u,t)*DS(a,r))))) +
*Transportation costs 5
  sum(t, sum(u, sum(f, sum(r, Q(f,r,u,t)*B(f,u)*Tc(u,t)*DS(f,r))))) +
  sum(t, sum(u, sum(w, sum(r, Q(w,r,u,t)*B(w,u)*Tc(u,t)*DS(w,r))))) +
  sum(t, sum(u, sum(r, sum(k, Q(r,k,u,t)*B(r,u)*Tc(u,t)*DS(r,k))))) +
*Transportation costs 6
  sum(t, sum(u, sum(c, sum(a, Q(c,a,u,t)*B(c,u)*Tc(u,t)*DS(c,a))))) +
  sum(t, sum(u, sum(w, sum(d, Q(w,d,u,t)*B(w,u)*Tc(u,t)*DS(w,d))))) +
  sum(t, sum(u, sum(a, sum(k, Q(a,k,u,t)*B(a,u)*Tc(u,t)*DS(a,k))))) +
*Inventory holding costs
  sum(w, sum(u, sum(t, R_(w,u,t)*WH(w,u,t)))) +
  sum(d, sum(u, sum(t, R_(d,u,t)*DH(d,u,t))))
;

Models model1 Closed-loop supply chain system / all /

Solve model1 minimizing UU using mip;
```

Chapter 8. Optimal scheduling of distributions to demanding nodes based on preferences

Abstract. Optimization issues of scheduling problems have long been examined for more than fifty years. Numerous studies are devoted to the scheduling issues in logistics and supply chain systems. Indeed, scheduling problems of logistics and supply chain systems have a combinatorial nature and thus computational complexity is usually very high. In this regard, this study reviews the existing literature and then examines an optimal scheduling problem of distributions to demanding nodes based on preferences. The scheduling problem is approached as a mixed integer program and an optimal solution is retrieved by using a solver methodology.

Keywords: scheduling, logistics and supply chain, mixed integer programming

1. Introduction

Scheduling has been extensively studied for over 50 years. This field has attracted the attention of researchers in management, industrial engineering, operations research and computer science (Jarboui *et al.* 2013). Indeed, the coordination of activities along different stages of the supply chain has received much attention in production management and operations research. The supply chain is a network of facilities and business entities (manufacturers, suppliers, warehouses, distributors and retailers), scattered over a wide geographical area, and in many cases worldwide (Khodr 2012).

It has been shown that the correct scheduling is a very difficult job. Standard operations research approaches such as mixed integer linear programming or dynamic programming are often of limited use because of their excessive calculation time (Jarboui *et al.* 2013). Because discrete decisions involved, these scheduling problems have a combinatorial nature and therefore challenging from computational complexity of the viewpoint. With the tools, scheduling studies can be performed to simulate the current conditions of the industry and to assist in the long-term planning of new facilities for a given production plan. The tools also provide an opportunity for the production engineer to do things such as optimizing the production process applying various mathematical optimization techniques (Khodr 2012). In this regard, this study reviews the existing literature and investigates an optimal scheduling problem of distributions to demanding nodes based on preferences. The scheduling problem is approached as a mixed integer program and an optimal solution is found by utilizing a solver method.

The remainder of this paper is organized as follows. Section 2 reviews the literature on studies in scheduling in general and scheduling algorithms in particular. Section 3 introduces the data and methods used for the scheduling optimization. Specifically, this section optimizes a logistics scheduling problem by applying an optimization technique. Section 4 presents and discusses the results. The study is concluded in Section 5.

2. Literature review

Many researchers have examined the scheduling issues in logistics. In recent years, more attention has been devoted to computer-aided scheduling problems in the process industries. Computer programs are now available so production engineer can develop a true idea of how a decision can be taken to make it easy to determine when, where and how to produce a set of products according to the requirements indicated in a period of specific time (Khodr 2012). For example, Wang *et al.* (2009) studied analysis and design of decision support system of disruption management in logistics scheduling. A vehicle monitoring and dispatching system was proposed to monitor and schedule the vehicles in logistics, and a decision support system to manage the disruption events. Agnetis *et al.* (2014) considered coordination of production and inter-stage batch delivery with outsourced distribution. Authors considered two transportation modes: regular transportation, for which delivery departure times were fixed at the beginning, and express transportation, for which delivery departure times were flexible.

Hall and Potts (2003) demonstrated supply chain scheduling of batching and delivery. Authors demonstrated that cooperation between a supplier and a manufacturer may reduce the total system cost by at least 20%, or 25%, or by up to 100%, depending upon the scheduling objective. Li *et al.* (2014) studied batch delivery scheduling with multiple decentralized manufacturers. The objective was to find a joint schedule of production and distribution to optimize the customer service level and delivery cost. Karimi and Davoudpour (2015) performed a branch and bound method for solving multi-factory supply chain scheduling with batch delivery. This study addressed the scheduling of supply chain with interrelated factories containing suppliers and manufacturers.

Chen and Lee (2008) studied logistics scheduling with batching and transportation. Their objective was to minimize the sum of weighted job delivery time and total transportation cost. Authors drew an overall picture of the problem complexity for various cases of problem parameters accompanied by polynomial algorithms for solvable cases. Selvarajah and Steiner (2006) studied batch scheduling in customer-centric supply chains. Authors studied batch arrival scheduling problems at the manufacturer in a multi-level customer-centric supply chain, where promised job due dates were considered constraints which must be satisfied. Averbakh and Baysan (2013) considered batching and delivery in semi-online distribution systems. Authors considered the semi-online environment where at any instant authors knew the orders that will be released in the next S time units, but had no information about the orders that would be released later. Agnetis *et al.* (2015) suggested two faster algorithms for coordination of production and batch delivery. Their note suggested faster algorithms for two integrated production/distribution problems studied earlier.

Qi (2005) discussed a logistics scheduling model for inventory cost reduction by batching. In their study, authors discussed a logistics scheduling model where the raw material was delivered to the shop in batches. They also observed some managerial insights. Selvarajah and Steiner (2009) studied approximation algorithms for the supplier's supply chain scheduling problem to minimize delivery and inventory holding costs. Authors studied the upstream supplier's batch scheduling problem in a supply chain. Selvarajah and Zhang (2014) studied supply chain scheduling at the manufacturer to minimize inventory holding and delivery costs. Since the problem with arbitrary processing times, release times and weights is strongly NP-hard, authors first analyzed some polynomially solvable special problems. Then authors developed a heuristic algorithm to solve the general problem.

Wang and Cheng (2009) studied logistics scheduling to minimize inventory and transport costs. Authors studied a logistics scheduling problem where a manufacturer receives raw materials from a supplier, manufactures products in a factory, and delivers the finished products to a customer. For the general problem, authors examined several special cases, identify their optimal properties, and developed polynomial-time algorithms to solve them optimally. Yeung *et al.* (2011) studied supply chain scheduling and coordination with dual delivery modes and inventory storage cost. Authors studied a two-echelon supply chain scheduling problem in which a manufacturer acquires supplies from an upstream supplier and processes orders from the downstream retailers. Authors developed two practically relevant and robust methods for the supply chain to achieve optimal profit-making performance through channel coordination. Gupta *et al.* (2009) performed sequencing deliveries to minimize inventory holding cost with dominant upstream supply chain partner. Authors developed algorithms for the distribution problem by exploiting its structural properties.

Badell *et al.* (2004) worked on planning, scheduling and budgeting value-added chains. The benefits of this work were shown through a case study that illustrates the modeling framework, the information flows and procedures necessary to implement a financial/supply chain scheduling methodology for the use of financial managers during planning and budgeting activities in process industries. Badell *et al.* (2005) studied optimal budget and cash flows during retrofitting periods in batch chemical process industries. The benefits of this work were demonstrated through a case study that illustrates the modeling framework, the information flows and procedures necessary to implement a financial/supply chain scheduling methodology to aid the high level staff during planning and budgeting activities.

Averbakh (2010) considered on-line integrated production-distribution scheduling problems with capacitated deliveries. Authors considered the capacitated case with an upper bound on the size of a batch. For several versions of the problem, authors presented efficient on-line algorithms, and used competitive analysis to study their worst-case performance. Averbakh and Baysan (2012) considered semi-online two-level supply chain scheduling problems. Averbakh and Baysan (2013) studied approximation algorithm for the on-line multi-customer two-level supply chain scheduling problem. Averbakh and Xue (2007) run on-line supply chain scheduling problems with preemption. Authors considered supply chain scheduling problems where customers release jobs to a manufacturer that has to process the jobs and deliver them to the customers. The objective was to minimize the total cost, which was the sum of the total flow time and the total delivery cost.

Tang *et al.* (2013) presented an improved ant colony optimization algorithm for three-tier supply chain scheduling based on networked manufacturing. This paper presented a study on supply chain scheduling from the perspective of networked manufacturing. The results obtained by applying the proposed algorithm to a real-life example showed that the presented scheduling optimization algorithm has better convergence, efficiency, and stability than conventional ant colony optimization. Pei *et al.* (2014) investigated the application of an effective modified gravitational search algorithm for the coordinated scheduling problem in a two-stage supply chain. This paper investigated a products and vehicles scheduling problem in a two-stage supply chain environment, where jobs first need to be processed on the serial batching machines of multiple manufacturers distributed in various geographic zones and then transported by vehicles to a customer for further processing. Terashima-Marin *et al.* (2005) studied scheduling transportation events with grouping genetic algorithms and the heuristic. Yimer and Demirli (2010) proposed a genetic approach to two-phase optimization of dynamic supply chain scheduling. In the proposed approach, the entire problem was first decomposed into two subsystems and evaluated sequentially. Naso *et al.* (2007) focused genetic algorithms for supply-chain scheduling as a case study in the distribution of ready-mixed concrete. This paper focused on the ready-mixed concrete delivery, in addition to the mentioned complexity, strict time-constraints forbid both earliness and lateness of the supply. A detailed case study derived from industrial data was used to illustrate the potential of the proposed approach.

Cakici *et al.* (2014) investigated scheduling parallel machines with single vehicle delivery. Authors investigated the integrated production and distribution scheduling problem in a supply chain. Bronja and Kudumovic (2012) presented a model of scheduling transportation vehicles in the function of improving transportation in supply chains. This paper presented a new, transformed, multi-criteria model for scheduling transportation vehicles in the process of goods transportation with maximal utilization of their capacities and minimal time needed for completing all jobs.

de Matta and Miller (2004) performed production and inter-facility transportation scheduling for a process industry. Using real and simulated data from a process industry firm, their computational study, which compared the production and transportation schedules obtained from coordinated scheduling and sequential scheduling, showed that coordinated schedules yield significant cost savings resulting from the modest use of the expensive fast transport mode, coordinated product changeovers between plants and reduced intermediate product inventories. Castelli *et al.* (2004) performed scheduling multimodal transportation systems. In this paper a Lagrangian based heuristic procedure for scheduling transportation networks was presented.

Furusho *et al.* (2008) proposed a distributed optimization method for simultaneous production scheduling and transportation routing in semiconductor fabrication bays. Authors proposed a decentralized optimization method for production scheduling, transportation routing for AGVs and motion planning for material handling robots simultaneously. Koc *et al.* (2013) examined a class of joint production and transportation planning problems under different delivery policies. Their paper examined a manufacturer's integrated planning problem for the production and the delivery of a set of orders. Pei *et al.* (2015) investigated coordination of production and transportation in supply chain scheduling. Their paper investigated a three-stage supply chain scheduling problem in the application area of aluminum production. The computational results showed the effectiveness of the proposed algorithms, especially for large-scale instances.

Sigurd *et al.* (2004) considered scheduling transportation of live animals to avoid the spread of diseases. Authors considered a variant of the VRP where the vehicles should deliver some goods between groups of customers. Stilgenbauer *et al.* (2001) studied scheduling transportation projects using a project planner as part of the software series in civil engineering technology independent learning experiment at a state college.

Wang and Lee (2005) performed production and transport logistics scheduling with two transport mode choices. This paper considered a new class of scheduling problems arising in logistics systems in which two different transportation modes were available at the stage of product delivery. Computational results showed that our branch and bound algorithm is more efficient than CPLEX. Delavar *et al.* (2010) studied genetic algorithms for coordinated scheduling of production and air transportation.

Gordon and Strusevich (2009) performed single machine scheduling and due date assignment with dependent processing times. Authors considered single machine scheduling and due date assignment problems in which the processing time of a job depends on its position in a processing sequence. Chang *et al.* (2013) studied applied column generation-based approach to solve supply chain scheduling problems. This paper studied a supply chain scheduling problem in which the production stage was modeled by an identical parallel machine scheduling problem and the distribution stage was modeled by a capacitated vehicle routing problem. The results of the computational experiments indicated that the proposed approach can solve the test problems to optimality.

Han *et al.* (2015) investigated on-line supply chain scheduling for single-machine and parallel-machine configurations with a single customer for minimizing the makespan and delivery cost. This paper investigated minimization of both the makespan and delivery costs in on-line supply chain scheduling for single-machine and parallel-machine configurations in a transportation system with a single customer. Lee *et al.* (2006) studied two-machine scheduling under disruptions with transportation considerations. Authors studied problems with different related costs.

Blocher and Chhajed (2008) studied minimizing customer order lead-time in a two-stage assembly supply chain. Authors looked at a two-stage assembly supply chain with the objective of minimizing the average customer order lead-time. Cakici *et al.* (2012) studied multi-objective analysis of an integrated supply chain scheduling problem. Choi *et al.* (2013) developed scheduling and co-ordination of multi-suppliers single-warehouse-operator single-manufacturer supply chains with variable production rates and storage costs. Authors then developed a theorem and two algorithms to solve the optimal scheduling problems in both the decentralized and centralized supply chains. Sawik (2009) studied coordinated supply chain scheduling. Numerical examples modeled after a real-world integrated scheduling in a customer driven supply chain in the electronics industry were presented and some computational results were reported.

Agnetis *et al.* (2006) performed supply chain scheduling for sequence coordination. Authors described efficient algorithms for all the supplier's and manufacturers' problems, as well as for a special case of the joint scheduling problem. Yeung *et al.* (2010) worked on optimal scheduling of a single-supplier single-manufacturer supply chain with common due windows. Authors studied a supply chain scheduling control problem involving a single supplier, a single manufacturer and multiple retailers, where the manufacturer with limited production capacity can only take some of the orders of the retailers. Ivanov *et al.* (2014) researched multi-stage supply chain scheduling with non-preemptive continuous operations and execution control. An integrated multi-stage scheduling and routing problem with alternative machines at each supply chain stage and non-preemptive operations were studied. Chen and Lee (2004) studied multi-objective optimization of multi-echelon supply chain networks with uncertain product demands and prices.

Fan and Lu (2015) studied supply chain scheduling problem in a hospital with periodic working time on a single machine. Their goal was to minimize sum of the total delivery time and the total delivery cost. Numerical simulation results showed that the approximation algorithm performs efficiently. Chang et al. (2014) proposed greedy-search-based multi-objective genetic algorithm for emergency logistics scheduling. To enable the immediate and efficient dispatch of relief to victims of disaster, this study proposed a greedy-search-based, multi-objective, genetic algorithm capable of regulating the distribution of available resources and automatically generating a variety of feasible emergency logistics schedules for decision-makers.

Chen and Hall (2007) studied supply chain scheduling by considering conflict and cooperation in assembly systems. Authors studied conflict and cooperation issues in supply chain manufacturing. Dawande et al. (2006) studied supply chain scheduling for distribution systems. Ekinci et al. (2015) proposed optimization of ATM cash replenishment with group-demand forecasts. This article proposed grouping ATMs into nearby-location clusters and also optimizing the aggregates of daily cash withdraws in the forecasting process. Gu et al. (2015) studied a mutualism quantum genetic algorithm to optimize the flow shop scheduling with pickup and delivery considerations. Li and Xiao (2004) researched lot streaming with supplier-manufacturer coordination. Authors developed and analyzed coordination mechanisms that enable different parties in the supply chain to coordinate their lot splitting decisions so as to achieve a system wide optimum.

Ivanov and Sokolov (2012) researched dynamic supply chain scheduling. Manoj et al. (2008) studied supply chain scheduling for just-in-time environment. Authors developed mathematical models for individual optimization goals of the two partners and compared the results of these models with the results obtained for a joint optimization model at the system level.

Moghaddam and Nof (2015) studied best-matching with interdependent preferences-implications for capacitated cluster formation and evolution. Mazdeh and Karamouzian (2014) deal with evaluating strategic issues in supply chain scheduling using game theory. This paper deal with the problem of scheduling and batch delivery of orders in a supply chain including a supplier, a manufacturer and a final customer. The numerical examples showed the superiority of integrated decisions over independent actions and also the importance of the sharing mechanism. Minguez et al. (2011) presented reliability and decomposition techniques to solve certain class of stochastic programming problems. This paper presented a new approach for solving a certain type of stochastic programming problems presenting the some characteristics.

Ng *et al.* (2008) studied a supply scheduling problem with non-monotone cost functions. Chauhan *et al.* presented a fully polynomial time approximation scheme for a supply scheduling problem, which was to minimize a total cost associated with the sizes of deliveries from several providers to one manufacturer.

Palander (2011) studied modeling renewable supply chain for electricity generation with forest, fossil, and wood-waste fuels. In this paper, a multiple objective model to large-scale and long-term industrial energy supply chain scheduling problems was considered.

Rasti-Barzoki and Hejazi (2013) studied minimizing the weighted number of tardy jobs with due date assignment and capacity-constrained deliveries for multiple customers in supply chains. In this paper, an integrated due date assignment and production and batch delivery scheduling problem for make-to-order production system and multiple customers was addressed. Computational tests were used to demonstrate the efficiency of the developed methods. Rasti-Barzoki and Hejazi (2015) studied pseudo-polynomial dynamic programming for an integrated due date assignment, resource allocation, production, and distribution scheduling model in supply chain scheduling. In this study, authors considered an integrated due date assignment, production, and batch delivery scheduling problem with controllable processing times for multiple customers in a supply chain.

Qi (2006) studied a logistics scheduling model for scheduling and transshipment for two processing centers. Authors study problems with different objective functions and constraints, and propose various algorithms to solve these problems. Ren *et al.* (2013) researched the complexity of two supply chain scheduling problems. Authors considered an assembling manufacture system where several suppliers provide component parts to a manufacturer, who assembles products from all the supplied component parts. Qi (2008) studied coordinated logistics scheduling for in-house production and outsourcing. In this paper, authors addressed a new scheduling model for a firm with an option of outsourcing.

Selvarajah and Zhang (2014) studied supply chain scheduling to minimize holding costs with outsourcing. This paper addressesed a scheduling problem in a flexible supply chain, in which the jobs can be either processed in house, or outsourced to a third-party supplier. Ruiz-Torres *et al.* (2006) studied generating Pareto schedules with outsource and internal parallel resources. This paper addressed a supply chain scheduling problem where both internal and external/outsourced parallel resources were available and the objectives were to minimize the number of late orders and the total outsource machine time.

Steiner and Zhang (2009) studied approximation algorithms for minimizing the total weighted number of late jobs with late deliveries in two-level supply chains. Authors studied a supply chain scheduling problem in which n jobs had to be scheduled on a single machine and delivered to m customers in batches.

Steiner and Zhang (2011) studied minimizing the weighted number of tardy jobs with due date assignment and capacity-constrained deliveries. Authors studied a supply chain scheduling problem, where a common due date was assigned to all jobs and the number of jobs in delivery batches was constrained by the batch size. Rasti-Barzoki *et al.* (2013) studied a branch and bound algorithm to minimize the total weighed number of tardy jobs and delivery costs. This paper addressed the production and delivery scheduling integration problem; a manufacturer receives n orders from one customer while the orders need to be processed on one or two machines and be sent to the customer in batches. Results of computational tests showed significant improvement over an existing dynamic programming method.

Steiner and Zhang (2011) presented revised delivery-time quotation in scheduling with tardiness penalties. Authors presented a model for the rescheduling of orders with simultaneous assignment of attainable revised due dates to minimize due date escalation and tardiness penalties for the supplier. Ruiz-Torres *et al.* (2008) studied minimizing the average tardiness as the case of outsource machines. This article deal with the problem of finding outsourcing strategies or solutions that consider trade-offs between outsourcing cost and average tardiness, an important measure of lost customer goodwill. The article presented lower bounds for the problem, which were used for comparisons.

Tang *et al.* (2014) investigated operations research that transforms Baosteel's operations. Surmann and Morales (2002) studied scheduling tasks to a team of autonomous mobile service robots in indoor environments. This paper presented a complete system for scheduling transportation orders to a fleet of autonomous mobile robots in service environments. One challenging key problem the multi robot cooperation was solved by the scheduling algorithms and by giving autonomy to the service robots.

Ullrich (2012) studied supply chain scheduling of makespan reduction potential. Ullrich (2013) studied integrated machine scheduling and vehicle routing with time windows. This paper integrated production and outbound distribution scheduling in order to minimize total tardiness. A genetic algorithm approach was introduced to solve the integrated problem as a whole. Tzur and Drezner (2011) introduced a look ahead partitioning heuristic for a new assignment and scheduling problem in a distribution system. Authors introduced a new assignment and scheduling problem in a distribution system, which authors referred to as the ASTV problem: Assigning and Scheduling transportation Tasks to Vehicles.

Yao (2011) studied supply chain scheduling optimization in mass customization based on dynamic profit preference and application case study. In this article, authors discussed the supply chain scheduling optimization in mass customization based on dynamic profit preference to solve these contradictions and bottlenecks, established a special optimization model to implement the scheduling. Yao (2013) performed scheduling optimization of co-operator selection and task allocation in mass customization supply chain based on collaborative benefits and risks. Wang *et al.* (2015) studied supply chain scheduling with receiving deadlines and non-linear penalty. Authors studied the operations scheduling problem with delivery deadlines in a three-stage supply chain process. Yao and Liu (2009) studied optimization analysis of supply chain scheduling in mass customization.

The present study adds another dimension to the existing literature. Specifically, it examines a scheduling model for distribution and customers.

3. A sample scheduling model

Table 1. Sets, constants, and variables

c	demand nodes
d	distributors
n	Distribution number
s	Timeslot
$x_{d,c,s}$	assign Distribution/demand nodes/slot
$xds_{d,s}$	assign Distribution/slot
$xdc_{d,c}$	Distribution assign Distribution /demand nodes
$\gamma_{c,d}$	0..10, 10 is highest priority
Q	Capacity1
K	Capacity2
P	Capacity3

Maximize

$$\sum_d \sum_c (\gamma_{c,d}) x d c_{d,c}$$

Subject to

$$xds_{d,s} \leq \sum_c x_{d,c,s} \qquad \forall d,s$$

$$xds_{d,s} \geq x_{d,c,s} \qquad \forall d,c,s$$

$$xdc_{d,c} = \sum_s x_{d,c,s} \qquad \forall d,c$$

$$\sum_s xds_{d,s} = Q_d \qquad \forall d$$

$$\sum_d xds_{d,s} \leq K \qquad \forall s$$

$$\sum_d x_{d,c,s} = 1 \qquad \forall c,s$$

$$\sum_c x_{d,c,s} \leq P \qquad \forall c,s$$

4. Optimization of the sample model

Fair Isaac Corporation (FICO) Xpress solver on GAMS environment is used to solve the problem with the size of 269 rows, 216 structural columns and 1044 non-zero elements (See Tables 2 through Table 5).

Table 2. Initial iterations summary

Its	Obj Value	S	Ninf	Nneg	Sum Inf	Time
0	222.000000	D	1	0	.0	0
100	189.000000	D	52	0	86.0	0
200	182.999999	D	27	0	9.0	0
223	183.000000	P	0	0	.0	0
223	183.000000	P	0	0	.0	0
223	183.000000	P	0	0	.0	0

Optimal solution found. LP relaxation solved: objective = 183. Solver is starting root cutting and heuristics.

Table 3. Final iterations summary. Number of integer feasible solutions found is 0. Best bound is 183.

Its	Type	BestSoln	BestBound	Sols	Add	Del	Gap	GInf	Time
+		165	183.000000		1			10.91%	0
+		171	183.000000	2			7.02%	0	0
1	K	171	181.842857	2	65	0	6.34%	109	0
2	K	171	180.850575	2	42	32	5.76%	108	0
3	K	171	180.376392	2	18	35	5.48%	121	0
4	K	171	180.009857	2	15	18	5.27%	114	0
5	K	171	179.884982	2	14	21	5.20%	120	0
6	K	171	179.820164	2	26	7	5.16%	123	0
7	K	171	179.811717	2	10	8	5.15%	123	0
8	K	171	179.811574	2	1	8	5.15%	125	0
9	K	171	179.811384	2	4	0	5.15%	123	0
10	K	171	179.811086	2	1	15	5.15%	125	0
11	K	171	179.811086	2	0	1	5.15%	125	0
+		178	179.811086	3			1.02%	0	0

Number of integer feasible solutions found is 3. Best integer solution found is178 and best bound is 179.764585. Presolved problem has 88 rows, 114 columns, and 426 non-zeros.

Table 4. Iterations summary

Its	Obj Value	S	Ninf	Nneg	Sum Inf	Time
0	178.000000	p	4	0	4.000000	0
51	178.000000	P	0	0	.000000	0
Uncrunching matrix						
51	178.000000	P	0	0	.000000	0

Optimal solution found and fixed LP solved successfully, objective = 178. Integer solution satisfies relative optimality tolerance of 0.1.

Table 5. Solution

MIP solution	178.000000	
Best possible	179.764585	
Absolute gap	1.764585	Optca: 0.00
Relative gap	0.009816	Optcr: 0.10

The purpose of supply chain scheduling is to optimize the short and medium term decisions in supply chains, given the trade-off between the tangible economic objectives such as reducing costs or profit maximization and less tangible goals such as customer satisfaction or customer service level (Sawik 2011).

Classical scheduling problems consist of finding a schedule of a set of tasks and also execution dates of these tasks, minimize or maximize objective function considered in a given set of constraints. They vary by type of workshop and production constraints (Jarboui et al 2013). In the short term scheduling supply chain is generally concerned about the distribution of tasks and resources in a single facility and sequence and timing of detailed decisions on a short-term horizon (e.g., a shift or day) to complete a number of jobs in such a way that one or more job completion time-related objectives are minimized (Sawik 2011). A typical single facility considered in a supply chain customer focus is a single-stage set of parallel machines or a multistage flow shop or job shop with single or parallel machines. However, planning for the medium term supply chain (also called planning) deals with the allocation of tasks and resources of one or more interconnected facilities on a longer time horizon (e.g. several shifts, a week, or month) to complete a number of customer orders for finished products so that the level of customer service is maximized and one or more cost objectives are minimized (Sawik 2011).

5. Conclusions

This study reviewed the literature and examined a sample scheduling problem of distributions to demanding nodes based on preferences. The scheduling problem was approached as a mixed integer program and an optimal solution was retrieved by using a solver methodology.

6. References

Agnetis, A., Aloulou, M. A., & Fu, L. L. (2014). Coordination of production and interstage batch delivery with outsourced distribution. *European Journal of Operational Research, 238*(1), 130-142. doi: 10.1016/j.ejor.2014.03.039

Agnetis, A., Aloulou, M. A., Fu, L. L., & Kovalyov, M. Y. (2015). Two faster algorithms for coordination of production and batch delivery: A note. *European Journal of Operational Research, 241*(3), 927-930. doi: 10.1016/j.ejor.2014.10.005

Agnetis, A., Hall, N. G., & Pacciarelli, D. (2006). Supply chain scheduling: Sequence coordination. *Discrete Applied Mathematics, 154*(15), 2044-2063. doi: 10.1016/j.dam.2005.04.019

Averbakh, I. (2010). On-line integrated production-distribution scheduling problems with capacitated deliveries. *European Journal of Operational Research, 200*(2), 377-384. doi: 10.1016/j.ejor.2008.12.030

Averbakh, I., & Baysan, M. (2012). Semi-online two-level supply chain scheduling problems. *Journal of Scheduling, 15*(3), 381-390. doi: 10.1007/s10951-011-0264-7

Averbakh, I., & Baysan, M. (2013). Approximation algorithm for the on-line multi-customer two-level supply chain scheduling problem. *Operations Research Letters, 41*(6), 710-714. doi: 10.1016/j.orl.2013.10.002

Averbakh, I., & Baysan, M. (2013). Batching and delivery in semi-online distribution systems. *Discrete Applied Mathematics, 161*(1-2), 28-42. doi: 10.1016/j.dam.2012.08.003

Averbakh, I., & Xue, Z. H. (2007). On-line supply chain scheduling problems with preemption. *European Journal of Operational Research, 181*(1), 500-504. doi: 10.1016/j.ejor.2006.06.004

Badell, M., Romero, J., Huertas, R., & Puigjaner, L. (2004). Planning, scheduling and budgeting value-added chains. *Computers & Chemical Engineering, 28*(1-2), 45-61. doi: 10.1016/s0098-1354(03)00163-7

Badell, M., Romero, J., & Puigjaner, L. (2005). Optimal budget and cash flows during retrofitting periods in batch chemical process industries. *International Journal of Production Economics, 95*(3), 359-372. doi: 10.1016/j.ijpe.2003.06.002

Blocher, J. D., & Chhajed, D. (2008). Minimizing customer order lead-time in a two-stage assembly supply chain. *Annals of Operations Research, 161*(1), 25-52. doi: 10.1007/s10479-007-0289-7

Bronja, H., & Kudumovic, D. (2012). The model of scheduling transportation vehicles in the function of improving transportation in supply chains. *Technics Technologies Education Management-Ttem, 7*(3), 1063-1071.

Cakici, E., Mason, S. J., Geismar, H. N., & Fowler, J. W. (2014). Scheduling parallel machines with single vehicle delivery. *Journal of Heuristics, 20*(5), 511-537. doi: 10.1007/s10732-014-9249-y

Cakici, E., Mason, S. J., & Kurz, M. E. (2012). Multi-objective analysis of an integrated supply chain scheduling problem. *International Journal of Production Research, 50*(10), 2624-2638. doi: 10.1080/00207543.2011.578162

Castelli, L., Pesenti, R., & Ukovich, W. (2004). Scheduling multimodal transportation systems. *European Journal of Operational Research, 155*(3), 603-615. doi: 10.1016/j.egor.2003.02.002

Chang, F. S., Wu, J. S., Lee, C. N., & Shen, H. C. (2014). Greedy-search-based multi-objective genetic algorithm for emergency logistics scheduling. *Expert Systems with Applications, 41*(6), 2947-2956. doi: 10.1016/j.eswa.2013.10.026

Chang, Y. C., Chang, K. H., & Chang, T. K. (2013). Applied column generation-based approach to solve supply chain scheduling problems. *International Journal of Production Research, 51*(13), 4070-4086. doi: 10.1080/00207543.2013.774476

Chen, B., & Lee, C. Y. (2008). Logistics scheduling with batching and transportation. *European Journal of Operational Research, 189*(3), 871-876. doi: 10.1016/j.ejor.2006.11.047

Chen, C. L., & Lee, W. C. (2004). Multi-objective optimization of multi-echelon supply chain networks with uncertain product demands and prices. *Computers & Chemical Engineering, 28*(6-7), 1131-1144. doi: 10.1016/j.compchemeng.2003.09.014

Chen, Z. L., & Hall, N. G. (2007). Supply chain scheduling: Conflict and cooperation in assembly systems. *Operations Research, 55*(6), 1072-1089. doi: 10.1287/opre.1070.0412

Choi, T. M., Yeung, W. K., & Cheng, T. C. E. (2013). Scheduling and co-ordination of multi-suppliers single-warehouse-operator single-manufacturer supply chains with variable production rates and storage costs. *International Journal of Production Research, 51*(9), 2593-2601. doi: 10.1080/00207543.2012.737949

Dawande, M., Geismar, H. N., Hall, N. G., & Sriskandarajah, C. (2006). Supply chain scheduling: Distribution systems. *Production and Operations Management, 15*(2), 243-261.

de Matta, R. T., & Miller, T. (2004). Production and inter-facility transportation scheduling for a process industry. *European Journal of Operational Research, 158*(1), 72-88. doi: 10.1016/s0377-2217(03)00358-8

Delavar, M. R., Hajiaghaei-Keshteli, M., & Molla-Alizadeh-Zavardehi, S. (2010). Genetic algorithms for coordinated scheduling of production and air transportation. *Expert Systems with Applications, 37*(12), 8255-8266. doi: 10.1016/j.eswa.2010.05.060

Ekinci, Y., Lu, J. C., & Duman, E. (2015). Optimization of ATM cash replenishment with group-demand forecasts. *Expert Systems with Applications, 42*(7), 3480-3490. doi: 10.1016/j.eswa.2014.12.011

Fan, J., & Lu, X. W. (2015). Supply chain scheduling problem in the hospital with periodic working time on a single machine. *Journal of Combinatorial Optimization, 30*(4), 892-905. doi: 10.1007/s10878-015-9857-y

Furusho, T., Nishi, T., & Konishi, M. (2008). Distributed optimization method for simultaneous production scheduling and transportation routing in semiconductor fabrication bays. *International Journal of Innovative Computing Information and Control, 4*(3), 559-575.

Gordon, V. S., & Strusevich, V. A. (2009). Single machine scheduling and due date assignment with positionally dependent processing times. *European Journal of Operational Research, 198*(1), 57-62. doi: 10.1016/j.ejor.2008.07.044

Gu, J. W., Gu, M. Z., & Gu, X. S. (2015). A Mutualism Quantum Genetic Algorithm to Optimize the Flow Shop Scheduling with Pickup and Delivery Considerations. *Mathematical Problems in Engineering,* 17. doi: 10.1155/2015/387082

Gupta, S., Vanajakumari, M., & Sriskandarajah, C. (2009). Sequencing deliveries to minimize inventory holding cost with dominant upstream supply chain partner. *Journal of Systems Science and Systems Engineering, 18*(2), 159-183. doi: 10.1007/s11518-009-5107-0

Hall, N. G., & Potts, C. N. (2003). Supply chain scheduling: Batching and delivery. *Operations Research, 51*(4), 566-584. doi: 10.1287/opre.51.4.566.16106

Han, B., Zhang, W. J., Lu, X. W., & Lin, Y. Z. (2015). On-line supply chain scheduling for single-machine and parallel-machine configurations with a single customer: Minimizing the makespan and delivery cost. *European Journal of Operational Research, 244*(3), 704-714. doi: 10.1016/j.ejor.2015.02.008

Ivanov, D., & Sokolov, B. (2012). Dynamic supply chain scheduling. *Journal of Scheduling, 15*(2), 201-216. doi: 10.1007/s10951-010-0189-6

Ivanov, D., Sokolov, B., & Dolgui, A. (2014). Multi-stage supply chain scheduling with non-preemptive continuous operations and execution control. *International Journal of Production Research, 52*(13), 4059-4077. doi: 10.1080/00207543.2013.793429

Jarboui, Bassem, Siarry, Patrick, and Teghem, Jacques, eds. Metaheuristics for Production Scheduling. Somerset, NJ, USA: John Wiley & Sons, 2013.

Karimi, N., & Davoudpour, H. (2015). A branch and bound method for solving multi-factory supply chain scheduling with batch delivery. *Expert Systems with Applications, 42*(1), 238-245. doi: 10.1016/j.eswa.2014.07.025

Khodr, Hussein M., ed. Computer Science, Technology and Applications: Scheduling Problems and Solutions. New York, NY, USA: Nova, 2012.

Koc, U., Toptal, A., & Sabuncuoglu, I. (2013). A class of joint production and transportation planning problems under different delivery policies. *Operations Research Letters, 41*(1), 54-60. doi: 10.1016/j.orl.2012.11.002

Lee, C. Y., Leung, J. Y. T., & Yu, G. (2006). Two machine scheduling under disruptions with transportation considerations. *Journal of Scheduling, 9*(1), 35-48. doi: 10.1007/s10951-006-5592-7

Li, C. L., & Xiao, W. Q. (2004). Lot streaming with supplier-manufacturer coordination. *Naval Research Logistics, 51*(4), 522-542. doi: 10.1002/nav.20013

Li, S., Zhong, X. L., Li, H., & Li, S. J. (2014). Batch Delivery Scheduling with Multiple Decentralized Manufacturers. *Mathematical Problems in Engineering*, 7. doi: 10.1155/2014/321513

Manoj, U. V., Gupta, J. N. D., Gupta, S. K., & Sriskandarajah, C. (2008). Supply chain scheduling: Just-in-time environment. *Annals of Operations Research, 161*(1), 53-86. doi: 10.1007/s10479-007-0290-1

Mazdeh, M. M., & Karamouzian, A. (2014). Evaluating strategic issues in supply chain scheduling using game theory. *International Journal of Production Research, 52*(23), 7100-7113. doi: 10.1080/00207543.2014.937880

Minguez, R., Conejo, A. J., & Garcia-Bertrand, R. (2011). Reliability and decomposition techniques to solve certain class of stochastic programming problems. *Reliability Engineering & System Safety, 96*(2), 314-323. doi: 10.1016/j.ress.2010.09.011

Moghaddam, M., & Nof, S. Y. (2015). Best-matching with interdependent preferences-implications for capacitated cluster formation and evolution. *Decision*

Support Systems, 79, 125-137. doi: 10.1016/j.dss.2015.08.005

Naso, D., Surico, M., Turchiano, B., & Kaymak, U. (2007). Genetic algorithms for supply-chain scheduling: A case study in the distribution of ready-mixed concrete. *European Journal of Operational Research, 177*(3), 2069-2099. doi: 10.1016/j.ejor.2005.12.019

Ng, C. T., Kovalyov, M. Y., & Cheng, T. C. E. (2008). An FPTAS for a supply scheduling problem with non-monotone cost functions. *Naval Research Logistics, 55*(3), 194-199. doi: 10.1002/nav.20276

Palander, T. (2011). Modelling renewable supply chain for electricity generation with forest, fossil, and wood-waste fuels. *Energy, 36*(10), 5984-5993. doi: 10.1016/j.energy.2011.08.017

Pei, J., Liu, X. B., Pardalos, P. M., Fan, W. J., Yang, S. L., & Wang, L. (2014). Application of an effective modified gravitational search algorithm for the coordinated scheduling problem in a two-stage supply chain. *International Journal of Advanced Manufacturing Technology, 70*(1-4), 335-348. doi: 10.1007/s00170-013-5263-8

Pei, J., Pardalos, P. M., Liu, X. B., Fan, W. J., Yang, S. L., & Wang, L. (2015). Coordination of Production and Transportation In Supply Chain Scheduling. *Journal of Industrial and Management Optimization, 11*(2), 399-419. doi: 10.3934/jimo.2015.11.399

Qi, X. T. (2005). A logistics scheduling model: Inventory cost reduction by batching. *Naval Research Logistics, 52*(4), 312-320. doi: 10.1002/nav.20078

Qi, X. T. (2006). A logistics scheduling model: scheduling and transshipment for two processing centers. *Iie Transactions, 38*(7), 609-618. doi: 10.1080/074081791009022

Qi, X. T. (2008). Coordinated logistics scheduling for in-house production and outsourcing. *Ieee Transactions on Automation Science and Engineering, 5*(1), 188-192. doi: 10.1109/tase.2006.887159

Rasti-Barzoki, M., & Hejazi, S. R. (2013). Minimizing the weighted number of tardy jobs with due date assignment and capacity-constrained deliveries for multiple customers in supply chains. *European Journal of Operational Research, 228*(2), 345-357. doi: 10.1016/j.ejor.2013.01.002

Rasti-Barzoki, M., & Hejazi, S. R. (2015). Pseudo-polynomial dynamic programming for an integrated due date assignment, resource allocation, production, and distribution scheduling model in supply chain scheduling. *Applied Mathematical*

Modelling, 39(12), 3280-3289. doi: 10.1016/j.apm.2014.11.031

Rasti-Barzoki, M., Hejazi, S. R., & Mazdeh, M. M. (2013). A branch and bound algorithm to minimize the total weighed number of tardy jobs and delivery costs. *Applied Mathematical Modelling, 37*(7), 4924-4937. doi: 10.1016/j.apm.2012.10.001

Ren, J. F., Du, D. L., & Xu, D. C. (2013). The complexity of two supply chain scheduling problems. *Information Processing Letters, 113*(17), 609-612. doi: 10.1016/j.ipl.2013.05.005

Ruiz-Torres, A. J., Ho, J. C., & Lopez, F. J. (2006). Generating Pareto schedules with outsource and internal parallel resources. *International Journal of Production Economics, 103*(2), 810-825. doi: 10.1016/j.ijpe.2005.11.010

Ruiz-Torres, A. J., Lopez, F. J., Ho, J. C., & Wojciechowski, P. J. (2008). Minimizing the average tardiness: the case of outsource machines. *International Journal of Production Research, 46*(13), 3615-3640. doi: 10.1080/00207540601158799

Sawik, T. (2009). Coordinated supply chain scheduling. *International Journal of Production Economics, 120*(2), 437-451. doi: 10.1016/j.ijpe.2008.08.059

Sawik, Tadeusz. Scheduling in Supply Chains Using Mixed Integer Programming. Hoboken, NJ, USA: John Wiley & Sons, 2011.

Selvarajah, E., & Steiner, G. (2006). Batch scheduling in customer-centric supply chains. *Journal of the Operations Research Society of Japan, 49*(3), 174-187.

Selvarajah, E., & Steiner, G. (2009). Approximation Algorithms for the Supplier's Supply Chain Scheduling Problem to Minimize Delivery and Inventory Holding Costs. *Operations Research, 57*(2), 426-438. doi: 10.1287/opre.1080.0622

Selvarajah, E., & Zhang, R. (2014). Supply chain scheduling at the manufacturer to minimize inventory holding and delivery costs. *International Journal of Production Economics, 147*, 117-124. doi: 10.1016/j.ijpe.2013.08.015

Selvarajah, E., & Zhang, R. (2014). Supply chain scheduling to minimize holding costs with outsourcing. *Annals of Operations Research, 217*(1), 479-490. doi: 10.1007/s10479-013-1522-1

Sigurd, M., Pisinger, D., & Sig, M. (2004). Scheduling transportation of live animals to avoid the spread of diseases. *Transportation Science, 38*(2), 197-209. doi: 10.1287/trsc.1030.0053

Steiner, G., & Zhang, R. (2009). Approximation algorithms for minimizing the total weighted number of late jobs with late deliveries in two-level supply chains. *Journal of Scheduling, 12*(6), 565-574. doi: 10.1007/s10951-009-0109-9

Steiner, G., & Zhang, R. (2011). Minimizing the weighted number of tardy jobs with due date assignment and capacity-constrained deliveries. *Annals of Operations Research, 191*(1), 171-181. doi: 10.1007/s10479-011-1000-6

Steiner, G., & Zhang, R. (2011). Revised Delivery-Time Quotation in Scheduling with Tardiness Penalties. *Operations Research, 59*(6), 1504-1511. doi: 10.1287/opre.1110.0948

Stilgenbauer, T. M., Nicholas, T., & Brizendine, A. L. (2001). Scheduling transportation projects using primavera project planner as part of the software series in civil engineering technology independent learning experiment at fairmont state college. *Journal of Engineering Technology, 18*(1), 24-30.

Surmann, H., & Morales, A. (2002). Scheduling tasks to a team of autonomous mobile service robots in indoor enviroments. *Journal of Universal Computer Science, 8*(8), 809-833.

Tang, L., Jing, K., & He, J. (2013). An improved ant colony optimisation algorithm for three-tier supply chain scheduling based on networked manufacturing. *International Journal of Production Research, 51*(13), 3945-3962. doi: 10.1080/00207543.2012.760853

Tang, L. X., Meng, Y., Wang, G. S., Chen, Z. L., Liu, J. Y., Hu, G. F., . . . Zhang, B. (2014). Operations Research Transforms Baosteel's Operations. *Interfaces, 44*(1), 22-38. doi: 10.1287/inte.2013.0719

Terashima-Marin, H., Tavernier-Deloya, J. M., & Valenzuela-Rendon, M. (2005). Scheduling transportation events with grouping genetic algorithms and the heuristic DJD. In A. Gelbukh, A. DeAlbornoz & H. TerashimaMarin (Eds.), *Micai 2005: Advances in Artificial Intelligence* (Vol. 3789, pp. 185-194). Berlin: Springer-Verlag Berlin.

Tzur, M., & Drezner, E. (2011). A lookahead partitioning heuristic for a new assignment and scheduling problem in a distribution system. *European Journal of Operational Research, 215*(2), 325-336. doi: 10.1016/j.ejor.2011.06.013

Ullrich, C. A. (2012). Supply chain scheduling: makespan reduction potential. *International Journal of Logistics-Research and Applications, 15*(5), 323-336. doi: 10.1080/13675567.2012.742045

Ullrich, C. A. (2013). Integrated machine scheduling and vehicle routing with time windows. *European Journal of Operational Research, 227*(1), 152-165. doi:

10.1016/j.ejor.2012.11.049

Wang, G., Lei, L., & Lee, K. (2015). Supply chain scheduling with receiving deadlines and non-linear penalty. *Journal of the Operational Research Society, 66*(3), 380-391. doi: 10.1057/jors.2014.2

Wang, H. Y., & Lee, C. Y. (2005). Production and transport logistics scheduling with two transport mode choices. *Naval Research Logistics, 52*(8), 796-809. doi: 10.1002/nav.20116

Wang, X. L., & Cheng, T. C. E. (2009). Logistics scheduling to minimize inventory and transport costs. *International Journal of Production Economics, 121*(1), 266-273. doi: 10.1016/j.ijpe.2009.05.007

Wang, X. P., Liang, A., Xu, C. L., & Yang, D. L. (2009). Analysis and Design of Decision Support System of Disruption Management in Logistics Scheduling. *International Journal of Innovative Computing Information and Control, 5*(6), 1559-1568.

Yao, J. M. (2011). Supply chain scheduling optimisation in mass customisation based on dynamic profit preference and application case study. *Production Planning & Control, 22*(7), 690-707. doi: 10.1080/09537287.2010.537577

Yao, J. M. (2013). Scheduling optimisation of co-operator selection and task allocation in mass customisation supply chain based on collaborative benefits and risks. *International Journal of Production Research, 51*(8), 2219-2239. doi: 10.1080/00207543.2012.709645

Yao, J. M., & Liu, L. W. (2009). Optimization analysis of supply chain scheduling in mass customization. *International Journal of Production Economics, 117*(1), 197-211. doi: 10.1016/j.ijpe.2008.10.008

Yeung, W. K., Choi, T. M., & Cheng, T. C. E. (2010). Optimal Scheduling of a Single-Supplier Single-Manufacturer Supply Chain With Common due Windows. *Ieee Transactions on Automatic Control, 55*(12), 2767-2777. doi: 10.1109/tac.2010.2049766

Yeung, W. K., Choi, T. M., & Cheng, T. C. E. (2011). Supply chain scheduling and coordination with dual delivery modes and inventory storage cost. *International Journal of Production Economics, 132*(2), 223-229. doi: 10.1016/j.ijpe.2011.04.012

Yimer, A. D., & Demirli, K. (2010). A genetic approach to two-phase optimization of dynamic supply chain scheduling. *Computers & Industrial Engineering, 58*(3), 411-422. doi: 10.1016/j.cie.2009.01.010

Appendix

Table A1. Priorities (0..10)

priority	d				
c	d1	d2	d3	d4	d5
c1	1	9	6	3	3
c2	2	3	9	0	5
c3	10	6	10	8	1
c4	7	1	2	7	4
c5	3	3	1	1	6
c6	9	2	7	8	3
c7	1	5	1	9	2
c8	3	6	7	6	5
c9	4	1	3	0	3
c10	2	7	6	8	3

Table A2. Optimal schedule

c	s	d				
		d1	**d2**	**d3**	**d4**	**d5**
c1	slot1		1			
	slot2			1		
	slot3				1	
c2	slot1		1			
	slot2			1		
	slot3					1
c3	slot1	1				
	slot2			1		
	slot3				1	
c4	slot1	1				
	slot2			1		
	slot3				1	
c5	slot1		1			
	slot2			1		
	slot3					1
c6	slot1	1				
	slot2			1		
	slot3				1	
c7	slot1		1			
	slot2			1		
	slot3				1	
c8	slot1		1			
	slot2			1		
	slot3				1	
c9	slot1	1				
	slot2			1		
	slot3					1
c10	slot1		1			
	slot2			1		
	slot3				1	

Part III – Logistics Impact Assessment of Trade Policies by CGE Modeling

Chapter 9. An overview of the computable general equilibrium (CGE) modeling

This chapter provides you an overview on the concepts of the computable general equilibrium modeling. It first introduces you shortly to the historical roots of CGE modeling. It then introduces you to the aim and the main mechanisms of models.

1.1. Historical roots of CGE modeling

Leon Walras (1874) formulated the first general equilibrium model, called the Walrasian model that could adapt to complex economic interactions. Based on the theoretical structure of the Walrasian general equilibrium, the CGE modeling approach is based on the work, which was completed in the 1950s by Kenneth Arrow and Gerard Debreu. CGE models are based on Arrow and Debreu's general equilibrium (GE) theories where agents interact in competitive markets by introducing optimum quantities and optimal prices that satisfy markets and agents' equilibrium conditions.

Since then, CGE models have been developed in the early 1960s to solve for both market prices and quantities simultaneously, simulating the operation of a competitive market economy. Eventually, interest in these models has increased since their inception in the 1960s as applied economists have recognized the benefits of their use for the counterfactual analysis and improvements in the computations that have allowed for analysis that is more detailed. The first CGE model, developed in Norway by Leif Johansen (Johansen, 1960), was designed to be used for policy analysis. The aim of a CGE model is to attempt to model the entire economy and the relations between economic agents in it. Later, CGE models have been commonly used to analyze the impact of macroeconomic policies and the impact of the allocation of development resources in developed countries since the early 1980s. CGE models are preferred in comparison to the partial equilibrium models because they include complex interdependencies in the analysis.

1.2. The aim of CGE models and its mechanisms

The aim of CGE models is to quantify the effects of policy on equilibrium allocations and relative prices using the standard theory of general equilibrium.

CGE models have their roots in the input-output theory. They are widely used for economic, social and environmental planning and evaluation, as it is effective to capture inter-sectoral linkages. Indeed, CGE models are based on complex hypothetical mathematical relationships between different sectors of the economy that reflect the behavior of the key players in the economy and provide more realistic assessments than those obtained from input-output models. However, the input-output models have limitations, which simulated the development of CGE models. These limitations are mainly reflected in the assumptions of the model, including fixed-price, indefinite supply factor and fixed share factor and intermediate inputs in the production process. Under these assumptions, the input-output model cannot show substitution between production inputs, and the responsive behavior of producers and consumers to changes in relative prices.

In sum, CGE models are adapted to analyze contemporary strategic issues in a competitive market economy because they have the price mechanism that plays an important role in the economy. The price mechanism is able to solve complex problems in an economy that cannot be performed by input-output models. In CGE models, economic agents make their own decisions on economic activities based on changes in the market price as given resource constraints and technology. In addition, the market balances supply and demand by adjusting prices. Because CGE models can quantify these market behavior and changes, they are widely used in various analyzes of policies. A CGE model describes an economy in equilibrium with prices and relative quantities, which are determined endogenously. While most empirical approaches examine the impacts of policies and effects in a *ceteris paribus* condition, a CGE model, which provides comparative scenarios *based on the baseline scenario* that incorporates the factor markets, the markets for goods and external trade markets.

Chapter 10. The standard CGE model with GAMS code

This chapter presents the main components of a CGE model. It provides shortly the main ingredients of a typical CGE model along with some of the advantages of CGE modeling. Appendix section provides the mathematical representation of a typical CGE model with general algebraic modeling system (GAMS) code. The aim of the optimization model is to maximize the utility function while satisfying all the constraints.

1.1 The standard CGE model

Computable general equilibrium (CGE) models are based on mathematical relationships (See Appendix section) between the different sectors of the economy that reflect the behavior of the key stakeholders in the economy and provide more realistic assessments than are obtained from input-output models. A CGE model provides an analytical framework in which the economy is represented as a complete system of interdependent components. All economic agents, that is, households, firms, government and external sectors are all linked by transactions on the markets and the price system. The CGE model takes into account the fact that an economic shock has subsequent effects on the overall economy, which is not the case in a partial equilibrium framework.

As the CGE model is a numerical simulation model, which takes into account the links between all markets, including links to many economic agents with optimal behavior and the links between economic agents and markets, the CGE model puts all economic agents and all markets in a unified framework, reflecting the universal links between each component of the real economic system. A CGE model is typically used to simulate policies. A base case is constructed to reflect the observed reality. Scenarios are then built by altering some exogenous variables or parameters of the model to reflect the intended or experienced changes.

The CGE model mainly comprises the following:

- Producers, which operate under full competition and maximize profit subject to given prices and current technology.

- Consumers, which have an initial endowment of factors and maximize utility subject to the budget constraint; the value of income must equal the total value of expenditures.

- The government, which obtains its income through taxes on capital, formal labor. Government revenue is the sum of revenue collected from tariffs, indirect taxes and export taxes as well as household and company income taxes.

- and Foreign trade

A typical CGE model specifies:

- Producers, households

- Profit maximization, utility maximization

- Technological constraints

- Input and output prices

- The institutional structure in which they operate

- System constraints that include supply–demand flow balance across all product and factor markets.

Some of the advantages of CGE modeling:

- Solving complex equilibrium problems and presenting more details.

- Describing the efficiency and stability of the market mechanism

- Providing the logical foundations of economic analyses

- Shedding light on real economic problems in many different areas and even to become the basis for formulating policies

- Analyzing complex economic relationships in a microeconomic environment based on a solid microeconomic foundation.

It should be noted that CGE modeling is a challenging arena. It requires mastery of economic theory, careful data preparation and familiarity with the underlying accounting conventions, knowledge of econometric methods, and an understanding of solution algorithms and associated software for solving large systems of equations.

Appendix

For more detailed information on the standard CGE model, refer to Hosoe, N., Gasawa, K., and Hashimoto, H., *Handbook of Computable General Equilibrium Modeling*, University of Tokyo Press, Tokyo, Japan, 2004

Table 2.1. Description of the variables used in GAMS

GAMS Variable	Description
$Y(j)$	composite factor
$F(h,j)$	the h-th factor input by the j-th firm
$X(i,j)$	intermediate input
$Z(j)$	output of the j-th good
$Xp(i)$	household consumption of the i-th good
$Xg(i)$	government consumption
$Xv(i)$	investment demand
$E(i)$	exports
$M(i)$	imports
$Q(i)$	Armington's composite good
$D(i)$	domestic good
$pf(h)$	the h-th factor price
$py(j)$	composite factor price
$pz(j)$	supply price of the i-th good
$pq(i)$	Armington's composite good price
$pe(i)$	export price in local currency
$pm(i)$	import price in local currency
$pd(i)$	the i-th domestic good price
epsilon	exchange rate
Sp	private saving
Sg	government saving
Td	direct tax
$Tz(j)$	production tax
$Tm(i)$	import tariff
UU	utility

Table 2.2. The domestic production functions

Composite factor aggregation function

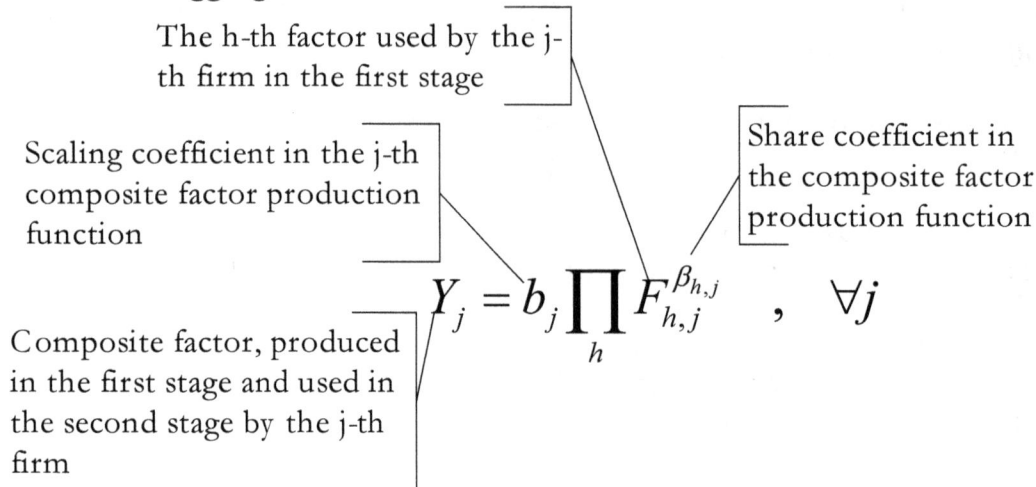

The h-th factor used by the j-th firm in the first stage

Scaling coefficient in the j-th composite factor production function

Share coefficient in the composite factor production function

Composite factor, produced in the first stage and used in the second stage by the j-th firm

$$Y_j = b_j \prod_h F_{h,j}^{\beta_{h,j}} \quad , \quad \forall j$$

```
eqpy(j)          composite factor agg. func.
eqpy(j)..        Y(j)    =e= b(j)*prod(h, F(h,j)**beta(h,j));
```

Factor demand function

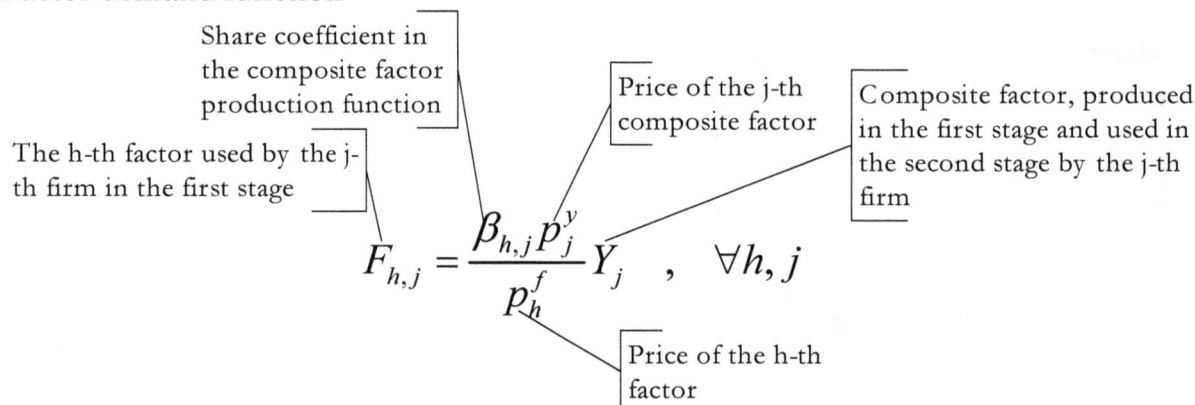

Share coefficient in the composite factor production function

The h-th factor used by the j-th firm in the first stage

Price of the j-th composite factor

Composite factor, produced in the first stage and used in the second stage by the j-th firm

$$F_{h,j} = \frac{\beta_{h,j} p_j^y}{p_h^f} Y_j \quad , \quad \forall h,j$$

Price of the h-th factor

```
eqF(h,j)         factor demand function
eqF(h,j)..       F(h,j)  =e= beta(h,j)*py(j)*Y(j)/pf(h);
```

282

Intermediate demand function

Intermediate input of the i-th good used by the j-th firm

Input requirement coefficient of the i-th intermediate input for a unit output of the j-th good

$$X_{i,j} = \alpha x_{i,j} Z_j \quad , \quad \forall i,j$$

Gross domestic output of the j-th firm

```
eqX(i,j)          intermediate demand function
eqX(i,j)..        X(i,j)   =e= ax(i,j)*Z(j);
```

Composite factor demand function

Composite factor, produced in the first stage and used in the second stage by the j-th firm

Input requirement coefficient of the j-th composite good for a unit output of the j-th good

$$Y_j = \alpha y_j z_j \quad , \quad \forall j$$

Gross domestic output of the j-th firm

```
eqY(j)            composite factor demand function
eqY(j)..          Y(j)     =e= ay(j)*Z(j);
```

Unit cost function

Input requirement coefficient of the j-th composite good for a unit output of the j-th good

Price of the j-th gross domestic output

$$p_j^z = \alpha y_j p_j^y + \sum_i \alpha x_{i,j} p_i^q \quad , \quad \forall j$$

Price of the j-th composite factor

Input requirement coefficient of the i-th intermediate input for a unit output of the j-th good

```
eqpzs(j)          unit cost function
eqpzs(j)..        pz(j)    =e= ay(j)*py(j) +sum(i, ax(i,j)*pq(i));
```

Table 2.3. The government tax revenue and demand functions

Direct tax revenue function

Direct tax

Direct tax rate

Endowments of the h-th factor for the household

$$T^d = \tau^d \sum_h p_h^f FF_h$$

Price of the h-th factor

```
eqTd          direct tax revenue function
eqTd..        Td      =e= taud*sum(h, pf(h)*FF(h));
```

Production tax revenue function

Production tax on the j-th good

Production tax rate on the j-th good

Gross domestic output of the j-th firm

$$T_j^z = \tau_j^z p_j^z z_j \quad , \quad \forall j$$

Price of the j-th gross domestic output

```
eqTz(j)       production tax revenue function
eqTz(j)..     Tz(j)   =e= tauz(j)*pz(j)*Z(j);
```

Import tariff revenue function

Import tariff on the i-th good

Import tariff rate on the i-th good

Imports of the i-th good

$$T_i^m = \tau_i^m p_i^m M_i \quad , \quad \forall i$$

Price of the i-th imported good

```
eqTm(i)       import tariff revenue function
eqTm(i)..     Tm(i)   =e= taum(i)*pm(i)*M(i);
```

Government demand function

Government consumption of the i-th good

Share of the i-th good in government expenditure

Production tax on the j-th good

Import tariff on the i-th good

$$X_i^g = \frac{\mu_i}{p_i^q}\left(T^d + \sum_j T_j^z + \sum_j T_j^m - S^g \right) \quad , \quad \forall i$$

Price of the i-th composite good

Direct tax

Government savings

```
eqXg(i)          government demand function
eqXg(i)..        Xg(i)   =e= mu(i)*(Td +sum(j, Tz(j)) +sum(j, Tm(j))-Sg)/pq(i);
```

Table 2.4. Household demand function

Household demand function

Share coefficient of for the i-th good good consumption in the utility function

Household consumption of the i-th good

Price of the h-th factor

Household savings

Direct tax

$$X_i^p = \frac{\alpha_i}{p_i^q}\left(\sum_h p_h^f FF_h - S^p - T^d \right) \quad , \quad \forall i$$

Price of the i-th composite good

Endowments of the h-th factor for the household

```
eqXp(i)          household demand function
eqXp(i)..        Xp(i)   =e= alpha(i)*(sum(h, pf(h)*FF(h)) -Sp -Td)/pq(i);
```

Table 2.5. Investment and savings functions

Investment demand function

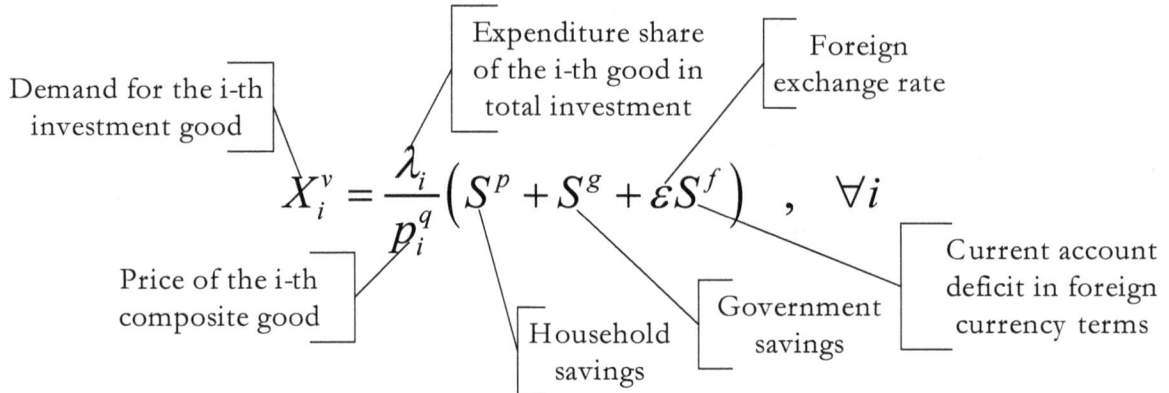

Demand for the i-th investment good

Expenditure share of the i-th good in total investment

Foreign exchange rate

$$X_i^v = \frac{\lambda_i}{p_i^q}\left(S^p + S^g + \varepsilon S^f\right) \quad , \quad \forall i$$

Price of the i-th composite good

Household savings

Government savings

Current account deficit in foreign currency terms

```
eqXv(i)           investment demand function
eqXv(i)..         Xv(i)   =e= lambda(i)*(Sp +Sg +epsilon*Sf)/pq(i);
```

Private saving function

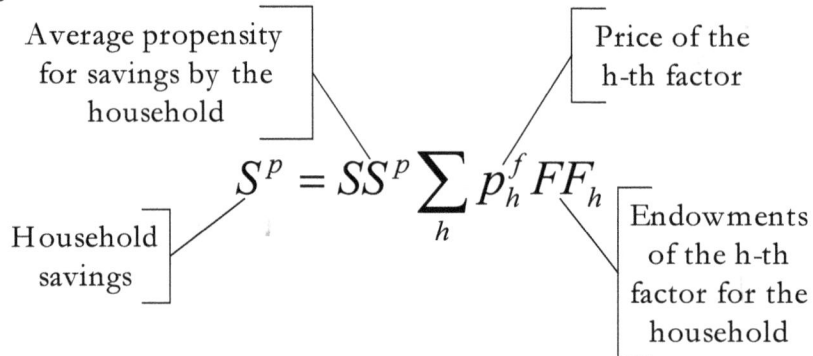

Average propensity for savings by the household

Price of the h-th factor

$$S^p = SS^p \sum_h p_h^f FF_h$$

Household savings

Endowments of the h-th factor for the household

```
eqSp              private saving function
eqSp..            Sp      =e= ssp*sum(h, pf(h)*FF(h));
```

Factor market clearing condition

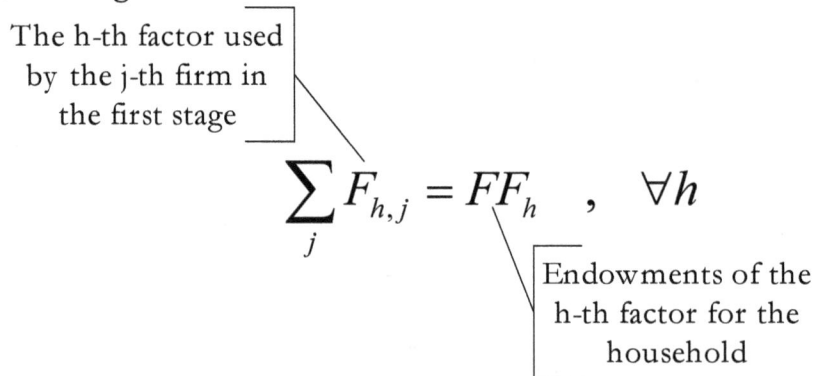

The h-th factor used by the j-th firm in the first stage

$$\sum_j F_{h,j} = FF_h \quad , \quad \forall h$$

Endowments of the h-th factor for the household

```
eqpf(h)           factor market clearing condition
eqpf(h)..         sum(j, F(h,j)) =e= FF(h);
```

286

Government saving function

Average propensity for savings by the household

Direct tax

Import tariff on the i-th good

$$S^g = SS^g \left(T^d + \sum_j T_j^z + \sum_j T_j^m \right)$$

Production tax on the j-th good

```
eqSg          government saving function
eqSg..        Sg      =e= ssg*(Td +sum(j, Tz(j))+sum(j, Tm(j)));
```

Table 2.6. Export and import prices and the balance of payments

World export price equation

Export price in terms of domestic currency

Export price in terms of foreign currency

$$p_i^e = \varepsilon\, p_i^{W_e} \quad , \quad \forall i$$

Foreign exchange rate

```
eqpe(i)       world export price equation
eqpe(i)..     pe(i)    =e= epsilon*pWe(i);
```

World import price equation

Import price in terms of domestic currency

Import price in terms of foreign currency

$$p_i^m = \varepsilon\, p_i^{W_m} \quad , \quad \forall i$$

Foreign exchange rate

```
eqpm(i)       world import price equation
eqpm(i)..     pm(i)    =e= epsilon*pWm(i);
```

Balance of payments

$$\sum_i p_i^{W_e} E_i + S^f = \sum_i p_i^{W_m} M_i$$

Export price in terms of foreign currency

Import price in terms of foreign currency

Exports of the i-th good

Imports of the i-th good

Current account deficit in terms of foreign currency

```
eqepsilon        balance of payments
eqepsilon..      sum(i, pWe(i)*E(i)) +Sf =e= sum(i, pWm(i)*M(I));
```

Table 2.7. Substitution between imports and domestic goods (Armington composite)

Armington function

$$Q_i = \gamma_i \sqrt[\eta_i]{(\delta m_i M_i^{\eta_i} + \delta d_i D_i^{\eta_i})} \quad , \quad \forall i$$

Scaling coefficient in the Armington composite good production function

Parameter defined by the elasticity of substitution

The i-th Armington composite good

Input share coefficient in the Armington composite good production function

The i-th imported good

The i-th domestic good

Input share coefficient in the Armington composite good production function

```
eqpqs(i)         Armington function
eqpqs(i)..       Q(i)     =e=
gamma(i)*(deltam(i)*M(i)**eta(i)+deltad(i)*D(i)**eta(i))**(1/eta(i));
```

Import demand function

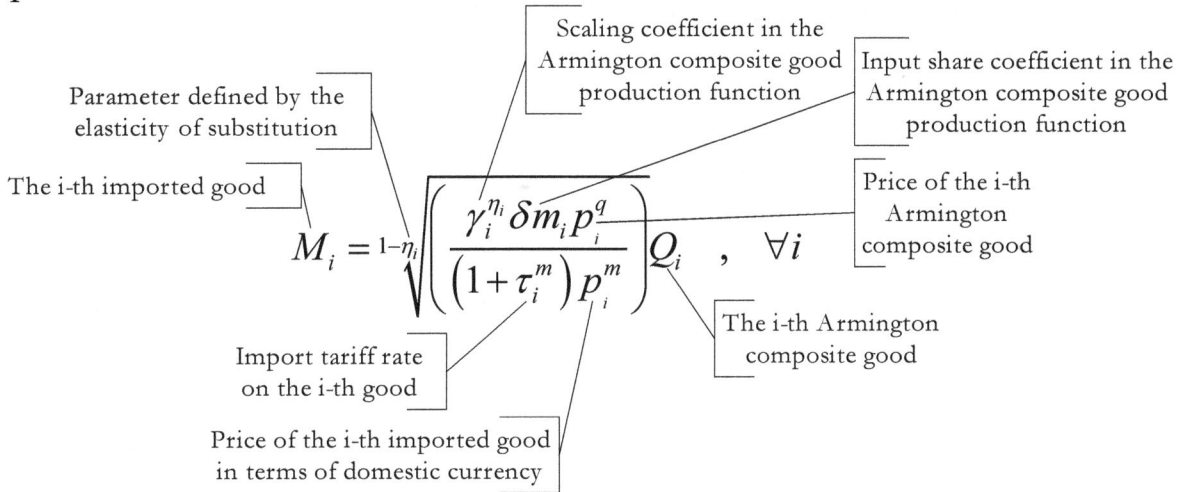

Parameter defined by the elasticity of substitution

The i-th imported good

Scaling coefficient in the Armington composite good production function

Input share coefficient in the Armington composite good production function

Price of the i-th Armington composite good

$$M_i = {}^{1-\eta_i}\!\!\sqrt{\left(\frac{\gamma_i^{\eta_i}\,\delta m_i\, p_i^q}{\left(1+\tau_i^m\right) p_i^m}\right) Q_i}\quad,\quad \forall i$$

Import tariff rate on the i-th good

Price of the i-th imported good in terms of domestic currency

The i-th Armington composite good

```
eqM(i)          import demand function
eqM(i)..        M(i)    =e= (gamma(i)**eta(i)*deltam(i)*pq(i)
/((1+taum(i))*pm(i)))**(1/(1-eta(i)))*Q(i);
```

Domestic good demand function

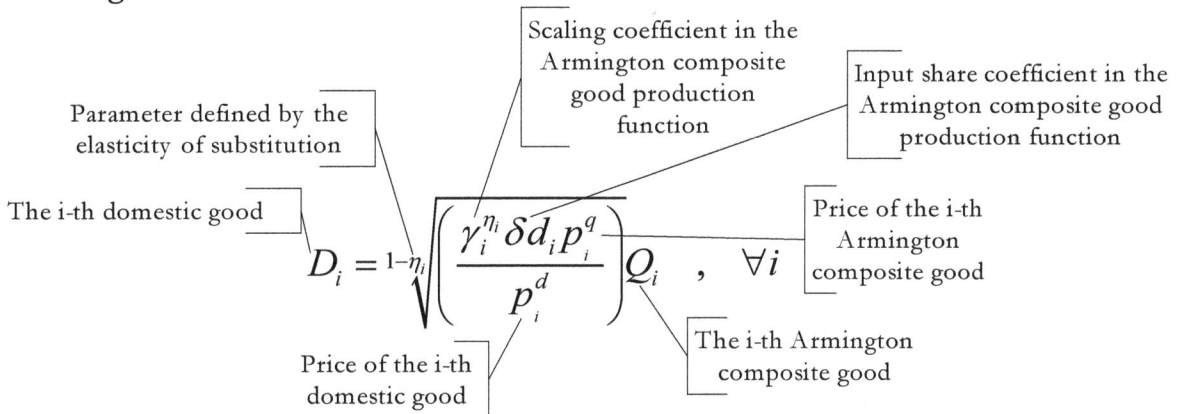

Parameter defined by the elasticity of substitution

The i-th domestic good

Scaling coefficient in the Armington composite good production function

Input share coefficient in the Armington composite good production function

Price of the i-th Armington composite good

$$D_i = {}^{1-\eta_i}\!\!\sqrt{\left(\frac{\gamma_i^{\eta_i}\,\delta d_i\, p_i^q}{p_i^d}\right) Q_i}\quad,\quad \forall i$$

Price of the i-th domestic good

The i-th Armington composite good

```
eqD(i)          domestic good demand function
eqD(i)..        D(i)    =e= (gamma(I)**eta(i)*deltad(i)*pq(i)/pd(i))**(1/(1-
eta(i)))*Q(i);
```

Table 2.8. Transformation between exports and domestic goods

Transformation function

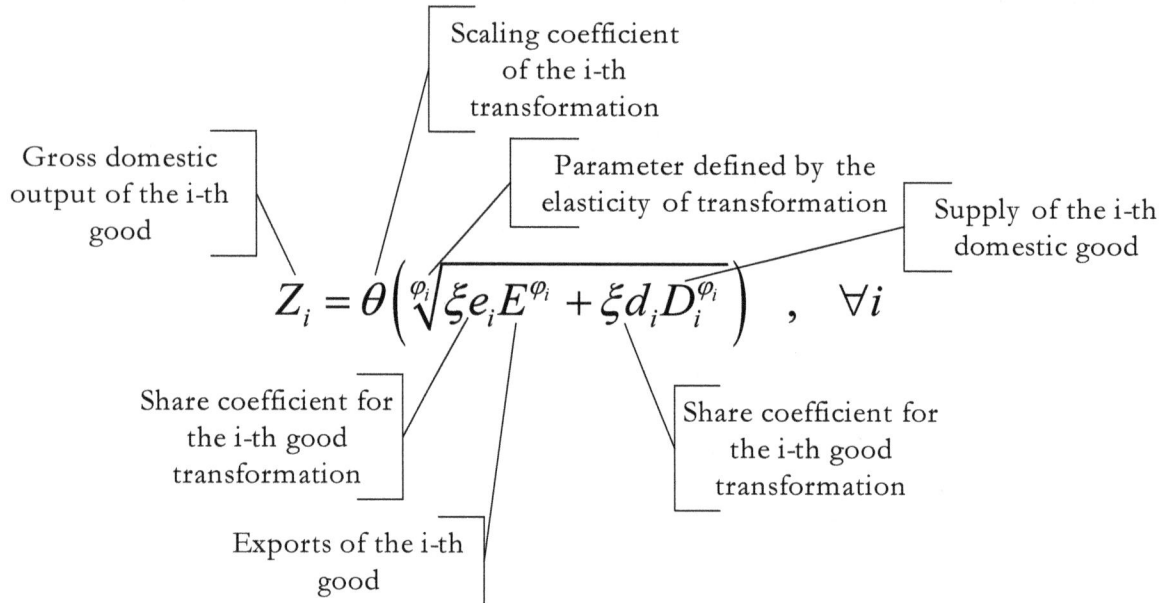

Scaling coefficient of the i-th transformation

Gross domestic output of the i-th good

Parameter defined by the elasticity of transformation

Supply of the i-th domestic good

$$Z_i = \theta \left(\sqrt[\varphi_i]{\xi e_i E^{\varphi_i} + \xi d_i D_i^{\varphi_i}} \right) \quad , \quad \forall i$$

Share coefficient for the i-th good transformation

Share coefficient for the i-th good transformation

Exports of the i-th good

```
eqpzd(i)        transformation function
eqpzd(i)..      Z(i)   =e=
theta(i)*(xie(i)*E(i)**phi(i)+xid(i)*D(i)**phi(i))**(1/phi(i));
```

Domestic good supply function

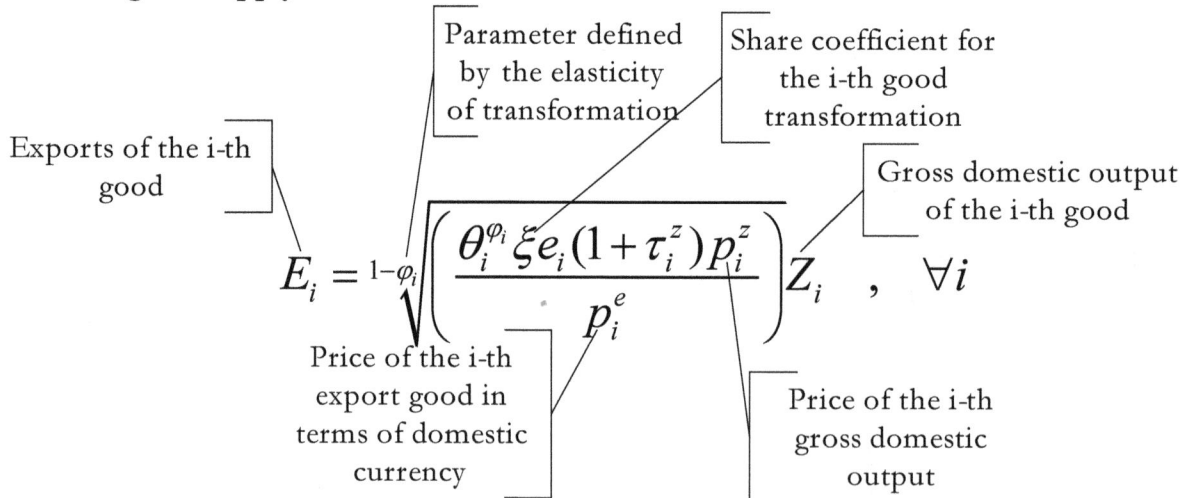

Parameter defined by the elasticity of transformation

Share coefficient for the i-th good transformation

Exports of the i-th good

Gross domestic output of the i-th good

$$E_i = \sqrt[1-\varphi_i]{\left(\frac{\theta_i^{\varphi_i} \xi e_i (1+\tau_i^z) p_i^z}{p_i^e} \right) Z_i} \quad , \quad \forall i$$

Price of the i-th export good in terms of domestic currency

Price of the i-th gross domestic output

```
eqDs(i)        domestic good supply function
eqDs(i)..      D(i)   =e=
(theta(i)**phi(i)*xid(i)*(1+tauz(i))*pz(i)/pd(i))**(1/(1-phi(i)))*Z(i);
```

Export supply function

Supply of the i-th domestic good

Parameter defined by the elasticity of transformation

Share coefficient for the i-th good transformation

Gross domestic output of the i-th good

Price of the i-th domestic good

Price of the i-th gross domestic output

$$D_i = \sqrt[1-\varphi_i]{\left(\frac{\theta_i^{\varphi_i} \xi d_i (1+\tau_i^z) p_i^z}{p_i^d}\right) Z_i} \quad , \quad \forall i$$

```
eqE(i)          export supply function
eqE(i)..        E(i)    =e=
(theta(i)**phi(i)*xie(i)*(1+tauz(i))*pz(i)/pe(i))**(1/(1-phi(i)))*Z(i);
```

Table 2.9. Market clearing conditions

Market clearing condition for composite good

Government consumption of the i-th good

Household consumption of the i-th good

Demand for the i-th investment good

The i-th Armington composite good

Intermediate input of of the i-th good used by the j-th firm

$$Q_i = X_i^p + X_i^g + X_i^v + \sum_j X_{i,j} \quad , \quad \forall i$$

```
eqpqd(i)          market clearing cond. for comp. good
eqpqd(i)..        Q(i)    =e= Xp(i) +Xg(i) +Xv(i) +sum(j, X(i,j));
```

Table 2.10. Objective function

Utility function

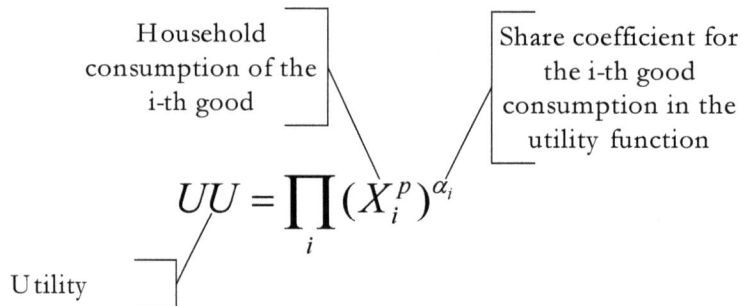

$$UU = \prod_{i}(X_i^p)^{\alpha_i}$$

Household consumption of the i-th good

Share coefficient for the i-th good consumption in the utility function

Utility

```
obj              utility function
obj..            UU        =e= prod(i, Xp(i)**alpha(i));
```

Part IV

Simulation 1

Chapter 11. The economic rise of the Eurasian trade and its implications for logistics services in Turkey

Abstract. Turkey considers additional potential trading partners other than the western economies; Turkey takes into account the enormous potential of Eurasia offerings. Indeed, Eurasia is the large landmass covering the continents of Europe and Asia. Eurasia offers tremendous opportunities for the Turkish economy and thus for logistics services. This research examines the effects on logistics services based on the scenario of a 10% growth in Eurasian economies. Based on this scenario, the results indicate that various logistics services to/from Eurasian economies exhibit increases. These results are reported using computable general equilibrium (CGE) simulations.

Keywords: Eurasia, Turkey, logistics, maritime transport, air transport, containerized products, computable general equilibrium, GTAP

1. Introduction

If Eurasia is to be defined as the large landmass covering the continents of Europe and Asia, then Turkey can easily be called a Eurasian country because it acts as a natural bridge with its territories in both continents (Ersen 2012). However, the term "Eurasian" has rarely been used in Turkey to define the geographical location of the country (Ersen 2012).

Eurasia is the landmass that shelters more than seventy percent of the population of the world and covers almost a tenth of the surface of the Earth (Ersen 2014). Geographically, it was named in the late nineteenth century to define the supercontinent including Europe and Asia, which until that time were usually considered as two separate continents (Ersen 2014).

As the center of global economic gravity continues to shift from West to East and that the international system is becoming more multipolar because of the rise of new powers, Eurasia seems to have become identified with the idea of a balance or a geopolitical alternative against the West (Ersen 2014). Turkey is a country that is most affected by this new geopolitical reality not only because it is geographically located in the center of the Eurasian supercontinent, but also because of its decision in particular in the last decade to develop its strategic relationship with the rise Eurasian powers (Ersen 2014).

After several years of political stability and high economic growth between 2002 and 2007, most Turks are optimistic about the status of their country as a power in the region and in the world (Gordon and Taspinar 2008).

In the last decade, however, Turkey's relations with Russia have improved significantly, especially in the economic field (Larrabee, 2010). Russia is the largest trading partner of Turkey and its largest supplier of natural gas. Russia is also an important market for the Turkish construction industry (Larrabee, 2010). Projects in Russia account for about one quarter of all projects undertaken by Turkish contractors worldwide (Larrabee, 2010).

The rediscovery of Turkey a greater Turkish world in the Caucasus and Central Asia has had a significant effect on the perception of Turkey's national interests (Larrabee and Lesser 2002). It was also a vehicle for a more active participation, and officially by the Turkish private sector (Larrabee and Lesser 2002).

Since the early 1990s, Central Asia and the Caucasus have emerged as important focal point of Turkish policy (Larrabee and Lesser 2002). This represents a significant shift in Turkish foreign policy (Larrabee and Lesser 2002). Under Atatürk, Turkey consciously avoided efforts to cultivate contacts with Turkic and Muslim populations beyond the borders of Turkey (Larrabee and Lesser 2002).

In this regard, this study investigates logistics services in case the Eurasian economies grow. What are the effects on logistics services in both directions, from Eurasia to Turkey and vice versa? This study looks into the effects on logistics services based on the scenario of a 10% growth in Eurasian economies. Based on this scenario, the results indicate that various logistics services from/to Eurasian economies exhibit changes. This study reports all these changes in logistics services by using Computable General Equilibrium (CGE) simulations.

The remainder of this study is organized as follows. Section 2 presents some highlighted studies. Section 3 presents the framework, methodology, results and short discussion. The study is concluded in Section 4.

2. Highlighted Studies about Turkey and Eurasia

Yildiz (2014) examined the logistics performance in developed and developing countries. Bilgin and Bilgic (2011) have highlighted the new policy of Turkey towards Eurasia. Aras and Fidan (2009) examined Turkey and Eurasia. They concluded that the renewed activism of Turkey has opened new horizons for relations in the region and that this new direction of foreign policy is linked to the reform and change in the internal landscape of Turkey. In the same vein, Cohen (2011) emphasized the central role of Turkey as a bridge between Europe and Eurasia.

Cinar (2013) analyzed the political, economic and cultural ties of Turkey with some of the countries of Eurasia. The study examined Turkey's relations with other regional and international powers. Ayhan (2009) investigated the oil, natural gas and integration from a critical perspective of European security of energy supply and Turkey. The author emphasized the role of Turkey, which could have a special role to balance the rising energy dependence of the EU.

Ersen (2013) focused on Turkish politics in its geopolitical perspectives to Russia and the Turkic republics of Central Asia and the Caucasus. Yeşiltaş (2013) studied the transformation of the geopolitical vision in Turkish foreign policy. The study explains how the different political actors spatialize geography of Turkey. Additionally, Yercan and Yildiz (2012) highlighted theoretical issues in global maritime trade and logistics.

3. Framework, methodology, results, brief discussion

In this study, Global Trade Analysis Project (GTAP) is used for performing trade simulations. GTAP and the database developed by the Global Trade Analysis Project at Purdue University. The general structure of the model is considered to be relatively simple. The demand side of the model based on the assumption of the existence of regional home that takes all expenditure decisions in the economy. Thus, the entity allocates expenditures to private consumption, government expenditures or savings. The structure of the regional preference of family is based on the function of multiply nested utility. At the top nest household decides on the allocation of expenses between private consumption, public consumption and savings, depending on the Cobb-Douglas utility function. The government consumption is a Cobb-Douglas composite of goods from different sectors. The demand for private consumption is governed by a constant difference of elasticity preferences to reflect the non-homothetic nature of consumer demand. Two levels of Armington CES aggregates are used to distinguish domestic from foreign products and to differentiate foreign products by country of origin.

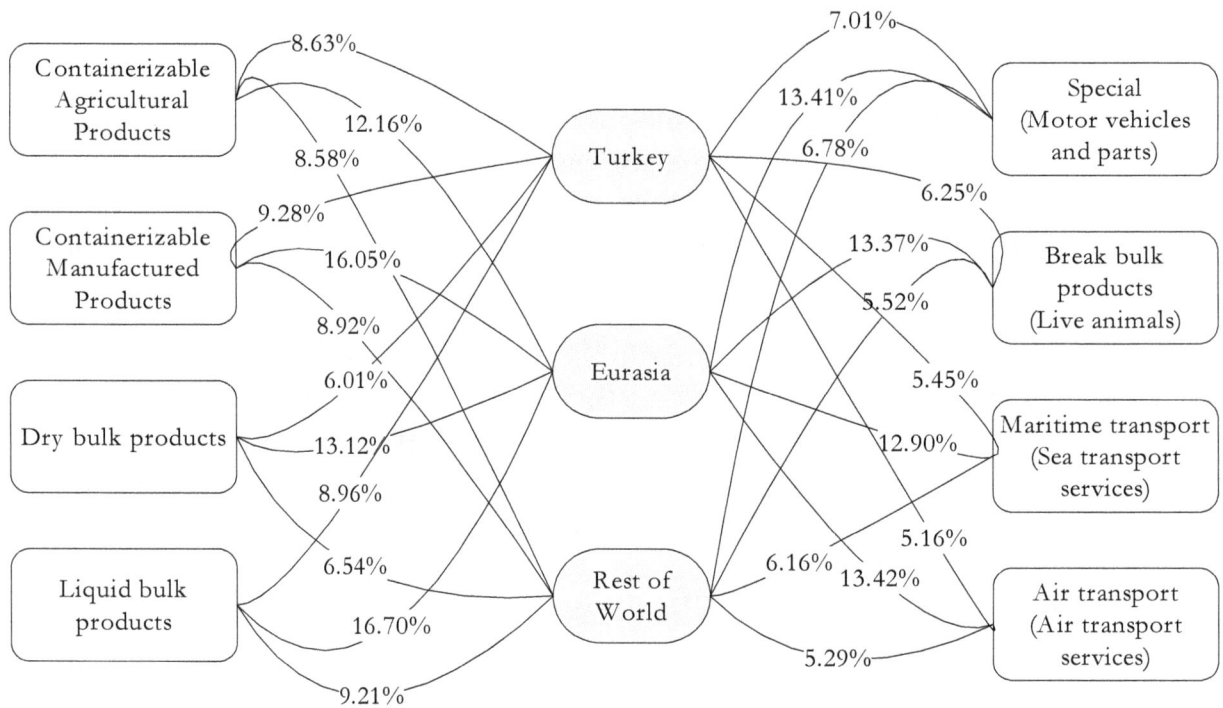

Figure 1. Percentage change of logistics services *from* Eurasia *to* Turkey and Rest of World

Firms produce using intermediate goods and primary factors purchased from the regional house. The primary sources are purely internal factors - it is assumed that the factors are strictly immobile internationally and mobile within a region (with the exception of land and natural resources). Intermediate goods can be either domestically produced or imported in the country.

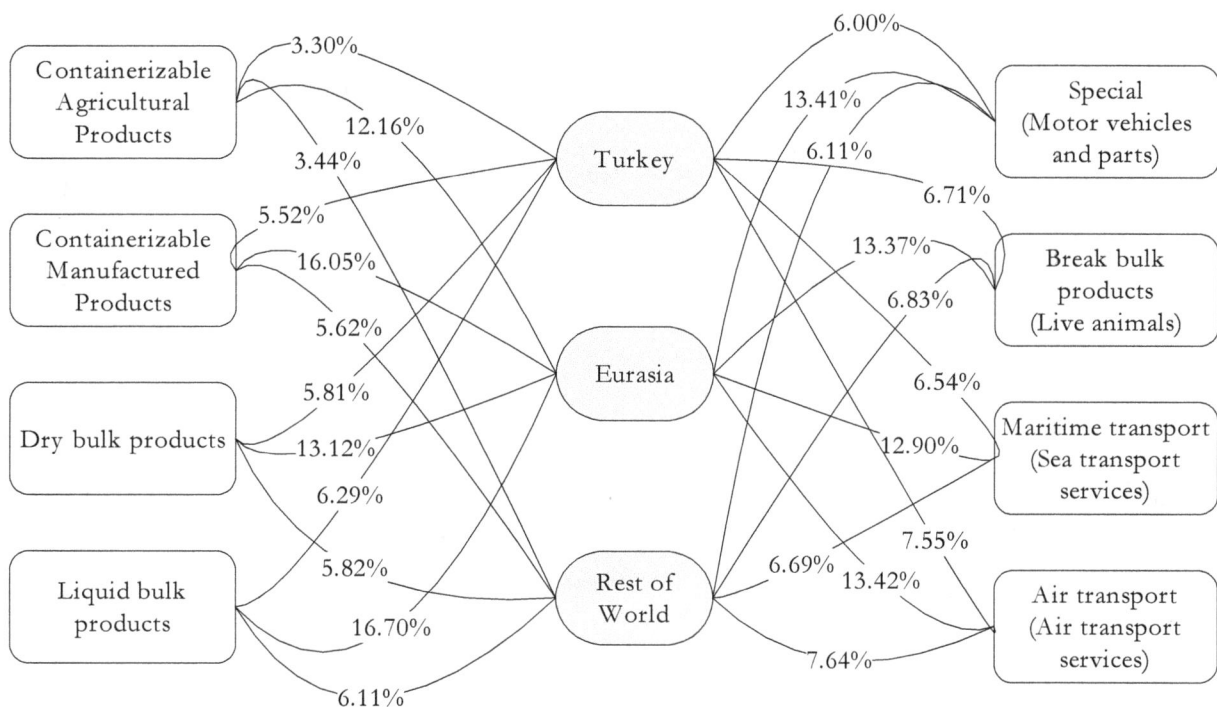

Figure 2. Percentage change of logistics services *to* Eurasia *from* Turkey and Rest of World

The GTAP database has information on 57 sectors, including services in all regions. These data include information on production volumes, sales in both domestic and international, through use and the use of primary factors. It also contains information on bilateral trade between the countries in both goods and services. Version 6 of the database uses 2001 as the base year.

In this study, the simulation looks into the possible effects on logistics services based on the scenario of a 10% growth in Eurasian economies. Based on this scenario, GTAP results are shown in Figure 1 and 2. Results indicate that various logistics services from/to Eurasian economies exhibit large increases.

Sea transport services	Maritime Transport
Air transport services	Air Transport
Oil, gas, petroleum, coal products, chemicals, rubber, plastic products	Liquid Bulk Products
Motor vehicles and parts	Special
Live animals	Break Bulk Products
Paddy rice, vegetables, fruits, animal products, raw milk, wool, silk-worm cocoons, fishery, cattle meat, sheep meat, horse meat, meat products nec, vegetable oils and fats, dairy products, processed rice, food products nec.	Containerazable Agricultural Products
Beverages, tobacco products, textiles, wearing apparel, leather products, wood products, paper products, metals nec, metal products, transport equipment nec, electronic equipment, machinery and equipment nec, manufactures nec.	Containerazable Manufactured Products
Wheat, cereal grains nec, oil seeds, sugar cane, sugar beet, plant-based fibers, crops nec, forestry, coal, minerals nec, sugar, mineral products nec, ferrous metals.	Dry Bulk Products
Electricity, gas and water, construction, trade, transport nec, communication, financial services nec, insurance, business services nec, recreational services, public administration and defense, health, education and dwellings	Other Services

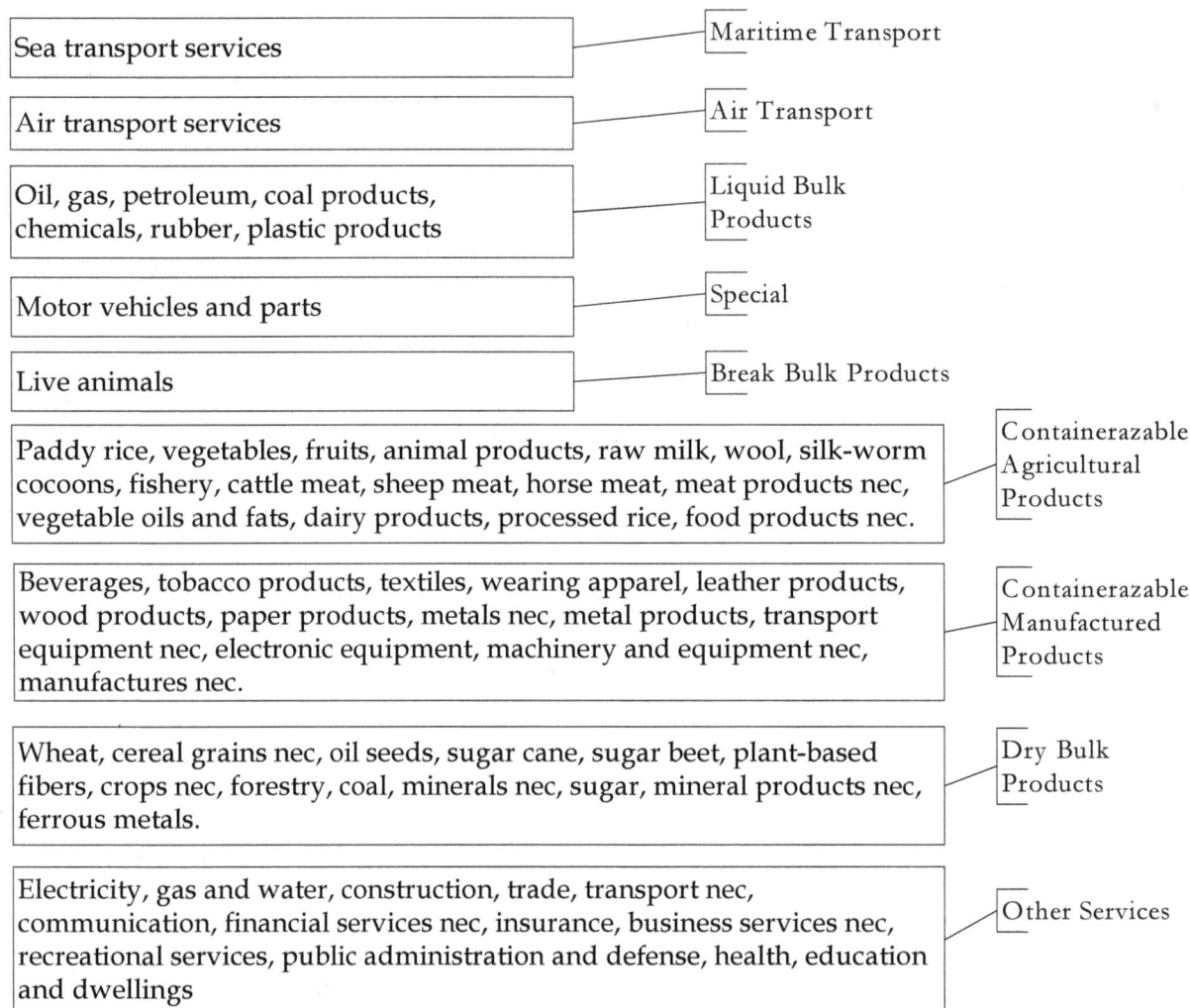

Figure 3. Commodity aggregation (Source: GTAP Database)

Figure 3 depicts the aggregation of the commodities to represent their corresponding logistics services.

The remarkable changes that have occurred in the geopolitical framework of Eurasia in the post-2007 period also had significant implications for Turkish foreign policy. First of all, mainly because of its impressive economic growth in the 2000s, Turkey has become one of the "new powers" of the world as also recognized by Western politicians. In this regard, it is frequently associated with the BRICS and other similar groups as a country, which should enhance its global economic influence in the coming decades. Second, the fact that it acts as a natural bridge with its territories in both Asia and Europe makes Turkey a truly Eurasian countries (Ersen 2014).

Turkish policy makers seem to have clearly anticipated the evolution of the meaning of Eurasia in world politics (Ersen 2014). Besides the fact that Turkey has recently labeled as a "rising Eurasian Tiger", its political and economic relations with other Eurasian powers such as Russia, China and India have also improved remarkably especially in the last few years (Ersen 2014).

Turkey may consider full membership in the Shanghai Cooperation Organization (SCO) as part of the Eurasia as a reaction to the neutral process of accession to the EU (Ersen 2014). Although it is currently difficult to imagine the SCO as a strong alternative to the EU or NATO in the Turkish foreign policy, it must be emphasized that Turkey is the only NATO member state to enjoy a privileged institutional relationship with the SCO (Ersen 2014).

In this study, after examining the increasing Eurasia relations, it should be highlighted that Turkey is not lost but it could be unless recent trends are reversed, and the Turks are given a reason to believe, as they have for more than 80 years, that their future is better served in the Western world (Gordon and Taspinar 2008).

4. Conclusion

This study looked into the effects on logistics services based on the scenario of a 10% growth in Eurasian economies. The results indicate that various logistics services from/to Eurasian economies exhibit large increases. These results are reported by using Computable General Equilibrium (CGE) simulations.

References

Aras, B. and H. Fidan (2009). Turkey and Eurasia: Frontiers of a new geographic imagination. *New Perspectives on Turkey* (40): 193-215.

Ayhan, V. (2009). European Energy Supply Security and Turkey: Oil, Natural Gas and Integration. *Uluslararasi Iliskiler-International Relations* 5(20): 155-178.

Bilgin, P. and A. Bilgic (2011). Turkey's "New" Foreign Policy toward Eurasia. Eurasian *Geography and Economics* 52(2): 173-195.

Cinar, K. (2013). Turkey and Turkic Nations: A Post-Cold War Analysis of Relations. *Turkish Studies* 14(2): 256-271.

Cohen, S. B. (2011). Turkey's Emergence as a Geopolitical Power Broker. Eurasian *Geography and Economics* 52(2): 217-227.

Erşen, E. (2012). The Evolution of 'Eurasia' as a Geopolitical Concept in Post–Cold War Turkey. *Geopolitics*, 18(1), 24-44. doi: 10.1080/14650045.2012.665106

Erşen, E. (2014). Rise of new centres of power in Eurasia: Implications for Turkish foreign policy. *Journal of Eurasian Studies*, 5(2), 184-191. doi:

http://dx.doi.org/10.1016/j.euras.2014.05.003

Gordon, Philip H., and Taspinar, Omer. (2008). *Winning Turkey: How America, Europe, and Turkey Can Revive a Fading Partnership*. Washington, DC, USA: Brookings Institution Press, 2008.

Larrabee, F. Stephan, and Lesser, Ian O. (2002). *Turkish Foreign Policy in an Age of Uncertainty*. Santa Monica, CA, USA: RAND Corporation, 2002.

Larrabee, F. Stephen. (2010). *Troubled Partnership: U.S.-Turkish Relations in an Era of Global Geopolitical Change*. Santa Monica, CA, USA: RAND Corporation, 2010.

Yercan, F. & Yildiz, T. (2012). International Maritime Trade and Logistics. In *Maritime Logistics: A Complete Guide to Effective Shipping and Port Management*, Edited by Dong-Wook Song, Photis M. Panayides, Kogan Page, pp 23-44

Yesiltas, M. (2013). The Transformation of the Geopolitical Vision in Turkish Foreign Policy. *Turkish Studies* 14(4): 661-687.

Yildiz, T. (2014). The Performances of Logistics Services in Developed and Developing Countries: A Review and Cluster Analysis. In *Business Logistics: Theory and Practice* (first edition). Charleston, SC, USA: CreateSpace, pp 43-73

Simulation 2

Chapter 12. The Eurasian trade and the effects on logistics services in the EU: An assessment by using CGE modeling

Abstract. Eurasia is a large landmass covering two continents: some parts of Europe and Asia. It is a unified region, which has been held together by political, economic and cultural layers. It offers enormous potential for regional economies and for the global economy. In this regard, this study examines a scenario taking into account a slowdown in Eurasian economies and the scale of the potential effects on European logistics services. The results indicate that some logistics services are much more sensitive to changes in these economies in Eurasia. These results are simulated and reported by using general equilibrium models.

Keywords: European Union, Eurasia, logistics, sea transport, air transport, containerized products

1. Introduction

As a unified region, political, economic and cultural layers have held Eurasia together (Tsygankov, 2012). Politically, the region consisted principalities or empires but rarely recognized with fixed boundaries, remaining relatively open inside and outside influences (Tsygankov, 2012).

Economically, the region has a wealth of resources, which it extracted and transported from one continent to another, mostly from East to West (Tsygankov, 2012). The Baltic Sea and the Black Sea have been particularly important for economic ties across continents (Tsygankov 2012). Politically, economically and culturally, the region has functioned as a unity in diversity, as a center of diverse influences and ensure the stability of European nations to the north and to the nations of Asia and the Middle East to the south (Tsygankov 2012).

In this regard, the Shanghai Cooperation Organization (SCO) has become the most influential organization in Central Asia, with a growing role in the Asian region at large. It was established in 2001 after more than five years of a process of strengthening the informal confidence-building process known as the "Shanghai Five" involving multilateral consultations on the demarcation of the China's border with Tajikistan, Kyrgyzstan and Kazakhstan. The founding members were China, Russia, Tajikistan, Kazakhstan, Kyrgyzstan and Uzbekistan. In 2005, India, Pakistan, Iran and Mongolia were admitted as observers: they all expressed interest in full membership (Strategic Survey of Russia and Eurasia 2006).

In this study, based on a scenario, answers to some questions are sought. In case of a slowdown in Eurasian economies, what will be the effects on European logistics services? How will be the logistics services affected in both directions from/to Europe? This study looks into a scenario by taking into account of a slowdown in Eurasian economies and its scale of the potential effects on European logistics services.

The remainder of this study is organized as follows. Section 2 presents some highlighted studies. Section 3 presents the framework, methodology, and results. The study is concluded in Section 4.

2. Highlighted Studies

Yildiz (2014) studied on the global analysis of logistics and business services, and examined the performance of logistics services in developed and developing countries. Gruere et al. (2009) explained the international differences in the labeling of genetically modified food policies focusing on regulations in Asia and Europe. Capannelli and Filippini (2010) compared the economic integration process of the European Union and the East Asian countries. They found that the two entities are currently facing difficult challenges to progress and growth. Chang (2009) explored the firm's investment strategies, as evidenced by the IT industry in Asia, European and US markets.

Plummer (2014) reported the integration of Asia and its implications for the EU. The study estimates the economic costs of excluding EU from economic cooperation initiatives in Asia. He found that the EU could face a significant diversion of trade in commodities. Bonham et al. (2007) focused on growth and the determinants of trade in information technology in the Asia-Pacific region. They found that the IT trade could be partly explained by the traditional income and the effects of relative prices, but also by inflows of foreign direct investment. Krapohl and Fink (2013) have highlighted the different paths of regional integration by investigating Europe, Southeast Asia and southern Africa.

3. Framework, methodology, and results

The Global Trade Analysis Project (GTAP) package covers much more than just a consistent set of bilateral international trade data by sector. The most important element, in the sense of being the most difficult to recreate in the construction of the model, is the set of social accounting matrix (SAM) for countries and regions included. These tables include all input-output (IO) relationships and income/payment flows.

Indeed, GTAP has the most detailed sectoral breakdowns and institutional details in agriculture. Manufacturing is well represented in its 57 sectors, but the distribution of services, particularly tradable services, is considered far from ideal.

The database GTAP 7 Links input-output tables with detailed coverage of gross bilateral trade, transport and protection data in 57 sectors and 113 regions. As such, multi-region CGE modeling has become the workhorse accepted in the quantitative analysis of the ex ante trade policy (Revoredo-Giha et al. 2013).

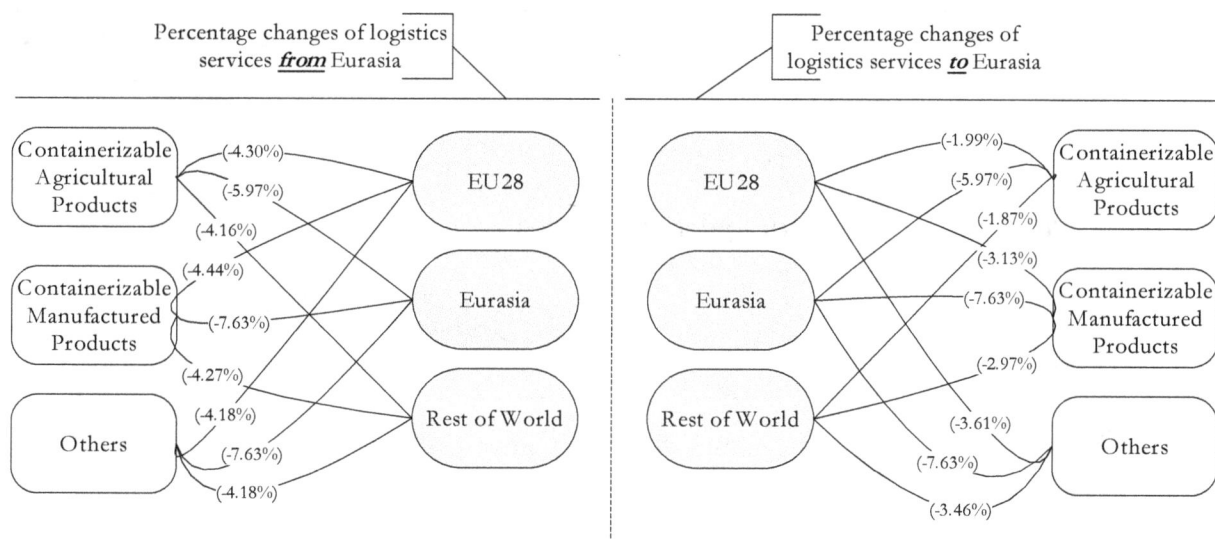

Figure 1. Percentage changes of logistics services

Based on a scenario of a 5% slowdown in Eurasian economies, their scales of the potential effects on European logistics services are examined. The results indicate that some logistics services are much more sensitive to the changes in Eurasia (See Figures 1 through 4).

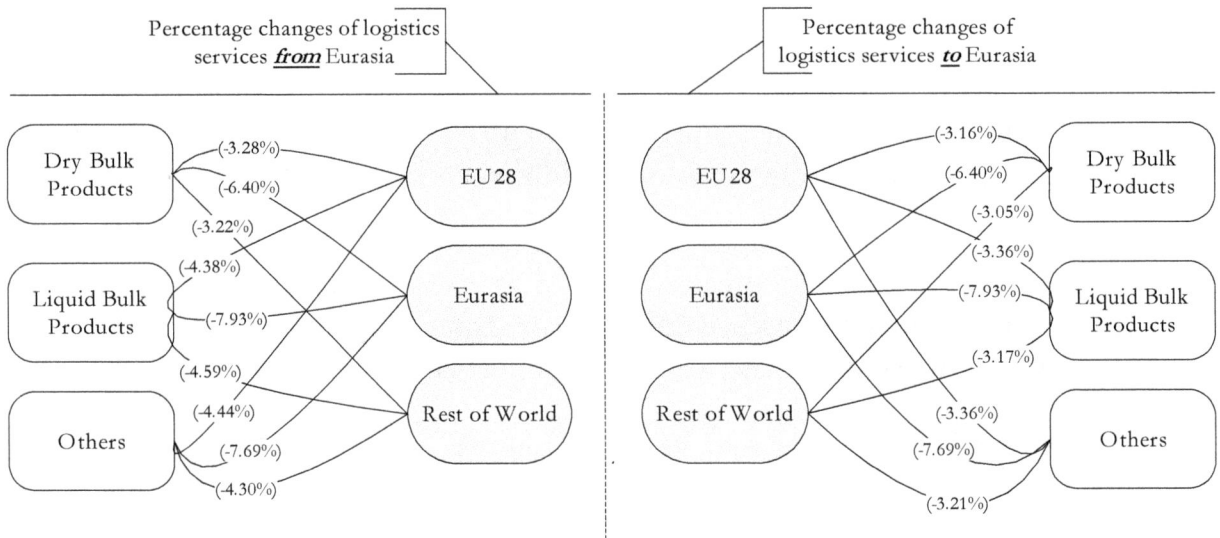

Figure 2. Percentage changes of logistics services

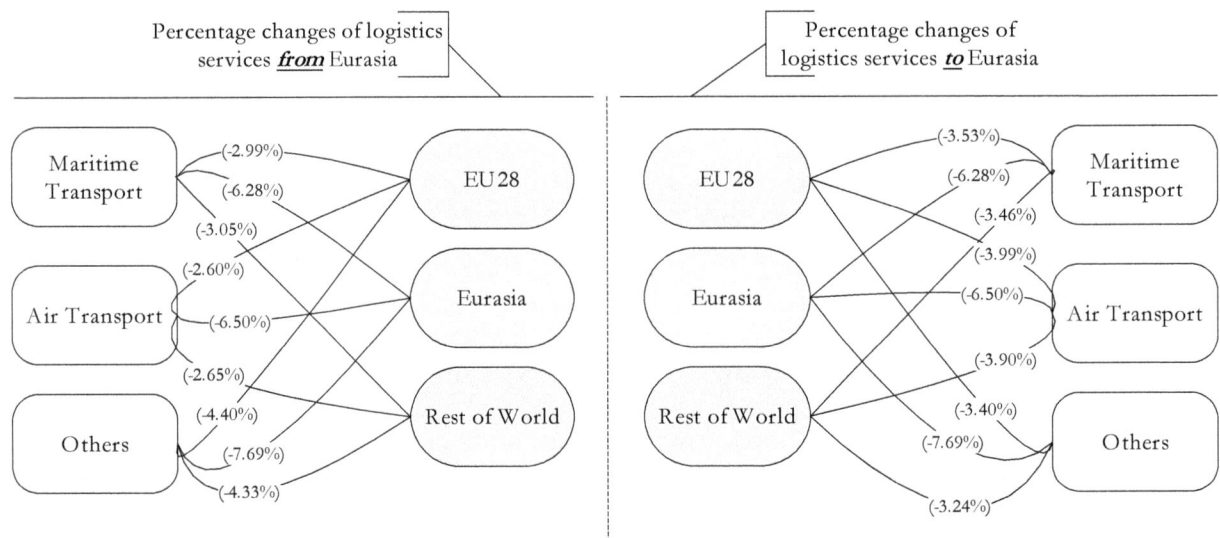

Figure 3. Percentage changes of logistics services

304

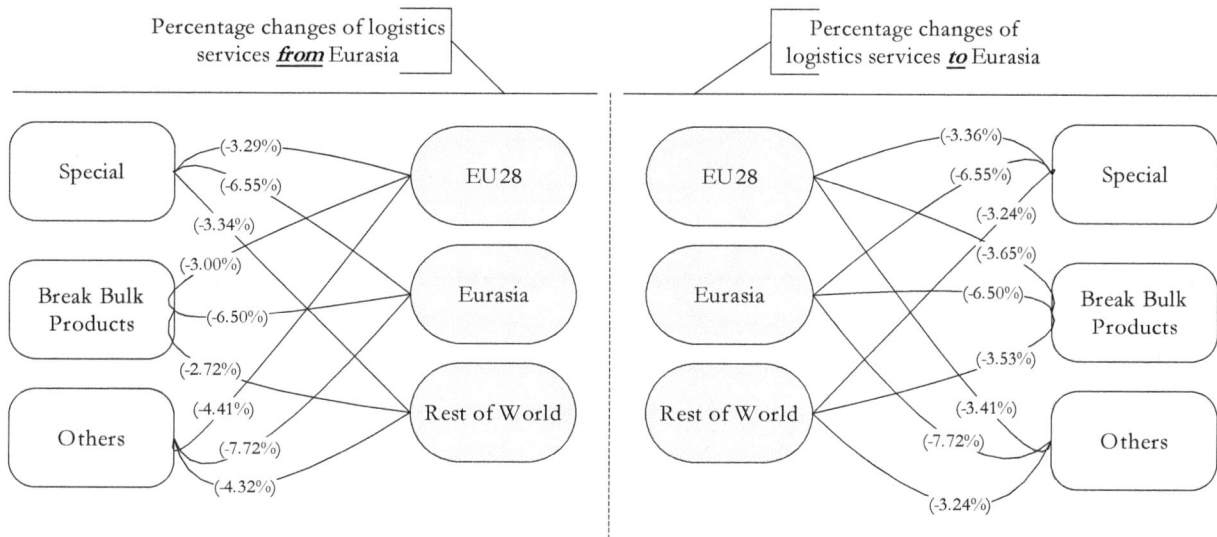

Figure 4. Percentage changes of logistics services

Figure 5 depicts the aggregation of the commodities to represent their corresponding logistics services.

Paddy rice, vegetables, fruits, animal products, raw milk, wool, silk-worm cocoons, fishery, cattle meat, sheep meat, horse meat, meat products nec, vegetable oils and fats, dairy products, processed rice, food products nec.	Containerazible Agricultural Products
Beverages, tobacco products, textiles, wearing apparel, leather products, wood products, paper products, metals nec, metal products, transport equipment nec, electronic equipment, machinery and equipment nec, manufactures nec.	Containerazible Manufactured Products
Wheat, cereal grains nec, oil seeds, sugar cane, sugar beet, plant-based fibers, crops nec, forestry, coal, minerals nec, sugar, mineral products nec, ferrous metals.	Dry Bulk Prducts
Oil, gas, petroleum, coal products, chemicals, rubber, plastic products	Liquid Bulk Prducts
Motor vehicles and parts	Special
Live animals	Break Bulk Products
Sea transport services	Maritime Transport
Air transport services	Air Transport
Electricity, gas and water, construction, trade, transport nec, communication, financial services nec, insurance, business services nec, recreational services, public administration and defense, health, education and dwellings	Other Services

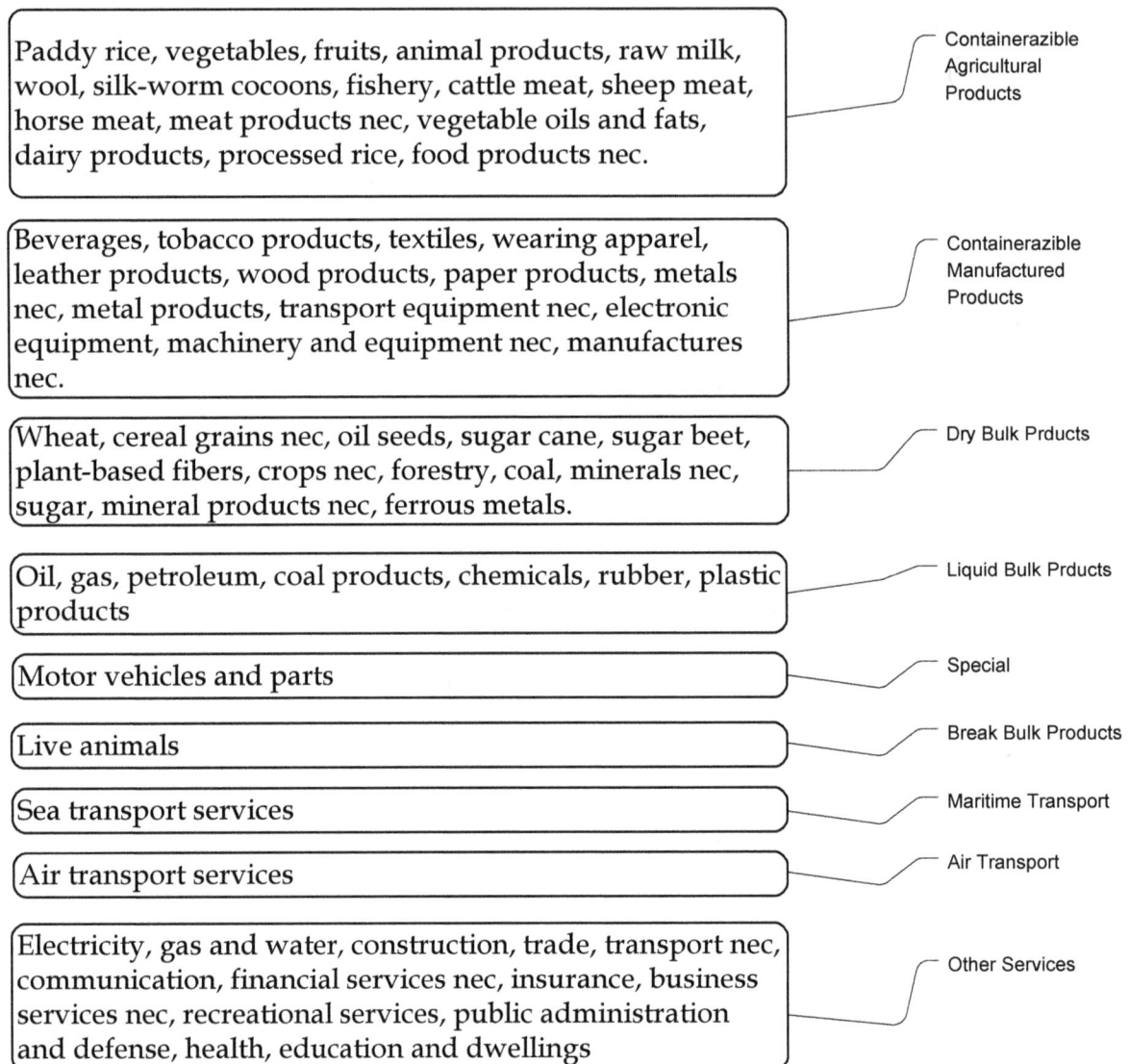

Figure 5. Commodity aggregation (Source: GTAP database)

4. Conclusion

This study looked into a scenario by taking into account of a slowdown in Eurasian economies and its scale of the potential effects on European logistics services. The results indicate that some logistics services are much more sensitive to the changes in these economies in Eurasia. These results are simulated and reported by using general equilibrium models.

References

Bonham, C. S., Gangnes, B., & Van Assche, A. (2007). Fragmentation and East Asia's information technology trade. *Applied Economics*, 39(2), 215-228. doi: 10.1080/00036840500427635

Capannelli, G. and C. Filippini (2010). Economic Integration In East Asia And Europe: Lessons From A Comparative Analysis. *Singapore Economic Review* 55(1): 163-184.

Chang, S. C. (2009). Horizontal and Vertical Intra-industry Trade and Firm's Investment Strategies: Evidence from the IT Industry in the Asian, EU and US Markets. *Global Economic Review* 38(1): 63-76.

Gruere, G. P., Carter, C. A., & Farzin, Y. H. (2009). Explaining International Differences in Genetically Modified Food Labeling Policies. *Review of International Economics*, 17(3), 393-408. doi: 10.1111/j.1467-9396.2008.00788.x

Krapohl, S., & Fink, S. (2013). Different Paths of Regional Integration: Trade Networks and Regional Institution-Building in Europe, Southeast Asia and Southern Africa. *Jcms-Journal of Common Market Studies*, 51(3), 472-488. doi: 10.1111/jcms.12012

Plummer, M. G. (2014). Asian Integration and its Implications for the EU "Post-Doha". *Asian Economic Papers*, 13(1), 53-82. doi: 10.1162/ASEP_a_00250

Revoredo-Giha, C., Philippidis, G., Toma, L., & Renwick, A. (2013). The Impact of EU Export Refunds on the African Continent: An Impact Assessment. *The Journal of Development Studies*, 49(12), 1651-1675. doi: 10.1080/00220388.2013.807500

Russia and Eurasia. (2006). *Strategic Survey, 106*(1), 177-196. doi: 10.1080/04597230600958007

Tsygankov, A. P. (2012). The heartland no more: Russia's weakness and Eurasia's meltdown. *Journal of Eurasian Studies, 3*(1), 1-9. doi: http://dx.doi.org/10.1016/j.euras.2011.10.001

Yildiz, T. (2014). A Global Analysis of Logistics and Business Services in Developed and Developing Countries. In *Business Logistics: Theory and Practice* (1st Edition). Charleston, SC, USA: CreateSpace, pp 9-42

Yildiz, T. (2014). The Performances of Logistics Services in Developed and Developing Countries: A Review and Cluster Analysis. In *Business Logistics: Theory and Practice* (1st Edition). Charleston, SC, USA: CreateSpace, pp 43-73

Simulation 3

Chapter 13. The rise of trade in NAFTA area and the logistics impact in the EU

Abstract. The EU accounts relatively large percentage of NAFTA's exports and imports. The commercial ties with NAFTA and the EU are strong. In addition, the trade between the EU and NAFTA is mostly unproblematic. This study looks into the logistics services as part of the global trade. It examines the growth of NAFTA area and its implications for the EU's logistics services. The results show that the logistics services from EU benefit highly in case of a growth in NAFTA area, but not all the logistics services exhibit large growths from the NAFTA area.

Keywords: NAFTA, European Union, logistics, maritime transport, air transport, containerized products, CGE

1. Introduction

The EU accounts for 35 percent of exports of NAFTA (excluding intra-North American trade) and 25 percent of its imports, and so is the largest trading partner of NAFTA (Aggarwal and Fogarty 2005). Together, on behalf of the EU and NAFTA for 35 percent of the world and more than 40 percent of world imports exports, making the transatlantic link not only the heart of the economy of each side, but central to the international economy as a whole (Aggarwal and Fogarty 2005).

Began in 1994, NAFTA was the largest free trade area in the world. It includes Canada, Mexico and the United States (Duina 2004). Its primary objectives are the elimination of tariffs and non-tariff barriers to trade in goods to facilitate the movement of services and capital (Duina 2004).

In this regard, based on a scenario, this study seeks answers to some questions. What will be the effects on logistics services in case the economy in NAFTA area grows? How will logistics services both in the NAFTA area and in Europe be affected by this growth? This study examines the growth of NAFTA area and its implications for the EU's logistics services.

The remainder of this study is organized as follows. Section 2 presents some highlighted studies. Section 3 presents the framework, methodology, and results used for the presented analysis. The study is concluded in Section 4.

2. Highlighted Studies

From a global perspective, Yildiz (2014) studied the complex relationship between logistics performance and global competitiveness. Aspinwall (2009) investigated the regionalization and adjustment of domestic policies in the economic zone of North America. Bems et al. (2010) highlighted the impact of demand and the collapse of trade in the global recession. They found that 20 to 30 percent of the decline in final demand in US and the final demand in EU has been supported by foreign countries, with the North American Free Trade Agreement (NAFTA) and the emerging Europe hit hardest. de Sousa, et al. (2012) studied the market access in the global and regional trade. They found that the EU, NAFTA, ASEAN and MERCOSUR agreements all tend to reduce the estimated degree of market fragmentation of the within these areas, with the expected ranking between their respective trade impacts.

Sbragia (2010) explored the EU, the US, and trade policy. The study revealed geo-economic competition between the EU and the United States is key to shaping the EU's trade policy.

More specifically, Andreff and Andreff (2009) studied the global trade of sporting goods. They found that the key negotiating areas are Asia, Europe and NAFTA. Asia, Eastern Europe and emerging countries have a surplus in the trade of sporting goods, while NAFTA and Europe are in deficit.

3. Framework, methodology, and results

Computable General Equilibrium (CGE) simulations in this study use Global Trade Analysis Project (GTAP), which is a multi-region model of comparative static general equilibrium. The model used for this analysis is a general equilibrium model of the world economy with several regions and sectors. The model was born in 1992 and is continuously updated by the GTAP team at Purdue University. The GTAP model runs on General Equilibrium Modeling Package (GEMPACK) software, which is designed for large complex nonlinear systems of equations. The standard GTAP model provides a detailed representation of the economy, including the links between the farming, agribusiness, industrial, and service sectors of the economy.

Trade is represented by bilateral matrices based on Armington assumption. Additional features of the standard GTAP model are perfect competition in all markets and profitability, and utility maximization behavior of producers and consumers. Price wedges represent all political interventions.

By using GTAP framework, this scenario examines the 5% growth of NAFTA area and its implications for the EU's logistics services. The results show that the logistics services from EU benefit highly in such growth in NAFTA area, but not all the logistics services from the NAFTA (See Figures 1 through 4).

Figure 1. Percentage changes of logistics services

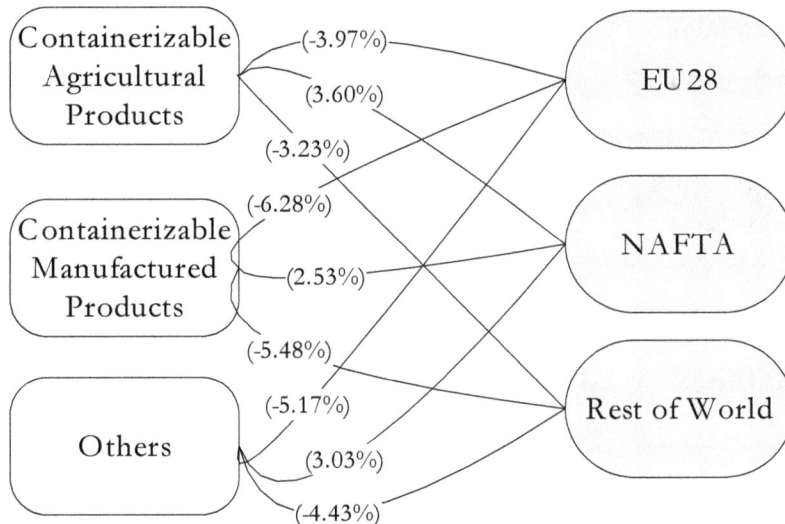

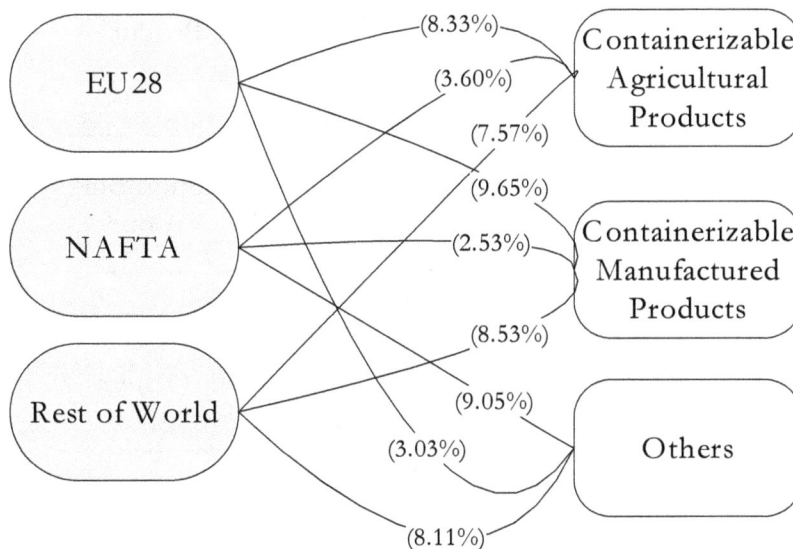

Dry Bulk Products

(-3.55%)

(3.37%)

(-2.97%)

(-4.33%)

Liquid Bulk Products

(1.98%)

(-3.65%)

(-5.93)

Others

(3.07%)

(-5.23%)

EU 28

NAFTA

Rest of World

Percentage changes of logistics
services ***to*** NAFTA

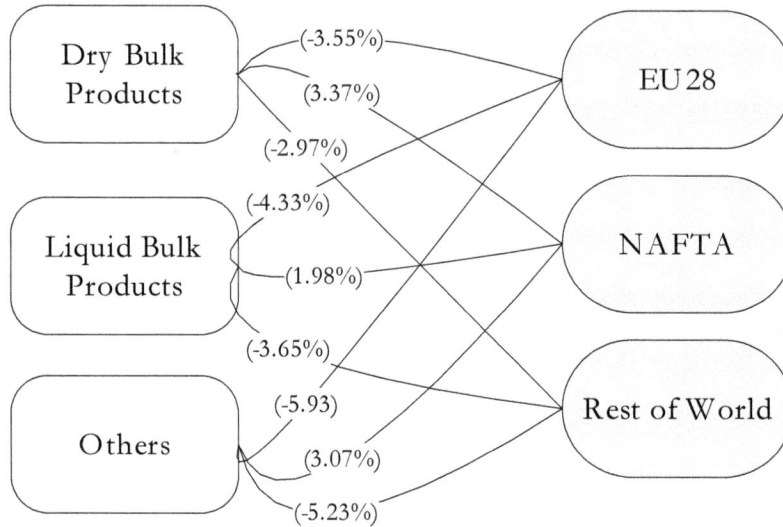

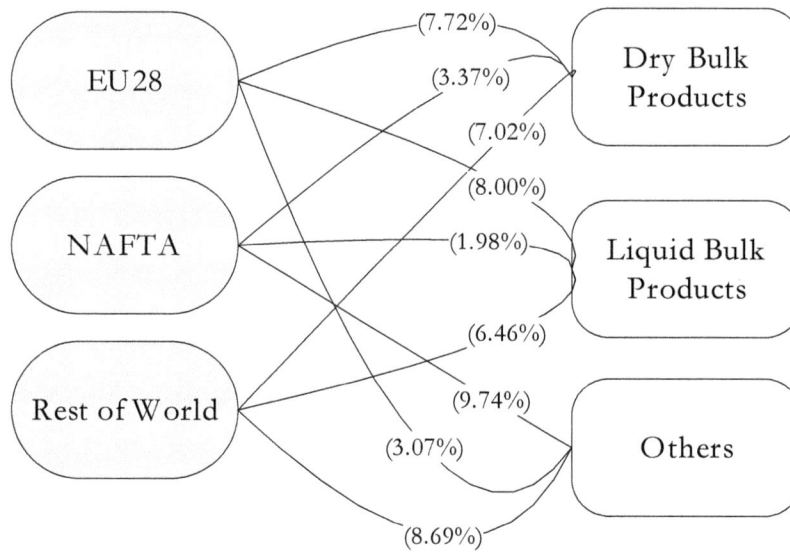

EU 28

(7.72%)

(3.37%)

(7.02%)

(8.00%)

NAFTA

(1.98%)

(6.46%)

(9.74%)

Rest of World

(3.07%)

(8.69%)

Dry Bulk Products

Liquid Bulk Products

Others

Figure 2. Percentage changes of logistics services

Figure 3. Percentage changes of logistics services

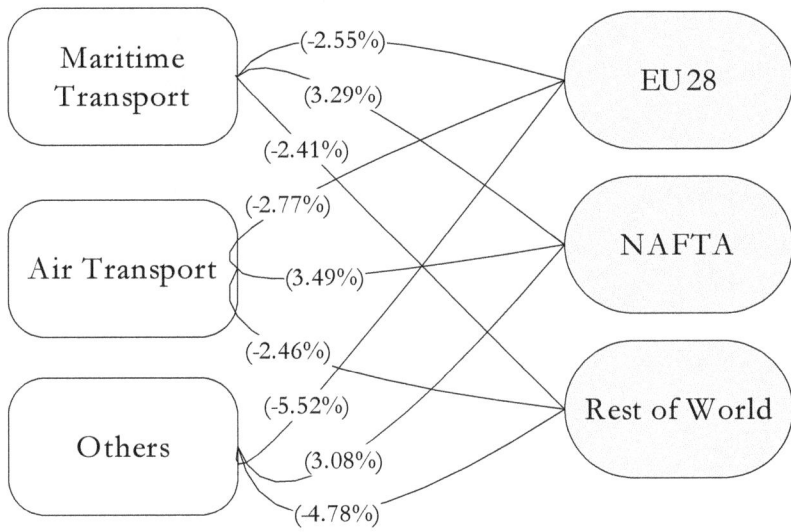

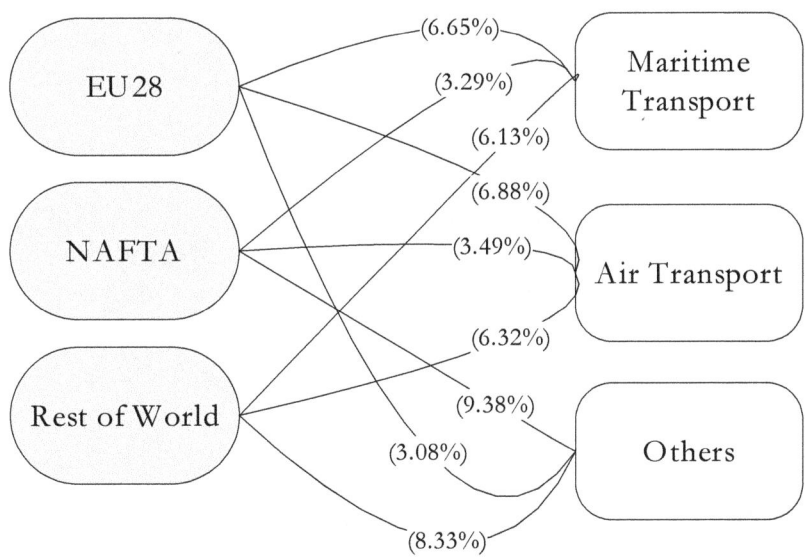

Figure 4. Percentage changes of logistics services

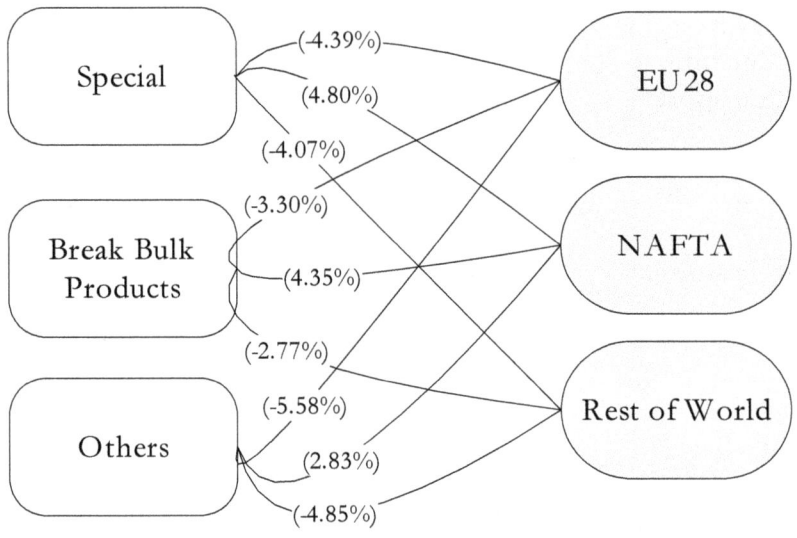

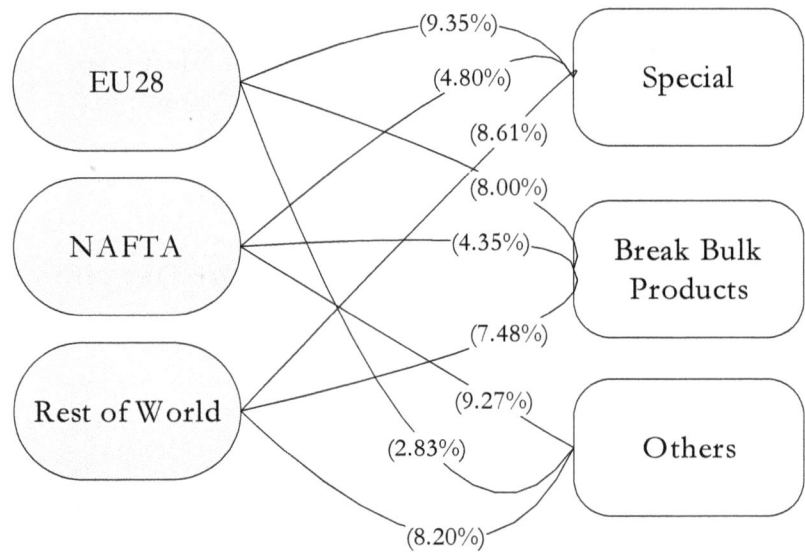

314

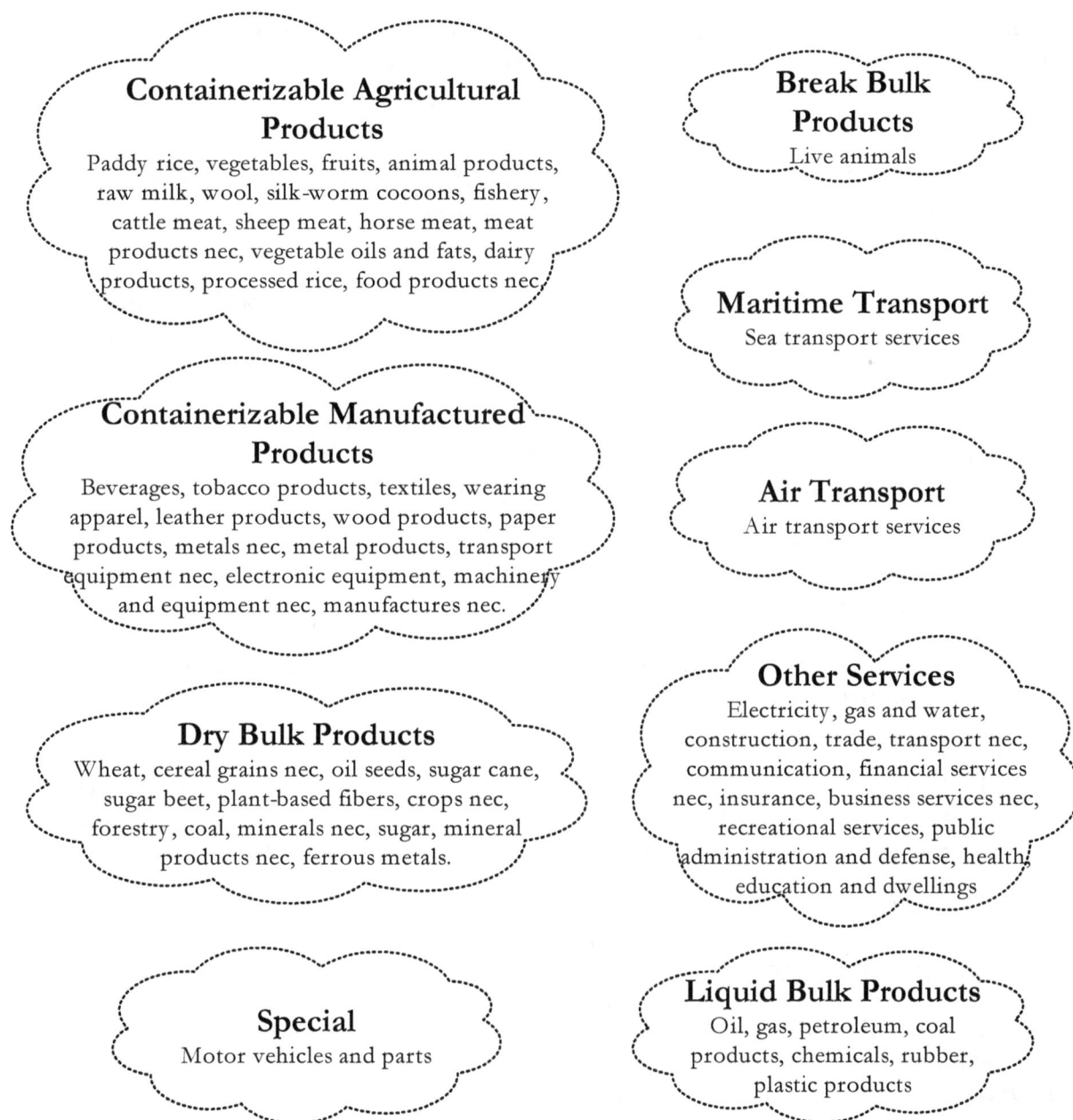

Figure 5. Commodity aggregation (Source: GTAP database)

Figure 5 shows the aggregation of the commodities to represent their corresponding logistics services.

4. Conclusion

This study looked into the logistics services as part of the global trade. It examined the growth of NAFTA area and its implications for the EU's logistics services. The results indicate that the logistics services from EU benefit highly in such growth in NAFTA area, but not all the logistics services from the NAFTA.

References

Aggarwal, V. K., & Fogarty, E. A. (2005). The Limits of Interregionalism: The EU and North America. *Journal of European Integration, 27*(3), 327-346. doi: 10.1080/07036330500190214

Andreff, M., & Andreff, W. (2009). Global Trade in Sports Goods: International Specialisation of Major Trading Countries. *European Sport Management Quarterly, 9*(3), 259-294. doi: 10.1080/16184740903024029

Aspinwall, M. (2009). NAFTA-ization: Regionalization and Domestic Political Adjustment in the North American Economic Area. *JCMS-Journal of Common Market Studies, 47*(1), 1-24. doi: 10.1111/j.1468-5965.2008.01831.x

Bems, R., Johnson, R. C., & Yi, K. M. (2010). Demand Spillovers and the Collapse of Trade in the Global Recession. *IMF Economic Review, 58*(2), 295-326. doi: 10.1057/imfer.2010.15

de Sousa, J., Mayer, T., & Zignago, S. (2012). Market access in global and regional trade. *Regional Science and Urban Economics, 42*(6), 1037-1052. doi: 10.1016/j.regsciurbeco.2012.07.011

Duina, F. (2004). Regional market building as a social process: an analysis of cognitive strategies in NAFTA, the European Union and Mercosur. *Economy and Society, 33*(3), 359-389. doi: 10.1080/0308514042000225707

Sbragia, A. (2010). The EU, the US, and trade policy: competitive interdependence in the management of globalization. *Journal of European Public Policy, 17*(3), 368-382. doi: 10.1080/13501761003662016

Yildiz, T. (2014). An Empirical Study on the Complex Relationships between Logistics Performances and Global Competitiveness. In *Business Logistics: Theory and Practice* (1st Edition). Charleston, SC, USA: CreateSpace, pp 75-107

Simulation 4

Chapter 14. The economic rise of the Sub-Saharan Africa and the impact on logistics services in Turkey

Abstract. As a growing economy with an expansionary foreign trade policy actively driven, Turkey has a focus in Africa. Over the past five years, Turkey's interest in seeking commercial partnerships with African economies is very well received on the continent. Africa offers significant trade potential. Turkey and Africa could greatly benefit from these relationships. In this regard, this study examines the potential for growth in the economies of Sub-Saharan Africa and their impact on logistics services.

Keywords: Sub-Saharan Africa, Turkey, logistics, transportation services, general equilibrium

1. Introduction

The focus of Turkey on trade and investment is an important element of its foreign markets and African policy are considered emerging opportunities (Wheeler 2011). Between 26 and 28 June 1998, the Turkish Foreign Ministry, having recognized that the level of Turkey's relations with Africa were inadequate, convened a series of meetings with stakeholders, including Turkish ambassadors in Africa, representatives of other ministries, non-governmental organizations (NGOs), the private sector and honorary consuls of African countries in Turkey to discuss "opening a door of Africa" (Wheeler 2011).

Turkey has used its historical ties with Africa as the basis of its recent policy, but the opportunity for change is the result of a combination of geopolitical factors (Wheeler 2011). Including changes that accompany the end of the Cold War, especially the end of the threat posed to Turkey by the Soviet Union; effective rejection by Europe of Turkey's longstanding ambitions to the European Union; and simultaneously flexing newly acquired economic muscle of Turkey at a time when its European neighbors are fighting against the impact of the global economic crisis (Wheeler 2011).

In this regard, based on a scenario, this study seeks some answers to questions. How will the growth in the Sub-Saharan African economies influence the logistics services in Turkey? This study examines the potential growth in the Sub-Saharan African economies and their impact on logistics services.

The remainder of this study is organized as follows. Section 2 presents some highlighted studies. Section 3 introduces the framework and methodology. The study is concluded in Section 4.

2. Highlighted Studies

Yildiz (2014) studied the global logistics performances in developed and developing countries, including African economies. Habiyaremye and Oguzlu (2014) studied the approach of Turkey in the context of growing East-West rivalry. They analyzed the main trends of the commitment of the global players with Africa, as well as shed light on how Turkey got involved in the continent. Bacik and Afacan (2013) studied the critical role of agents in the construction of Turkish foreign policy on Sub-Saharan Africa. They concluded that the foreign policy on sub-Saharan Africa that has emerged so far would suffer constant adjustment.

The opening of Turkey in Africa is a result of both the inner transformation of Turkey and a change in the global political economy (Ozkan 2012). The inner transformation of Turkey challenged the traditional Turkish partners in the economy and to diversify its trade opportunities in line with the change in the power of the global political economic configuration (Ozkan 2012).

3. Framework, methodology and brief results

The computable general equilibrium (CGE) model is economy wide in the sense that it includes all sectors. These models have gained increasingly widespread recognition in terms of policy evaluation. It provides a systematic analysis at the provincial level of external shocks, and allows monitoring of the effects of these changes on the various players in the economy. It is possible to distinguish the implications of various policies and external price plans for their effect on several variables of interest: the macroeconomic variables, sectoral output, employment, income and welfare.

Based on a scenario, this study examines the potential 5% growth in the Sub-Saharan African economies and their impact on logistics services. The results indicate both sides gain from trade. Logistics services grow higher that those from SSA (See Figures 1 through 4).

Figure 1. Percentage changes of logistics services

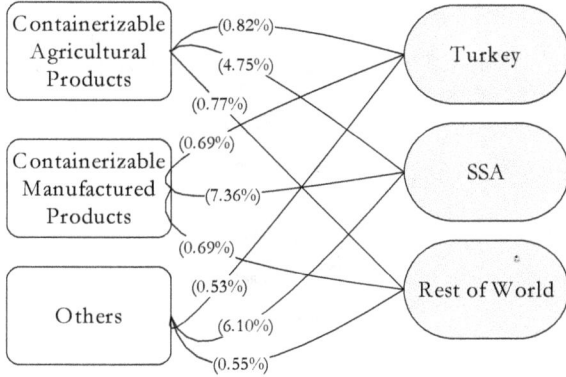

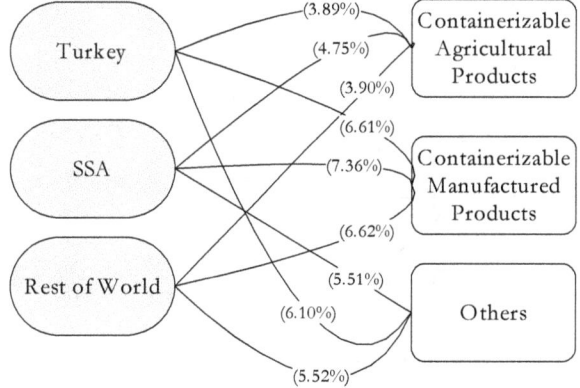

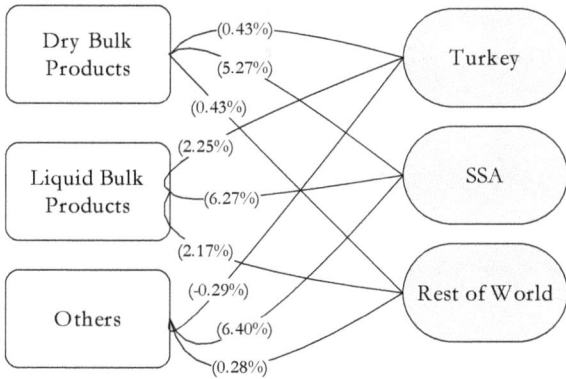

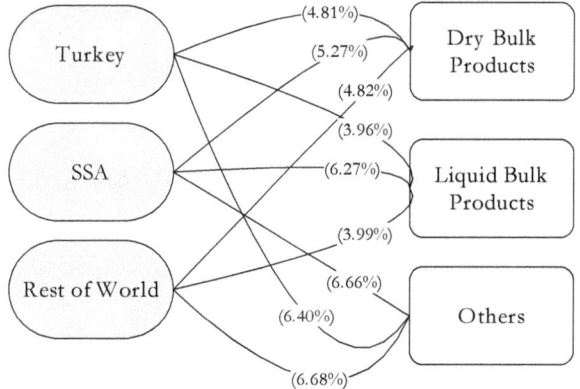

Figure 2. Percentage changes of logistics services

Percentage changes of logistics services ***from*** SSA

Maritime Transport		Turkey
(0.32%)		
(5.25%)		
(0.34%)		
(0.31%)		
Air Transport	(5.58%)	SSA
(0.30%)		
(0.61%)		Rest of World
Others	(6.51%)	
(0.60%)		

Percentage changes of logistics services ***to*** SSA

Turkey	(4.92%)	Maritime Transport
	(5.25%)	
	(4.93%)	
	(5.24%)	
SSA	(5.58%)	Air Transport
	(5.25%)	
Rest of World	(5.86%)	Others
	(6.51%)	
	(5.87%)	

Figure 3. Percentage changes of logistics services

Percentage changes of logistics services ***from*** SSA

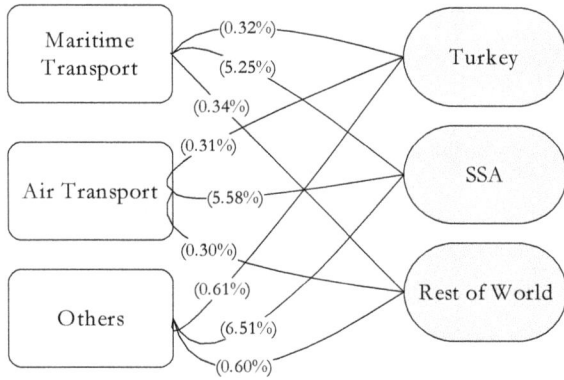

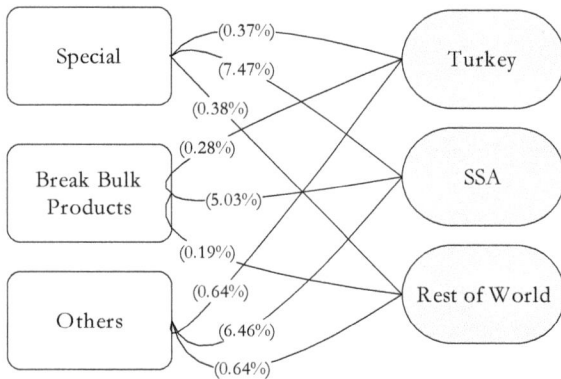

Special	(0.37%)	Turkey
	(7.47%)	
	(0.38%)	
	(0.28%)	
Break Bulk Products	(5.03%)	SSA
	(0.19%)	
	(0.64%)	Rest of World
Others	(6.46%)	
	(0.64%)	

Percentage changes of logistics services ***to*** SSA

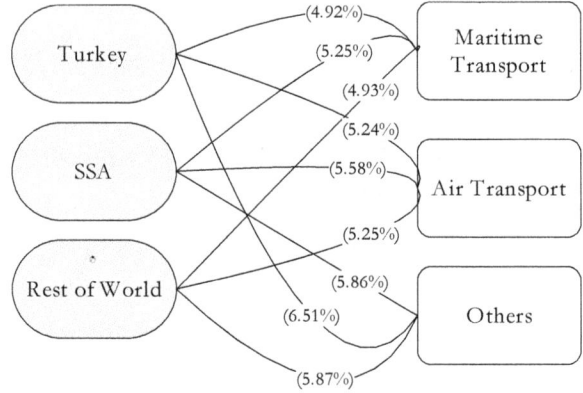

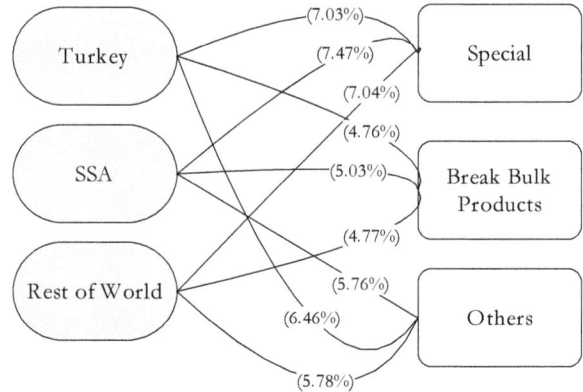

Turkey	(7.03%)	Special
	(7.47%)	
	(7.04%)	
	(4.76%)	
SSA	(5.03%)	Break Bulk Products
	(4.77%)	
Rest of World	(5.76%)	Others
	(6.46%)	
	(5.78%)	

Figure 4. Percentage changes of logistics services

Sea transport services	Maritime Transport
Air transport services	Air Transport
Oil, gas, petroleum, coal products, chemicals, rubber, plastic products	Liquid Bulk Products
Motor vehicles and parts	Special
Live animals	Break Bulk Products
Paddy rice, vegetables, fruits, animal products, raw milk, wool, silk-worm cocoons, fishery, cattle meat, sheep meat, horse meat, meat products nec, vegetable oils and fats, dairy products, processed rice, food products nec.	Containerazable Agricultural Products
Beverages, tobacco products, textiles, wearing apparel, leather products, wood products, paper products, metals nec, metal products, transport equipment nec, electronic equipment, machinery and equipment nec, manufactures nec.	Containerazable Manufactured Products
Wheat, cereal grains nec, oil seeds, sugar cane, sugar beet, plant-based fibers, crops nec, forestry, coal, minerals nec, sugar, mineral products nec, ferrous metals.	Dry Bulk Products
Electricity, gas and water, construction, trade, transport nec, communication, financial services nec, insurance, business services nec, recreational services, public administration and defense, health, education and dwellings	Other Services

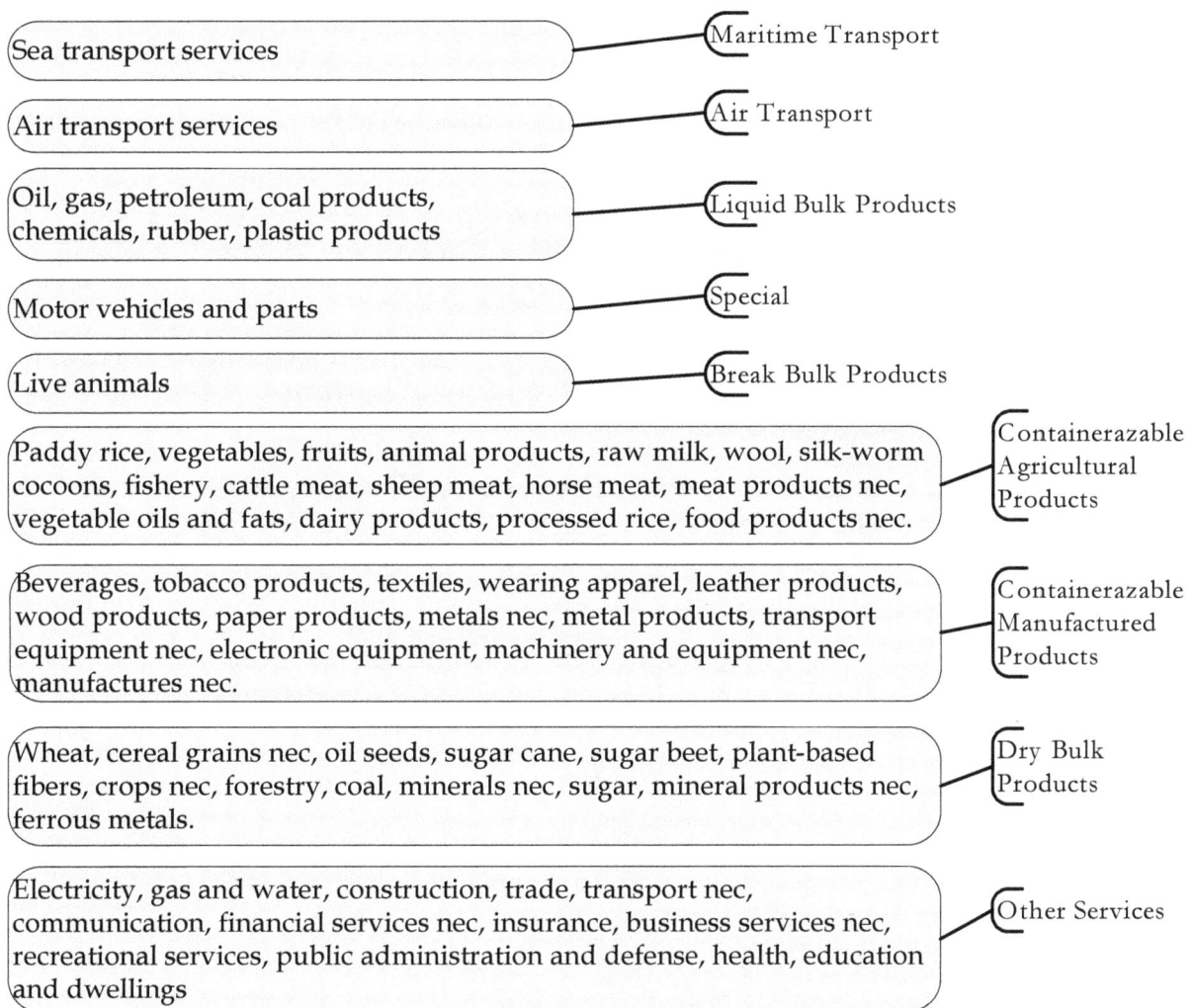

Figure 5. Commodity aggregation (Source: GTAP database)

Figure 5 shows the aggregation of the commodities to represent their corresponding logistics services.

4. Conclusion

This study examined the potential growth in the Sub-Saharan African economies and their impact on logistics services in Turkey and in the Sub-Saharan Africa. Results indicate that Turkey's logistics services will show more increase than the increase in the Sub-Saharan African logistics services. All estimated results are reported in the study.

References

Bacik, G., & Afacan, I. (2013). Turkey Discovers Sub-Saharan Africa: The Critical Role of Agents in the Construction of Turkish Foreign-Policy Discourse. *Turkish Studies*, 14(3), 483-502. doi: 10.1080/14683849.2013.832040

Habiyaremye, A., & Oguzlu, T. (2014). Engagement with Africa: Making Sense of Turkey's Approach in the Context of Growing East-West Rivalry. *Uluslararasi Iliskiler-International Relations*, 11(41), 65-85.

Ozkan, M. (2012). A New Actor or Passer-By? The Political Economy of Turkey's Engagement with Africa. *Journal of Balkan and Near Eastern Studies, 14*(1), 113-133. doi: 10.1080/19448953.2012.656968

Wheeler, A. T. (2011). Ankara to Africa: Turkey's outreach since 2005. *South African Journal of International Affairs, 18*(1), 43-62. doi: 10.1080/10220461.2011.564426

Yildiz, T. (2014). The Performances of Logistics Services in Developed and Developing Countries: A Review and Cluster Analysis. In *Business Logistics: Theory and Practice* (1st Edition). Charleston, SC, USA: CreateSpace, pp 43-73

Simulation 5

Chapter 15. Rise of the trade between the Sub-Saharan Africa and the EU: Implications for logistics services

Abstract. Africa is often referred to as the forgotten continent in which most international traders do not have a strong interest because of various reasons. Although the Sub-Saharan part is the poorest and least developed region in the world, Europe has long been shown interest in Africa. This study, based on a scenario, takes into account the growth in Sub-Saharan Africa and its possible effects on logistics services. The results indicate that the logistics services in the EU greatly benefit from growth in Africa.

Keywords: Sub-Saharan Africa, European Union, logistics, maritime transport, air transport, computable general equilibrium, GTAP

1. Introduction

Africa is often referred to as the "forgotten continent" in which most international actors do not have a vested interest (and Scheipers Sicurelli 2008). Its sub-Saharan region is the poorest and least developed region in the world. In addition, there are conflicts in the region. The role of the EU in Africa is remarkable: the EU is the largest export market for African products and the largest development aid donor. Since the adoption of the Lomé Convention in 1975, the EU has a formal relationship with Sub-Saharan Africa (SSA) countries (Scheipers and Sicurelli 2008). Relations between Europe and Africa were ruled by the Lomé Conventions and since 2001 have been formalized by the Cotonou Partnership Agreement (CPA) between the European Union (EU) and the African, Caribbean and Pacific (ACP) group. (Helly 2013). The Cotonou Partnership Agreement focuses primarily on aid and trade (Helly 2013)

Indeed, the future of Africa and Africans will be determined largely by the trade relationship with Europe (Goodison 2007). Before dismissing this as a serious over-statement, examine the past of Africa: the slave trade, colonial protectionism and post-colonial hostility to economic nationalism, the imposition of market forces while the Europe has strengthened its common agricultural policy (Goodison 2007).

Based on a scenario, this study seeks answers to some questions. How will the growth in Sub-Saharan Africa affect logistics services in the EU? Which logistics services will grow in the Sub-Saharan Africa and the EU? This study, based on a scenario, takes into account of the growth in Sub-Saharan Africa and its possible effects on logistics services.

The remainder of this study is organized as follows. Section 2 presents some highlighted studies. Section 3 presents the framework and methodology. The study is concluded in Section 4.

2. Highlighted Studies

Yildiz (2014) investigated the global logistics services in both developed and developing economies. Buzdugan (2013) studied the EU's external involvement in regionalism in Southern Africa. The study sheds new light on theories of regionalism, arguing that international actors can have a direct and significant impact on the dynamics of regionalism, particularly in Sub-Saharan Africa, examining the influence of the European Union (EU) on the development of the Southern African Community (SADC).

de Melo and Portugal-Perez (2014) studied the design of the preferential market access. Evidence and lessons learned from African clothing exports reported.

Jordaan and Kanda (2011) studied the trade effects of preferential trade agreements of the EU-SA and SADC, which South Africa is a member. The study also recommends that trade policy in South Africa should be increasingly oriented towards broad-based multilateral liberalization. In addition, South Africa should promote regional economic stability and development by supporting regional initiatives of trade agreements.

Langan (2012) studied Normative Power Europe and the moral economy of the EU-Africa ties. Langan (2012) argues that the moral outlook of the economy is at the center of a decisive shift in the concept of normative power to the evaluation of the differences between EU norms and EU policy outcomes.

McDonald and Walmsley (2008) studied the impact of the Republic of South Africa EU free trade agreement on Botswana. Their analyzes indicated that structural adjustments for African economies reported by the FTA are significant, which implies that there will be significant economic costs associated with the FTA.

3. Framework, methodology and brief results

In this study, Computable General Equilibrium (CGE) simulations use Global Trade Analysis Project (GTAP), which is a multi-region model of comparative static general equilibrium. The model used for this analysis is a general equilibrium model of the world economy with several regions and sectors.

Based on a scenario, 5% growth in Sub-Saharan Africa and its possible effects on logistics services are simulated. Results indicate that logistics services in the EU highly benefit from the growth in Africa (See Figures 1 through 4).

Figure 1. Percentage changes of logistics services

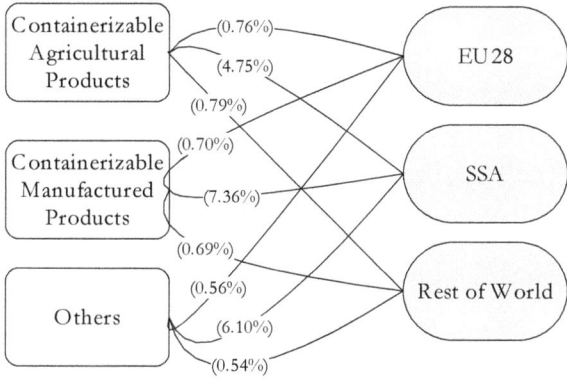

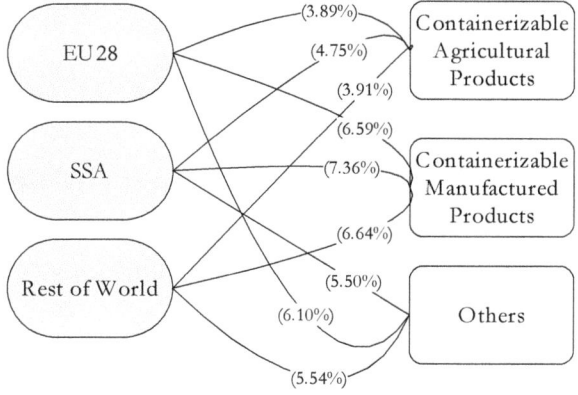

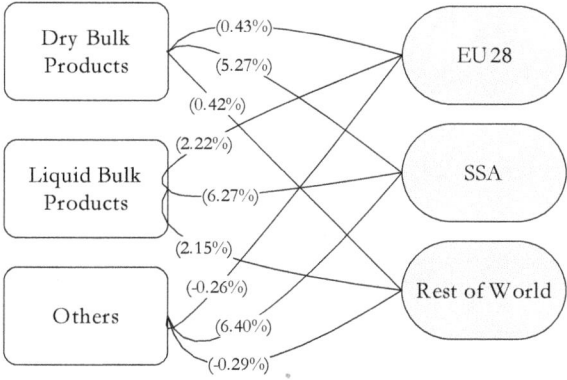

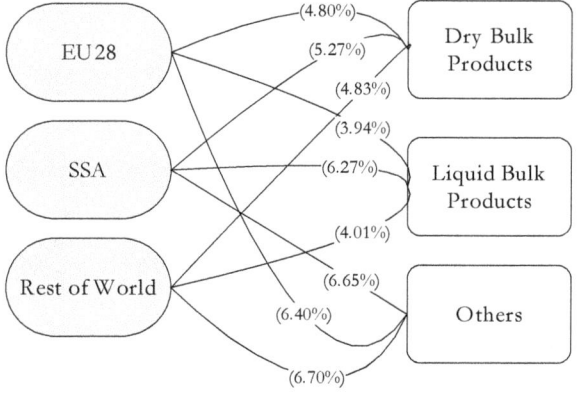

Figure 2. Percentage changes of logistics services

Figure 3. Percentage changes of logistics services

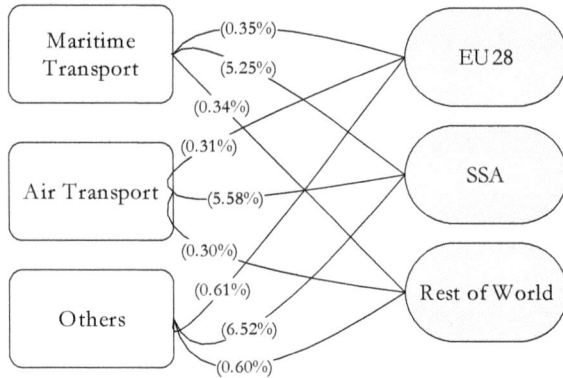

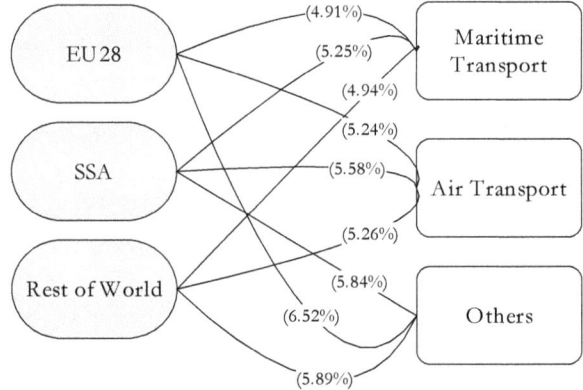

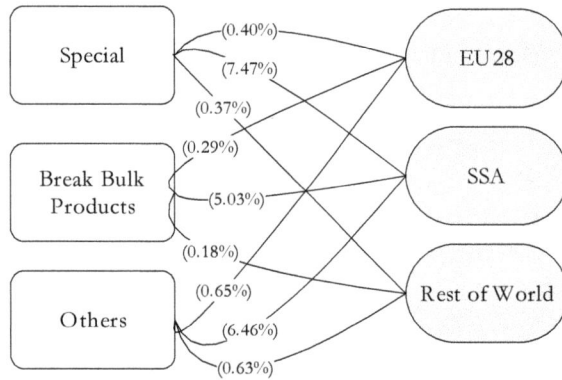

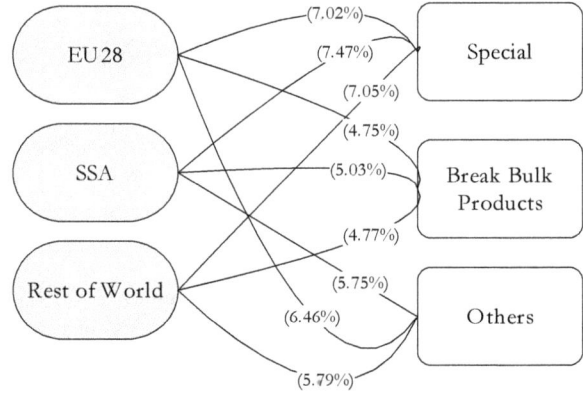

Figure 4. Percentage changes of logistics services

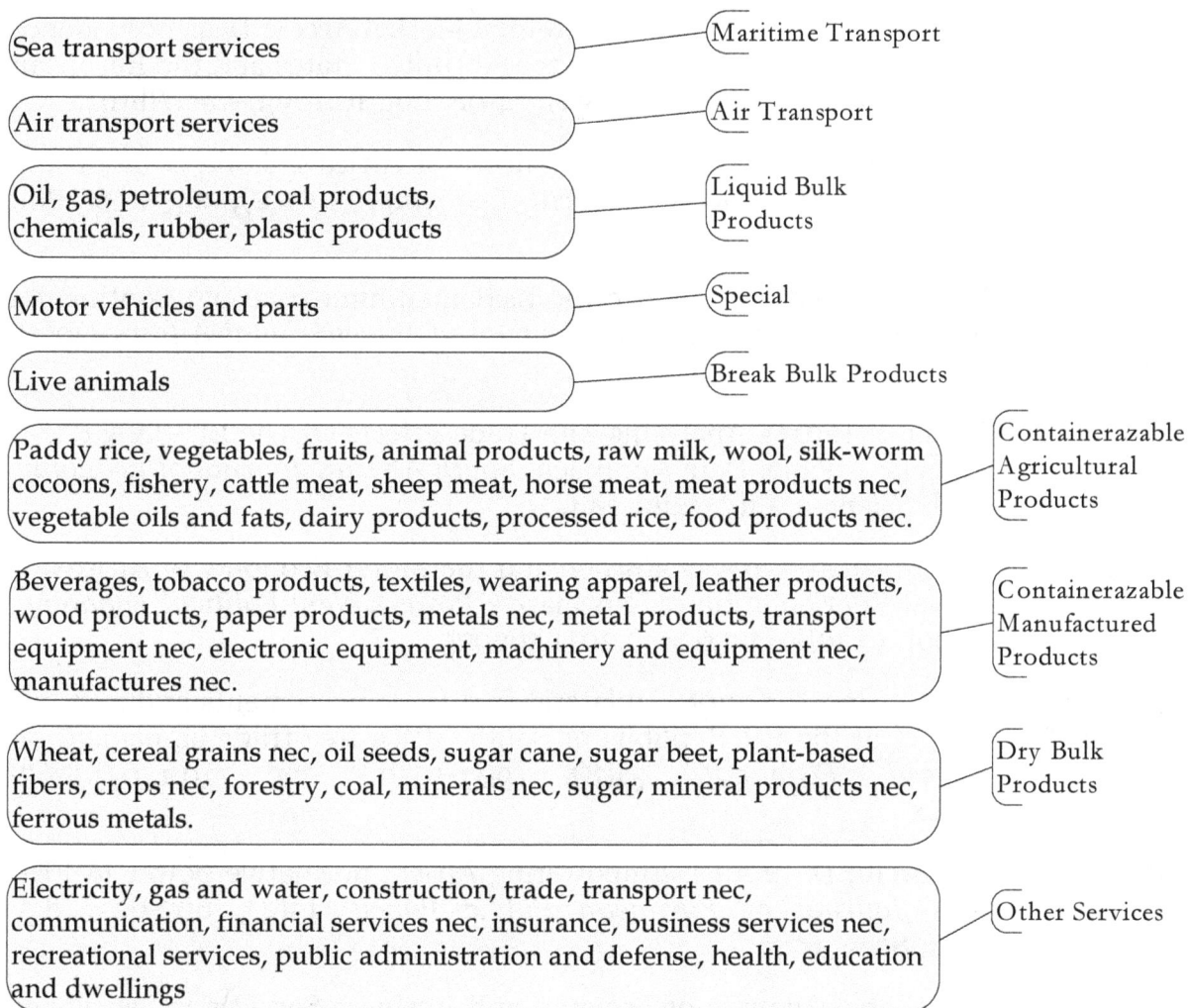

Figure 5. Commodity aggregation (Source: GTAP database)

Figure 5 depicts the aggregation of the commodities to represent their corresponding logistics services.

4. Conclusion

This study, based on a scenario, took into account of the growth in Sub-Saharan Africa and its possible effects on logistics services. Results indicate that logistics services in the EU highly benefit from the growth in Africa.

References

Buzdugan, S. R. (2013). Regionalism from without: External involvement of the EU in regionalism in southern Africa. *Review of International Political Economy*, 20(4), 917-946. doi: 10.1080/09692290.2012.747102 de Melo, J., & Portugal-Perez, A. (2014). Preferential Market Access Design: Evidence and Lessons from African Apparel Exports to the United States and the European Union. *World Bank Economic Review*, 28(1), 74-98. doi: 10.1093/wber/lht012

Goodison, P. (2007). EU Trade Policy & the Future of Africa's Trade Relationship with the EU. *Review of African Political Economy, 34*(112), 247-266. doi: 10.1080/03056240701449646

Helly, D. (2013). The EU and Africa since the Lisbon summit of 2007: Continental drift or widening cracks? *South African Journal of International Affairs, 20*(1), 137-157. doi: 10.1080/10220461.2013.780323

Jordaan, A., & Kanda, P. (2011). Analysing The Trade Effects Of The EU-SA & SADC Trading Agreements: A Panel Data Approach. *South African Journal of Economic and Management Sciences*, 14(2), 229-244.

Langan, M. (2012). Normative Power Europe and the Moral Economy of Africa-EU Ties: A Conceptual Reorientation of 'Normative Power'. *New Political Economy*, 17(3), 243-270. doi: 10.1080/13563467.2011.562975

McDonald, S., & Walmsley, T. (2008). Bilateral free trade agreements and customs unions: The impact of the EU Republic of South Africa free trade agreement on Botswana. *World Economy*, 31(8), 993-1029. doi: 10.1111/j.1467-9701.2008.01112.x

Scheipers, S., & Sicurelli, D. (2008). Empowering Africa: normative power in EU–Africa relations. *Journal of European Public Policy, 15*(4), 607-623. doi: 10.1080/13501760801996774

Yildiz, T. (2014). A Global Analysis of Logistics and Business Services in Developed and Developing Countries. In *Business Logistics: Theory and Practice* (1st Edition). Charleston, SC, USA: CreateSpace, pp 9-42

Yildiz, T. (2014). The Performances of Logistics Services in Developed and Developing Countries: A Review and Cluster Analysis. In *Business Logistics: Theory and Practice* (1st Edition). Charleston, SC, USA: CreateSpace, pp 43-73

Part V – Notes on Estimation of Logistics Impact of Trade Policies – A Recursively Dynamic Applied General Equilibrium Approach

Chapter 16. An overview of the computable general equilibrium (CGE) modeling and the *dynamic* CGE model

This chapter provides you an overview on the basics of the computable general equilibrium modeling (CGE) and the dynamic CGE model. It first introduces you shortly to the historical roots of CGE modeling. It then introduces you to the aim and the main mechanisms of models.

1.1. History of CGE modeling

Walras (1874) formulated the first general equilibrium model, called the Walrasian model that could adapt to complex economic interactions. Based on the theoretical structure of the Walrasian general equilibrium, the CGE modeling approach is based on the work, which was completed in the 1950s by Kenneth Arrow and Gerard Debreu. CGE models are based on general equilibrium (GE) theories of Arrow and Debreu, where agents interact in competitive markets by introducing optimal prices and optimal amounts that satisfy markets and balance agents' equilibrium conditions.

Since then, CGE models have been developed in the early 1960s to solve for both market prices and quantities simultaneously, simulating the operation of a competitive market economy.

Finally, interest in these models has increased since their creation in the 1960s as applied economists have recognized the benefits of their use for the counterfactual analysis and improvement in the computations that allowed for detailed analysis. The first CGE model developed in Norway by Leif Johansen (Johansen, 1960), was designed to be used for policy analysis. The purpose of a CGE model is to try to model the entire economy and the relations between economic agents in it. Later, CGE models have been widely used to analyze the impact of macroeconomic policies and the impact of the allocation of development resources in developed countries since the early 1980s CGE models are preferred over the partial equilibrium models because they include the complex interdependencies in the analysis.

1.2. The aim of CGE models and its mechanisms

The aim of CGE models is to quantify the effects of the policy on equilibrium allocations and relative prices using the standard theory of general equilibrium.

CGE models have their roots in the input-output theory. They are widely used for economic, social and environmental planning, as it is effective for capturing inter-sectoral linkages. Indeed, CGE models are based on complex hypothetical mathematical relationships between different sectors of the economy that reflect the behavior of the key players in the economy and provide more realistic estimates than those obtained from input-models. However, input-output models have limitations, which simulated the development of CGE models. These limitations are reflected mainly in the assumptions of the model, including fixed price, the factor of the unlimited supply and fixed share factor and intermediate inputs in the production process. Under these assumptions, the input-output model cannot show substitution between production inputs, and reactive behavior of producers and consumers to changes in relative prices.

In short, CGE models are adapted to analyze contemporary strategic issues in a competitive market economy, as they have the price mechanism that plays an important role in the economy. The price mechanism is able to solve complex problems in an economy that cannot be achieved by input-output models. In general equilibrium models, economic agents make their own decisions on economic activities based on changes in the market price as indicated resource constraints and technology. In addition, the market balances supply and demand by adjusting prices. Because CGE models can quantify market behavior and changes, they are widely used in various policy analyzes. A CGE model describes an economy in equilibrium with prices and relative quantities, which are determined endogenously. While most empirical approaches to examine the impacts of policies and effects in a *ceteris paribus* condition, a CGE model, which provides comparative scenarios based on the baseline scenario incorporates the factor markets, markets for goods and foreign trade markets.

1.3. The dynamic CGE model

Analytical processing of overall economic growth has its origin in the work of early theorists such as Ramsey (1928), Solow (1956) and Koopmans (1965). However, because of their heavy computing requirements, real dynamic extensions of computable general equilibrium models are a recent development.

The extension of a static CGE model to a dynamic process is simple. Although computationally more complex, a dynamic CGE model differs from its static counterpart by including a driving force to move the economy from period to period. In most dynamic models, this force is provided by the growth of underlying labor and/or a change in the level of technology in one or more sectors of the economy. These changes are facilitated by new investments and growth in the capital stock in the economy.

As with the static model, real output for each sector in a particular reference year is reproduced through the calibration procedure. In addition, the economy is now expected to grow, and the initial benchmark must be run with all sectors, the quantities and production factors, each of which are needed to grow at the same steady state rate.

When counterfactual shock is given to a dynamic CGE model two things happen.

1. The affected prices and quantities traverse to a new growth path in the years following the shock.

2. The new growth path itself returns to a stable state, but with economic variables at a different level than they would have been in the benchmark case.

In general, the interest of these dynamic models is on this new path and how much higher or lower, it is than the initial benchmark path.

1.4. The GDyn

The Dynamic GTAP (Global Trade Analysis Project) model known as GDyn (see Ianchovichina and McDougall, 2012). Being a general equilibrium global trade model with dynamic elements (see Ianchovichina and McDougall, 2012 for details) the GDyn is an extension of the widely used GTAP model (Hertel, 1997).

It can be applied to analyze various issues, such as trade policy, regional economic integration and climate change. The GTAP database model is a comparative static global model of general equilibrium, linking bilateral trade flows between all countries or regions, and explicitly models the consumption and production for all commodities of each national or regional economy.

As with other neoclassical CGE or applied general equilibrium (AGE) model in the GTAP, the producers are assumed to maximize profits and consumers are assumed to maximize utility (in GTAP). Product and factor market clearing requires that supply equals demand in each market (see Hertel, 1997 for details).

The GDyn is an extension of GTAP and retains its basic characteristics. It also provides an improved long-term treatment in the context of the GTAP data modeling.

It is a recursive model, generating a sequence of static equilibria based on the investment theory of adaptive expectations, and is bound by a number of dynamic characteristics.

The main features of the model include (Ianchovichina and McDougall, 2012, p 5.)

- the treatment of time; the distinction between physical and financial assets, and between domestic and foreign financial assets

- the treatment of capital and asset accumulation, assets and liabilities of firms and households, income from financial assets, and

- the investment theory of adaptive expectations

GTAP-Dyn is a recursive dynamic model applied general equilibrium (AGE) of the global economy. It extends the standard GTAP model (Hertel, 1997) to include:

- international capital mobility,

- capital accumulation

- adaptive expectations theory of investment.

Standard GTAP (Hertel and Tsigas, 1997) is a comparative static AGE model of the world economy, developed as a vehicle for teaching multi-country AGE modeling and to complement the GTAP multi-country AGE data base (Gehlhar, Gray, Hertel et al., 1997).

In general, it aims to provide a simple presentation of AGE modeling techniques widely used. It does, however, include some special features, notably an extensive decomposition of welfare results.

Chapter 17. Theoretical structure of dynamic GTAP and the recursive dynamic model

The Dynamic GTAP model (GTAP-Dyn) is a recursively dynamic applied general equilibrium (AGE) model of the world economy. It extends the standard GTAP model (Hertel, 1997) to include:

- International capital mobility

- Capital accumulation

- An adaptive expectations theory of investment

2.1. Theoretical structure of dynamic GTAP

A salient technical feature of the new extension is the treatment of *time*. Many dynamic models treat time as an index, so that each of the variables in the model has a time index.

In GTAP-Dyn, time itself is a variable, subject to exogenous change with the usual policy, technology, and demographic variables.

The differences between standard GTAP and GTAP-Dyn model can be generalized as follows (Ianchovichina and McDougall, 2001; Walmsley and Strutt, 2010):

1. Compared to the standard model of GTAP, GTAP-Dyn provides *a better long-term analysis*. Because the dynamic model needs to build the baseline scenario as well as take the accumulative effects of variable factors into consideration.

2. In the standard GTAP model, capitals are allowed to move between industries in a region, *but not between regions*. While in GTAP-Dyn, *capitals can move across regions*, which allow the investment allocation and endowment to respond to region-specific rates of return on capital.

3. The adjustment for the rate of return needs time. The standard GTAP model assumes that the rate of return adjustment in all countries is instantaneous without any delay. While in GTAP-Dyn we describe *a lagged adjustment*, which is more realistic.

4. GTAP-Dyn pulls in *the adaptive expectations theory of investment*. The investment movements depend on the changes of investors' expected rates of rates other than the actual rates. Their expectations of rates of return may be in error in the short term, but remain consistent with long-term real rates.

5. GTAP-Dyn includes *the capitals and gains of financial assets* to achieve the dynamic links across years.

Dynamic general equilibrium models can be classified as truly dynamic ("*intertemporal*") or sequential dynamic ("*recursive*") models. Truly dynamic models are based on the theory of optimal growth where the behavior of economic agents is characterized by perfect foresights. They know everything about the future and react to future changes in prices.

- Households maximize their intertemporal utility function under a wealth constraint to determine their schedule of consumption over time.

- Investment decisions by firms are the result of the maximization of cash flow over the entire time horizon.

A recursive dynamic model is essentially a series of static CGE models, which are linked between periods by an exogenous and endogenous variable updating procedure.

The static CGE model used to develop a recursive dynamic process is based on several standard assumptions:

- constant returns to scale,

- perfect competition and price taking behavior,

- market-clearing conditions hold for commodities and primary factors,

- and zero profit conditions hold, implying that price equals marginal cost.

Recursive-dynamic models: multi-period CGE models in which results are computed one-period-at-a-time. In contrast, for intertemporal models, results are computed simultaneously for all periods.

Intertemporal models: multi-period CGE models in which results are computed simultaneously for all periods. In contrast, for recursive-dynamic models, results are computed one-period-at-a-time.

RunDynam and GEMPACK

RunDynam allows you to build a reference scenario (which may be a forecast) and policy deviations from the base case with a model, which was implemented using GEMPACK. The model is solved on a year to year basis (that is, recursively) over a number of years, from initial data. For each year thereafter, the input data is the data updated by the previous simulation.

1. First, you solve the base case, and then carry out the policy deviation. You can choose a group of input data files for the model; these are the starting points for your base case.

2. You specify the closures and the impact on text files, using the syntax required in GEMPACK command files.

3. You can choose names for the output files of the base case and of policy runs.

4. You can choose from several methods to solve the model.

5. You can view the results of the base case or the deviation of the policy on the screen or export them to other programs.

6. RunDynam can produce graphics of selected variables over time.

7. You can view or copy either the initial model database or any of the updated data files produced during the base case or policy deviation.

References

Farmer, K., & Wendner, R. (2004). Dynamic multi-sector CGE modeling and the specification of capital. *Structural Change and Economic Dynamics*, 15(4), 469-492. doi: http://dx.doi.org/10.1016/ j.strueco.2003.12.002

Auerbach, A.J., Kotlikoff, L.J., 1987. *Dynamic Fiscal Policy*. Cambridge University Press, Cambridge.

P.A. Diamond, National debt in a neoclassical growth model, *American Economic Review*, 55 (1965), pp. 1135–1150

Chapter 18. Global Trade Analysis Project

The GTAP Modeling Framework

Recursive Dynamic GTAP

The main objective of GTAP-Dyn is to provide better long-term treatment in the GTAP framework. In standard GTAP, capital can move between industries in a region, *but not between regions*. This hinders analysis of policy shocks and other developments diversely affecting incentives to invest in different regions. For a good long-term treatment, we need *international capital mobility*.

The main distinctive features of GDyn are its specification investment income flows associated with financial assets. The model distinguishes between *physical* and *financial* assets, and in the latter between *domestic* and *foreign*.

The model allows to determine the accumulation of capital and assets of each national economy, and the assets and liabilities of businesses and households in each region. The theory of investment in each region is characterized by *adaptive expectations*, in which the differences between *actual* and *expected rates of return* are corrected over time by displacing investment and international capital mobility.

The GDyn uses a simplified and unified treatment of the mobility of capital and investment in the context of a global CGE model. This specification endogenously captures the overall effects of the accumulation of capital and wealth in the country, and the effects of income from foreign ownership of assets.

Final demand: private consumption, saving and government

Final demand in each region is represented by an aggregate called "*Regional Household*," which is a Cobb-Douglas combination of private household consumption, saving and government spending.

Private consumption optimizer is represented by an agent governed by a function of spending CDE (constant difference of elasticity).

Government consumption follows a Cobb–Douglas function, which implies a constant share of public spending on goods and services. The savings is a residual element of the country's income and determines the net investment in the economy.

GDyn

The standard version of GDyn is a recursive-dynamic extension of the standard GTAP (Hertel, 1997), developed for better treatment of medium and long-term simulations, as it strengthens the investment side of the modeling framework to enable international capital mobility (Ianchovichina & Walmsley, 2012).

GDyn extends the standard, comparative static version of the GTAP model by introducing

- international capital mobility,

- endogenous capital accumulation and

- adaptive expectations theory of investment in a recursive dynamics setting.

GDyn is a real assets model, i.e. investment is associated with equity: the regional households (shareholders) own equity in the firm equal to the value of physical capital and earn income (dividends) corresponding to their ownership share - there are no financial markets and no differentiation between debt and equity.

The model keeps track of gross ownership positions and income flows associated with them and thus compared to the comparative static version the GTAP model is augmented to improve the representation of balance of payments relationships.

Despite the advantages offered by perfect foresight models, the solution procedure chosen for the GDyn is a recursive one in which investors are allowed to have errors in their expectations, i.e. a novel adaptive expectations specification of investors' behavior. Compared to perfect foresight models GDyn offers greater empirical realism, flexibility in data specification and lower computational complexity.

GDyn inherits the treatment of savings of the comparative static GTAP model. As implemented by, the representative household allocates regional income that would maximize the per capita utility based on a Cobb–Douglas utility function complemented with non-homothetic preferences on the private consumption side. Real saving is a single commodity that is defined as savings deflated by the price of savings. The Cobb–Douglas specification keeps the budget shares constant, implicitly assuming a constant marginal propensity to save of the household.

Capital goods are a production sector and their supply is determined by a Leontieff type production technology. On the other hand, capital is a value added component and is a direct input into production of all goods (except capital goods) governed by a CES type allocation. Capital is assumed to be perfectly mobile across sectors determining a single rental rate across sectors that clears the market.

As in most recursive dynamic models, each period's equilibrium determines the level of global savings and implicitly the aggregate amount of investment expenditure available in that specific period. International capital mobility is modeled using a disequilibrium approach that reconciles investment theory with empirical findings.

The disequilibrium approach adopted here is described by two mechanisms in the model:

1. There is a gradual convergence of the expected rate of return leading to the equalization of expected rates of return on the long run; and

2. Errors in expectations with respect to the actual rate of return are eliminated over time.

Investors are assumed to respond to expected rates of return as opposed to actual rates of return when making investment decisions allowing for errors in expectations. For instance, when investment in the base data is low despite high actual rates of return it is assumed to be due to errors in expectations; investors are assumed to behave adaptively and over time these errors are eliminated and the expected rate of return will converge toward the observed rate of return.

The GDyn model in its current form does not make use of portfolio allocation theory in determining gross ownership positions, i.e. investors reactions are based only on (expected) rates of return and hence the GDyn model is an investment demand driven model.

Moreover, domestic firms hold equity directly in domestic firms, the lack of availability of bilateral data on foreign assets, precludes the representative household from holding equity directly in foreign firms. This lack of bilateral data on foreign assets and liabilities compels many CGE modelers to employ a somewhat artificial representation of foreign investment. The GDyn model overcomes this problem through the adoption of a fictional entity called the global trust. The global trust collects the saving of all the regional households and allocates this to regional investment on their behalf.

The mechanism that the GDyn model uses to determine the composition of the cross-ownership matrix over time is cross-entropy minimization. The choice of the cross-entropy allocation of wealth is motivated by the fact that this type of specification is able to reproduce some of the empirical findings of the investment literature such as the home bias of puzzle of investment.

Part VI

Chapter 19. The long-term assessment of exports of food and industrial products and its logistics impacts

(Simulation 1)

1. Introduction

This study examines the fictitious long-term trade among North America (NAM), European Union (EUN) and the rest of the world (ROW) and it effects on the logistics services (food, mnfc, serv). Based on the scenario, the results indicate that various logistics services to/from economies exhibit increases or decreases. Results are reported using *dynamic* computable general equilibrium (CGE) simulations.

The remainder of this study is organized as follows. Section 2 presents some highlighted studies. Section 3 presents the framework, methodology, results and short discussion. The study is concluded in Section 4.

2. Highlighted studies

Feraboli (2007) studied preferential trade liberalization, fiscal policy responses and welfare for Jordan. Lucke *et al.* (2007) assessed economic and fiscal reforms in Lebanon with debt constraints. Radulescu and Stimmelmayr (2010) explored the impact of the 2008 German corporate tax reform. Lu *et al.* (2010) studied the impacts of carbon tax and complementary policies on Chinese economy. Fehr (2000) analyzed consumption taxation. Cho *et al.* (2010) researched allocation and banking in Korean permits trading.

Espinosa *et al.* (2014) performed ex-ante analysis of the Regional Impacts of the common agricultural policy. Xu *et al.* (2011) explored impacts of agricultural public spending on Chinese food economy. Dogruel *et al.* (2003) researched macroeconomics of Turkey's agricultural reforms. Femenia (2010) investigated impacts of stockholding behavior on agricultural market volatility.

Bruvoll and Foehn (2006) analyzed trans-boundary effects of environmental policy. Bretschger *et al.* (2011) investigated growth effects of carbon policies. Chi *et al.* (2014) performed scenarios analysis of the energies' consumption and carbon emissions in China. Doumax *et al.* (2014) investigated biofuels, tax policies and oil prices in France. He *et al.* (2014) performed low-carbon-oriented dynamic optimization of residential energy pricing for China. Takeda (2007) investigated the double dividend from carbon regulations in Japan. Rive *et al.* (2006) investigated climate agreements based on responsibility for global warming. Schenker (2013) investigated exchanging goods and damages from the perspective of the role of trade on the distribution of climate change costs. Markandya *et al.* (2015) analyzed trade-offs in international climate policy options. Mori *et al.* (2006) studied integrated assessments of global warming issues. Okagawa *et al.* (2012) assessed GHG emission reduction pathways in a society without carbon capture and nuclear technologies. O'Ryan *et al.* (2011) studied the socioeconomic and environmental effects of free trade agreements for Chile. Loisel (2009) explored environmental climate instruments in Romania. Kishimoto *et al.* (2014) modeled regional transportation demand in China and explored the impacts of a national carbon policy. Fujino *et al.* (2006) performed multi-gas mitigation analysis on stabilization scenarios using aim global model. Xie *et al.* (2015) studied disaster risk decision of regional mitigation Investment.

Georges (2008) analyzed liberalizing NAFTA Rules of Origin. Vellinga (2008) commented on dynamic general-equilibrium model of an open economy. Brocker and Korzhenevych (2013) explored forward-looking dynamics in spatial CGE modelling. Giesecke (2002) explained regional economic performance. Deepak *et al.* (2001) studied local government portfolios and regional growth. Hubler (2011) investigated technology diffusion under contraction and convergence of China.

Zhang (2001) performed iterative method for finding the balanced growth solution of the non-linear dynamic input-output model. Kristkova (2012) explored impact of R&D investment on economic growth of the Czech Republic. Lay *et al.* (2008) studied shocks, policy reforms and pro-poor growth in Bolivia. Ihori *et al.* (2011) studied health insurance reform and economic growth for Japan. Breisinger *et al.* (2011) explored impacts of the triple global crisis on growth and poverty for Yemen. Breisinger *et al.* (2009) modeled growth options and structural change to reach middle-income country status for Ghana.

Wu and Xiao (2014) run dynamic CGE model and performed simulation analysis on the impact of citizenization of Rural migrant workers on the labor and capital Markets in China. AlShehabi (2013) modeled energy and labor linkages for Iran.

Loisel (2010) explored quota allocation rules in Romania which are assessed by a dynamic CGE model. Wittwer (2009) explored the economic impacts of a new dam in South-East Queensland.

Ozdemir and Bayar (2009) analyzed the peace dividend effect of Turkish convergence to the EU using a multi-region dynamic CGE model for Greece and Turkey. Aydin and Acar (2011) studied economic impact of oil price shocks on the Turkish economy in the coming decades. Barkhordar and Saboohi (2013) assessed alternative options for allocating oil revenue in Iran.

Dixon *et al.* (2011) explored the economic costs to the U.S. of closing its borders. Cardenete and Delgado (2015) simulated the impact of withdrawal of European funds on Andalusian economy.

Fougere *et al.* (2007) performed a sectoral and occupational analysis on ageing population in Canada. Dixon and Rimmer (2010) validated a detailed, dynamic CGE Model of the USA. Dixon *et al.* (2005) studied rational expectations for large CGE models.

3. Framework, methodology, and results

The GDyn is an extension of GTAP and retains its basic characteristics. It also provides an improved long-term treatment in the context of the GTAP data modeling. It is a recursive model, generating a sequence of static equilibria based on the investment theory of adaptive expectations, and is bound by a number of dynamic characteristics. GTAP-Dyn provides a better long-term analysis. A recursive dynamic model is essentially a series of static CGE models, which are linked between periods by an exogenous and endogenous variable updating procedure.

This study examines the fictitious long-term trade among North America (NAM), European Union (EUN) and the rest of the world (ROW) and it effects on the logistics services. This study looks into the effects on logistics services based on the scenario of an economy wide technology shock in ROW economies. The results indicate that various logistics services from/to economies exhibit increases or decreases (See Tables 1 through 4).

Policy Shock

ashock afereg(*"row"*) = -5 ;

afereg (REG): Economy wide afe shock

afe (ENDW_COMM, PROD_COMM, REG) : Primary factor i augmenting tech change by j of r

Table 1. Food, Mnfc, Serv 2007 (percentage changes)

qxs[food**](D)	1 NAM	2 EUN	3 ROW
1 NAM	-0.013	-0.693	-2.309
2 EUN	1.231	0.440	-1.186
3 ROW	-0.212	-0.952	-2.592

qxs[mnfc**](D)	1 NAM	2 EUN	3 ROW
1 NAM	-0.444	-1.467	-6.705
2 EUN	1.365	0.283	-5.068
3 ROW	3.575	2.441	-3.044

qxs[serv**](D)	1 NAM	2 EUN	3 ROW
1 NAM	-0.647	-1.066	-5.821
2 EUN	0.702	0.304	-4.523
3 ROW	2.286	1.899	-3.035

Table 2. Food, Mnfc, Serv 2012 (percentage changes)

qxs[food**](D)	1 NAM	2 EUN	3 ROW
1 NAM	0.926	0.723	0.393
2 EUN	1.640	1.509	1.128
3 ROW	-3.950	-4.098	-4.425

qxs[mnfc**](D)	1 NAM	2 EUN	3 ROW
1 NAM	1.735	0.998	-2.829
2 EUN	2.622	1.924	-1.975
3 ROW	-1.027	-1.764	-5.478

qxs[serv**](D)	1 NAM	2 EUN	3 ROW
1 NAM	0.848	0.374	-2.995
2 EUN	1.890	1.437	-1.970
3 ROW	-1.535	-1.957	-5.282

Table 3. Food, Mnfc, Serv 2017 (percentage changes)

qxs[food**](D)	1 NAM	2 EUN	3 ROW
1 NAM	1.578	1.773	2.558
2 EUN	1.829	2.226	2.923
3 ROW	-6.683	-6.385	-5.665

qxs[mnfc**](D)	1 NAM	2 EUN	3 ROW
1 NAM	2.908	2.504	0.599
2 EUN	2.973	2.685	0.719
3 ROW	-4.738	-5.092	-6.811

qxs[serv**](D)	1 NAM	2 EUN	3 ROW
1 NAM	1.810	1.272	-0.565
2 EUN	2.417	1.902	0.050
3 ROW	-4.389	-4.854	-6.615

Table 4. Food, Mnfc, Serv 2020 (percentage changes)

qxs[food**](D)	1 NAM	2 EUN	3 ROW
1 NAM	1.723	2.104	3.049
2 EUN	1.705	2.311	3.165
3 ROW	-7.238	-6.757	-5.892

qxs[mnfc**](D)	1 NAM	2 EUN	3 ROW
1 NAM	3.087	2.883	1.516
2 EUN	2.754	2.682	1.260
3 ROW	-5.581	-5.738	-6.940

qxs[serv**](D)	1 NAM	2 EUN	3 ROW
1 NAM	2.010	1.521	0.033
2 EUN	2.305	1.840	0.345
3 ROW	-4.971	-5.388	-6.808

Sea transport services — Maritime Transport

Air transport services — Air Transport

Oil, gas, petroleum, coal products, chemicals, rubber, plastic products — Liquid Bulk Products

Motor vehicles and parts — Special

Live animals — Break Bulk Products

Paddy rice, vegetables, fruits, animal products, raw milk, wool, silk-worm cocoons, fishery, cattle meat, sheep meat, horse meat, meat products nec, vegetable oils and fats, dairy products, processed rice, food products nec. — Containerazable Agricultural Products

Beverages, tobacco products, textiles, wearing apparel, leather products, wood products, paper products, metals nec, metal products, transport equipment nec, electronic equipment, machinery and equipment nec, manufactures nec. — Containerazable Manufactured Products

Wheat, cereal grains nec, oil seeds, sugar cane, sugar beet, plant-based fibers, crops nec, forestry, coal, minerals nec, sugar, mineral products nec, ferrous metals. — Dry Bulk Products

Electricity, gas and water, construction, trade, transport nec, communication, financial services nec, insurance, business services nec, recreational services, public administration and defense, health, education and dwellings — Other Services

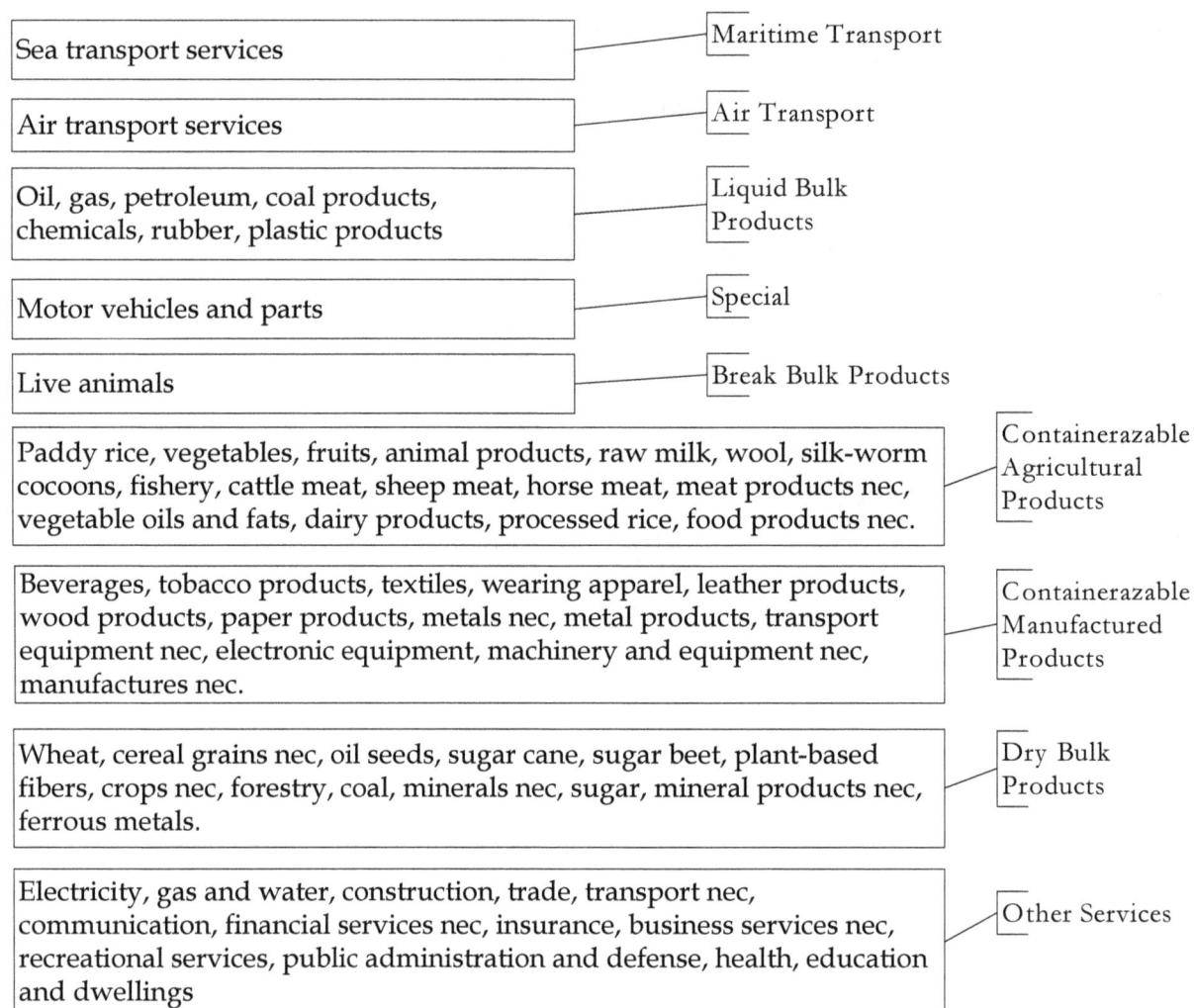

Figure 1. Commodities and sample aggregation (Source: GTAP Database)

Figure 1 depicts a sample aggregation of the commodities to represent their corresponding logistics services.

4. Conclusion

This study looked into the effects on logistics services based on the scenario of an economy wide technology shock in ROW economies. The results indicate that various logistics services from/to economies exhibit increases or decreases. These results are reported by using dynamic (CGE) simulations.

References

AlShehabi, O. H. (2013). Modelling energy and labour linkages: A CGE approach with an application to Iran. *Economic Modelling, 35,* 88-98. doi: 10.1016/j.econmod.2013.06.047

Aydin, L., & Acar, M. (2011). Economic impact of oil price shocks on the Turkish economy in the coming decades: A dynamic CGE analysis. *Energy Policy, 39*(3), 1722-1731. doi: 10.1016/j.enpol.2010.12.051

Barkhordar, Z. A., & Saboohi, Y. (2013). Assessing alternative options for allocating oil revenue in Iran. *Energy Policy, 63,* 1207-1216. doi: 10.1016/j.enpol.2013.08.099

Breisinger, C., Diao, X. S., Collion, M. H., & Rondot, P. (2011). Impacts of the Triple Global Crisis on Growth and Poverty: The Case of Yemen. *Development Policy Review, 29*(2), 155-184. doi: 10.1111/j.1467-7679.2011.00530.x

Breisinger, C., Diao, X. S., & Thurlow, J. (2009). Modeling growth options and structural change to reach middle income country status: The case of Ghana. *Economic Modelling, 26*(2), 514-525. doi: 10.1016/j.econmod.2008.10.007

Bretschger, L., Ramer, R., & Schwark, F. (2011). Growth effects of carbon policies: Applying a fully dynamic CGE model with heterogeneous capital. *Resource and Energy Economics, 33*(4), 963-980. doi: 10.1016/j.reseneeco.2011.06.004

Brocker, J., & Korzhenevych, A. (2013). Forward looking dynamics in spatial CGE modelling. *Economic Modelling, 31,* 389-400. doi: 10.1016/j.econmod.2012.11.031

Bruvoll, A., & Foehn, T. (2006). Transboundary effects of environmental policy: Markets and emission leakages. *Ecological Economics, 59*(4), 499-510. doi: 10.1016/j.ecolecon.2005.11.015

Cardenete, M. A., & Delgado, M. C. (2015). A simulation of impact of withdrawal European funds on Andalusian economy using a dynamic CGE model: 2014-20. *Economic Modelling, 45,* 83-92. doi: 10.1016/j.econmod.2014.09.021

Chi, Y. Y., Guo, Z. Q., Zheng, Y. H., & Zhang, X. P. (2014). Scenarios Analysis of the Energies' Consumption and Carbon Emissions in China Based on a Dynamic CGE Model. *Sustainability, 6*(2), 487-512. doi: 10.3390/su6020487

Cho, G. L., Kim, H. S., & Kim, Y. D. (2010). Allocation and banking in Korean permits trading. *Resources Policy, 35*(1), 36-46. doi: 10.1016/j.resourpol.2009.10.001

Deepak, M. S., West, C. T., & Spreen, T. H. (2001). Local government portfolios and regional growth: Some combined dynamic CGE/optimal control results. *Journal of Regional Science, 41*(2), 219-254. doi: 10.1111/0022-4146.00215

347

Dixon, P. B., Giesecke, J. A., Rimmer, M. T., & Rose, A. (2011). The economic costs to the U.S. Of closing its borders: a computable general equilibrium analysis. *Defence and Peace Economics, 22*(1), 85-97. doi: 10.1080/10242694.2010.491658

Dixon, P. B., Pearson, K. R., Picton, M. R., & Rimmer, M. T. (2005). Rational expectations for large CGE models: A practical algorithm and a policy application. *Economic Modelling, 22*(6), 1001-1019. doi: 10.1016/j.econmod.2005.06.007

Dixon, P. B., & Rimmer, M. T. (2010). Validating a Detailed, Dynamic CGE Model of the USA*. *Economic Record, 86*, 22-34. doi: 10.1111/j.1475-4932.2010.00656.x

Dogruel, F., Dogruel, A. S., & Yeldan, E. (2003). Macroeconomics of Turkey's agricultural reforms: an intertemporal computable general equilibrium analysis. *Journal of Policy Modeling, 25*(6-7), 617-637. doi: 10.1016/s0161-8938(03)00056-5

Doumax, V., Philip, J. M., & Sarasa, C. (2014). Biofuels, tax policies and oil prices in France: Insights from a dynamic CGE model. *Energy Policy, 66*, 603-614. doi: 10.1016/j.enpol.2013.11.027

Espinosa, M., Psaltopoulos, D., Santini, F., Phimister, E., Roberts, D., Mary, S., . . . Paloma, S. G. Y. (2014). Ex-Ante Analysis of the Regional Impacts of the Common Agricultural Policy: A Rural-Urban Recursive Dynamic CGE Model Approach. *European Planning Studies, 22*(7), 1342-1367. doi: 10.1080/09654313.2013.786683

Fehr, H. (2000). From destination- to origin-based consumption taxation: A dynamic CGE analysis. *International Tax and Public Finance, 7*(1), 43-61. doi: 10.1023/a:1008754029145

Femenia, F. (2010). Impacts of Stockholding Behaviour on Agricultural Market Volatility: A Dynamic Computable General Equilibrium Approach. *German Journal of Agricultural Economics, 59*(3), 187-201.

Feraboli, O. (2007). Preferential trade liberalisation, fiscal policy responses and welfare: A dynamic CGE model for Jordan. *Jahrbucher Fur Nationalokonomie Und Statistik, 227*(4), 335-357.

Fougere, M., Mercenier, J., & Merette, M. (2007). A sectoral and occupational analysis of population ageing in Canada using a dynamic CGE overlapping generations model. *Economic Modelling, 24*(4), 690-711. doi: 10.1016/j.econmod.2007.01.001

Fujino, J., Nair, R., Kainuma, M., Masui, T., & Matsuoka, Y. (2006). Multi-gas mitigation analysis on stabilization scenarios using aim global model. *Energy Journal*, 343-353.

Georges, P. (2008). Liberalizing NAFTA Rules of Origin: A Dynamic CGE Analysis. *Review of International Economics, 16*(4), 672-691. doi: 10.1111/j.1467-9396.2008.00771.x

Giesecke, J. (2002). Explaining regional economic performance: An historical application of a dynamic multi-regional CGE model. *Papers in Regional Science, 81*(2), 247-278. doi: 10.1007/s101100100100

He, Y. X., Liu, Y. Y., Wang, J. H., Xia, T., & Zhao, Y. S. (2014). Low-carbon-oriented dynamic optimization of residential energy pricing in China. *Energy, 66*, 610-623. doi: 10.1016/j.energy.2014.01.051

Hubler, M. (2011). Technology diffusion under contraction and convergence: A CGE analysis of China. *Energy Economics, 33*(1), 131-142. doi: 10.1016/j.eneco.2010.09.002

Ihori, T., Kato, R. R., Kawade, M., & Bessho, S. (2011). Health insurance reform and economic growth: Simulation analysis in Japan. *Japan and the World Economy, 23*(4), 227-239. doi: 10.1016/j.japwor.2011.07.003

Kishimoto, P. N., Zhang, D., Zhang, X. L., & Karplus, V. J. (2014). Modeling Regional Transportation Demand in China and the Impacts of a National Carbon Policy. *Transportation Research Record*(2454), 1-11. doi: 10.3141/2454-01

Kristkova, Z. (2012). Impact of R&D Investment on Economic Growth of The Czech Republic - A Recursively Dynamic CGE Approach. *Prague Economic Papers, 21*(4), 412-433.

Lay, J., Thiele, R., & Wiebelt, M. (2008). Shocks, policy reforms and pro-poor growth in Bolivia: A simulation analysis. *Review of Development Economics, 12*(1), 37-56. doi: 10.1111/j.1467-9361.2007.00394.x

Loisel, R. (2009). Environmental climate instruments in Romania: A comparative approach using dynamic CGE modelling. *Energy Policy, 37*(6), 2190-2204. doi: 10.1016/j.enpol.2009.02.001

Loisel, R. (2010). Quota allocation rules in Romania assessed by a dynamic CGE model. *Climate Policy, 10*(1), 87-102. doi: 10.3763/cpol.2008.0557

Lu, C. Y., Tong, Q., & Liu, X. M. (2010). The impacts of carbon tax and complementary policies on Chinese economy. *Energy Policy, 38*(11), 7278-7285. doi: 10.1016/j.enpol.2010.07.055

Lucke, B., Soto, B. G., & Zotti, J. (2007). Assessing economic and fiscal reforms in Lebanon - A dynamic CGE analysis with debt constraints. *Emerging Markets Finance and Trade, 43*(1), 35-63. doi: 10.2753/ree1540-496x430102

Markandya, A., Antimiani, A., Costantini, V., Martini, C., Palma, A., & Tommasino, M. C. (2015). Analyzing Trade-offs in International Climate Policy Options: The Case of the Green Climate Fund. *World Development, 74,* 93-107. doi: 10.1016/j.worlddev.2015.04.013

Mori, S., Akimoto, K., Homma, T., Sano, F., Oda, J., Hayashi, A., . . . Tomoda, T. (2006). Integrated assessments of global warming issues and an overview of project PHOENIX - A comprehensive approach. *Ieej Transactions on Electrical and Electronic Engineering, 1*(4), 383-396. doi: 10.1002/tee.20081

Okagawa, A., Masui, T., Akashi, O., Hijioka, Y., Matsumoto, K., & Kainuma, M. (2012). Assessment of GHG emission reduction pathways in a society without carbon capture and nuclear technologies. *Energy Economics, 34,* S391-S398. doi: 10.1016/j.eneco.2012.07.011

O'Ryan, R., De Miguel, C. J., Miller, S., & Pereira, M. (2011). The Socioeconomic and environmental effects of free trade agreements: a dynamic CGE analysis for Chile. *Environment and Development Economics, 16,* 305-327. doi: 10.1017/s1355770x10000227

Ozdemir, D., & Bayar, A. (2009). The Peace Dividend Effect of Turkish Convergence to The EU: A Multi-Region Dynamic CGE Model Analysis For Greece And Turkey. *Defence and Peace Economics, 20*(1), 69-78. doi: 10.1080/10242690701833217

Radulescu, D., & Stimmelmayr, M. (2010). The impact of the 2008 German corporate tax reform: A dynamic CGE analysis. *Economic Modelling, 27*(1), 454-467. doi: 10.1016/j.econmod.2009.10.012

Rive, N., Torvanger, A., & Fuglestvedt, J. S. (2006). Climate agreements based on responsibility for global warming: Periodic updating, policy choices, and regional costs. *Global Environmental Change-Human and Policy Dimensions, 16*(2), 182-194. doi: 10.1016/j.gloenvcha.2006.01.002

Schenker, O. (2013). Exchanging Goods and Damages: The Role of Trade on the Distribution of Climate Change Costs. *Environmental & Resource Economics, 54*(2), 261-282. doi: 10.1007/s10640-012-9593-z

Takeda, S. (2007). The double dividend from carbon regulations in Japan. *Journal of the Japanese and International Economies, 21*(3), 336-364. doi: 10.1016/j.jjie.2006.01.002

Vellinga, N. (2008). Dynamic general-equilibrium model of an open economy: A comment. *Journal of Policy Modeling, 30*(6), 993-997. doi: 10.1016/j.jpolmod.2007.04.009

Wittwer, G. (2009). The Economic Impacts of a New Dam in South-East Queensland. *Australian Economic Review, 42*(1), 12-23. doi: 10.1111/j.1467-8462.2009.00506.x

Wu, Q., & Xiao, H. (2014). Dynamic CGE Model and Simulation Analysis on the Impact of Citizenization of Rural Migrant Workers on the Labor and Capital Markets in China. *Discrete Dynamics in Nature and Society*, 8. doi: 10.1155/2014/351947

Xie, W., Li, N., Wu, J. D., & Hao, X. L. (2015). Disaster Risk Decision: A Dynamic Computable General Equilibrium Analysis of Regional Mitigation Investment. *Human and Ecological Risk Assessment, 21*(1), 81-99. doi: 10.1080/10807039.2013.871997

Xu, S. W., Zhang, Y. M., Diao, X. S., & Chen, K. Z. (2011). Impacts of agricultural public spending on Chinese food economy A general equilibrium approach. *China Agricultural Economic Review, 3*(4), 518-534. doi: 10.1108/17561371111192365

Zhang, J. S. (2001). Iterative method for finding the balanced growth solution of the non-linear dynamic input-output model and the dynamic CGE model. *Economic Modelling, 18*(1), 117-132. doi: 10.1016/s0264-9993(00)00031-6

Chapter 20. Assessment of exports of commodities and implication for logistics services

(Simulation 2)

1. Introduction

In this study, a fictitious trade and its 3-year-term potential effects on logistics services are simulated. The results indicate that some logistics services (food, mnfcs, svces) are much more sensitive to changes in some economies. These results are simulated and reported by using dynamic general equilibrium models. Based on the scenario, the results indicate that various logistics services to/from economies exhibit increases or decreases.

The remainder of this study is organized as follows. Section 2 presents some highlighted studies. Section 3 presents the framework, methodology, results and short discussion. The study is concluded in Section 4.

2. Highlighted studies

Parrado and De Cian (2014) explored technology spillovers embodied in international trade. Philip *et al.* (2014) investigated technological change in irrigated agriculture in a semiarid region of Spain.

Schenker (2013) investigated exchanging goods and damages as a role of trade on the distribution of climate change costs. Qin *et al.* (2011) assessed economic impacts of China's water-pollution-mitigation-measures through a dynamic computable general equilibrium analysis.

Xie *et al.* (2015) studied disaster risk decision by analyzing regional mitigation investment. Xie *et al.* (2014) modeled the economic costs of disasters and recovery. Gohin and Rault (2013) assessed the economic costs of a foot and mouth disease outbreak on Brittany by using a dynamic computable general equilibrium.

Berrittella and Zhang (2015) investigated fiscal sustainability in the EU. Bhattarai and Dixon (2014) explored equilibrium unemployment in a general equilibrium model with taxes. Mabugu *et al.* (2015) analyzed pro-poor tax policy changes in South Africa and investigated potentials and limitations.

Bhattarai (2015) investigated financial deepening and economic growth. Cheong and Tongzon (2013) compared the economic impact of the Trans-Pacific Partnership and the regional comprehensive economic partnership. Decreux and Fontagne (2015) investigated multilateral trade talks. Jiang and Mai (2015) studied social welfare housing project and its effects in China. Matovu (2012) investigated trade reforms and horizontal inequalities for Uganda. Seung and Kraybill (2001) explored the effects of infrastructure investment for Ohio. Verikios *et al.* (2015) studied improving health in an advanced economy for Australia. Lakatos and Walmsley (2012) studied investment creation and diversion effects of the ASEAN-China free trade agreement. Boccanfuso *et al.* (2014) performed a comparative analysis of funding schemes for public infrastructure spending in Quebec.

Barkhordar and Saboohi (2013) assessed alternative options for allocating oil revenue in Iran. Breisinger *et al.* (2012) investigated leveraging fuel subsidy reform for transition in Yemen. Dai *et al.* (2012) explored the impacts of China's household consumption expenditure patterns on energy demand and carbon emissions towards 2050. Faehn and Bruvoll (2009) investigated richer and cleaner issues at others expense. Hosoe (2014) investigated Japanese manufacturing facing post-Fukushima power crisis. Liang *et al.* (2014) studied platform for China energy & environmental policy analysis. Ruamsuke *et al.* (2015) explored energy and economic impacts of the global climate change policy on Southeast Asian countries. Wu *et al.* (2014) analyzed the future vehicle-energy-demand in China.

Femenia (2010) explored impacts of stockholding behaviour on agricultural market volatility. Femenia and Gohin (2013) studied optimal implementation of agricultural policy reforms. Furuya *et al.* (2015) studied economic evaluation of agricultural mitigation and adaptation technologies for Climate Change. Mariano *et al.* (2015) studied the effects of domestic rice market interventions outside business-as-usual conditions for imported rice prices. Akune *et al.* (2015) studied economic evaluation of dissemination of high temperature-tolerant rice in Japan. Bourne *et al.* (2012) measured the impact of the global financial crisis on Spanish agriculture. Breisinger and Ecker (2014) simulated economic growth effects on food and nutrition security in Yemen. Philippidis and Hubbard (2005) wrote a note on the ban of UK beef exports.

Arndt *et al.* (2012) studied biofuels and economic development for Tanzania. Asafu-Adjaye and Mahadevan (2013) studied implications of CO2 reduction policies for a high carbon emitting economy. Bao *et al.* (2013) studied impacts of border carbon adjustments on China's sectoral emissions. Chi *et al.* (2014) performed scenarios analysis of the energies' Consumption and carbon emissions in China. Doumax *et al.* (2014) studied biofuels, tax policies and oil prices in France. Lanzi *et al.* (2012) studied alternative approaches for levelling carbon prices in a world with fragmented carbon markets. Liang *et al.* (2013) assessed the distributional impacts of carbon tax among households across different income groups of China. Liu and Lu (2015) studied the economic impact of different carbon tax revenue recycling schemes in China. Ricci (2012) studied providing adequate economic incentives for bioenergies with CO2 capture and geological storage. Saveyn *et al.* (2012) performed economic analysis of a low carbon path to 2050 as a case for China, India and Japan.

Wittwer and Banerjee (2015) studied investing in irrigation development in North West Queensland, Australia. Li *et al.* (2015) explored economic impacts of total water use control in the Heihe River Basin in Northwestern China. He *et al.* (2007) constructed a dynamic computable general equilibrium model and performed sensitivity analysis for shadow price of water resource in China.

Mai *et al.* (2014) studied the economic effects of facilitating the flow of rural workers to urban employment in China. Cockburn *et al.* (2014) studied impacts of the global economic crisis and national policy responses on children in Cameroon.

3. *Framework, methodology, and results*

The standard version of GDyn is a recursive-dynamic extension of the standard GTAP (Hertel, 1997), developed for better treatment of medium and long-term simulations, as it strengthens the investment side of the modeling framework to enable international capital mobility (Ianchovichina & Walmsley, 2012).

GDyn extends the standard, comparative static version of the GTAP model by introducing

- international capital mobility,

- endogenous capital accumulation and

- adaptive expectations theory of investment in a recursive dynamics setting.

The model determines the global markets for products, so that the balance is determined by the conditions of all countries' supply and demand. Demand for imports of a country is determined by its demand for imported inputs and goods consumed by final demand.

In this study, a fictitious trade and its 3-year-term potential effects on logistics services are simulated. The results indicate that some logistics services are much more sensitive to changes in some economies. The results indicate that various logistics services to/from economies exhibit increases or decreases (See Tables 1 through 9).

355

Table 1. Food – 2015 (percentage changes)

qxs[Food**]	1 Australia	2 NZ	3 China	4 Japan	5 Korea	6 SA	7 Canada	8 US	9 SAM	10 Austria	11 Denmark	12 Finland	13 France	14 UK	15 Ireland	16 Italy	17 Netherl	18 Portugal	19 Sweden	20 Eur	21 Turkey	22 ROW
1 Australia	1.8	1.8	4.3	3.9	2.7	4.2	0.7	0.7	0.9	1.9	1.9	2.1	1.3	1.2	2.1	1.0	1.4	2.4	1.8	1.8	1.8	2.6
2 NZ	2.3	2.3	4.7	4.3	3.1	4.6	1.1	1.1	1.4	2.4	2.3	2.5	1.7	1.6	2.5	1.4	1.8	2.8	2.2	2.2	2.3	3.0
3 China	4.2	4.2	6.8	6.3	4.9	6.5	3.0	3.0	3.3	4.5	4.4	4.6	3.7	3.6	4.6	3.4	3.8	4.9	4.3	4.2	4.2	5.0
4 Japan	-5.2	-5.2	-2.8	-3.6	-4.3	-2.8	-6.2	-6.2	-6.1	-5.4	-5.4	-5.2	-5.9	-5.9	-5.1	-6.2	-5.6	-4.8	-5.4	-5.4	-5.2	-4.5
5 Korea	-1.0	-1.0	1.6	1.1	-0.3	1.4	-2.1	-2.1	-1.9	-1.0	-1.0	-0.7	-1.6	-1.7	-0.7	-1.8	-1.4	-0.4	-1.1	-1.0	-1.1	-0.2
6 SA	-0.1	-0.1	2.4	2.0	0.9	2.3	-1.2	-1.2	-0.9	0.0	0.0	0.2	-0.6	-0.7	0.2	-0.9	-0.4	0.5	-0.1	-0.1	-0.1	0.7
7 Canada	1.9	1.9	4.3	3.9	2.7	4.2	0.7	0.7	1.0	2.0	1.9	2.2	1.3	1.2	2.1	1.0	1.5	2.5	1.9	1.8	1.9	2.6
8 US	3.4	3.5	6.0	5.5	4.3	5.8	2.4	2.5	2.6	3.6	3.5	3.8	2.9	2.8	3.6	2.6	3.1	4.0	3.5	3.4	3.4	4.2
9 SAM	4.1	4.1	6.5	6.2	4.9	6.4	2.9	2.9	3.1	4.3	4.0	4.4	3.4	3.4	4.2	3.1	3.6	4.6	4.0	4.0	4.0	4.8
10 Austria	1.4	1.4	3.9	3.5	2.3	3.8	0.3	0.3	0.5	1.5	1.5	1.7	0.9	0.8	1.7	0.6	1.0	2.0	1.4	1.4	1.4	2.2
11 Denmark	0.7	0.7	3.1	2.7	1.5	3.0	-0.5	-0.5	-0.2	0.7	0.7	0.9	0.1	0.0	0.9	-0.2	0.2	1.2	0.6	0.6	0.6	1.4
12 Finland	0.9	0.9	3.4	2.9	1.8	3.2	-0.2	-0.2	0.1	1.0	0.9	1.2	0.3	0.3	1.2	0.1	0.5	1.5	0.9	0.9	0.9	1.7
13 France	2.1	2.1	4.5	4.1	3.0	4.4	0.9	0.9	1.2	2.2	2.1	2.4	1.6	1.5	2.4	1.3	1.7	2.7	2.1	2.1	2.1	2.8
14 UK	1.6	1.6	4.1	3.6	2.4	3.9	0.4	0.4	0.7	1.7	1.6	1.9	1.0	1.0	1.9	0.8	1.2	2.2	1.6	1.5	1.6	2.3
15 Ireland	3.6	3.6	6.0	5.6	4.5	5.9	2.4	2.4	2.6	3.7	3.7	3.9	3.1	3.0	4.0	2.8	3.3	4.2	3.6	3.6	3.5	4.3
16 Italy	2.7	2.7	5.2	4.8	3.6	5.0	1.5	1.5	1.8	2.8	2.8	3.0	2.2	2.1	3.0	1.9	2.3	3.3	2.7	2.7	2.7	3.4
17 Netherl	0.6	0.6	3.1	2.7	1.5	2.9	-0.5	-0.5	-0.3	0.7	0.6	0.9	0.0	-0.1	0.9	-0.3	0.2	1.2	0.6	0.6	0.6	1.4
18 Portugal	-0.1	-0.1	2.4	2.0	0.8	2.2	-1.2	-1.2	-0.9	0.0	0.0	0.2	-0.6	-0.7	0.2	-0.9	-0.5	0.5	-0.1	-0.1	-0.1	0.7
19 Sweden	0.4	0.4	2.8	2.4	1.3	2.7	-0.8	-0.8	-0.5	0.5	0.4	0.7	-0.2	-0.3	0.6	-0.5	0.0	0.9	0.3	0.4	0.4	1.2
20 Eur	0.8	0.8	3.3	2.9	1.7	3.2	-0.3	-0.3	0.0	1.0	0.9	1.1	0.3	0.2	1.1	0.0	0.5	1.4	0.8	0.8	0.8	1.6
21 Turkey	3.1	3.0	5.7	5.2	4.0	5.5	2.0	2.0	2.2	3.2	3.2	3.4	2.5	2.4	3.4	2.3	2.7	3.7	3.1	3.0	3.2	3.8
22 ROW	1.6	1.6	4.1	3.7	2.5	4.0	0.5	0.5	0.7	1.7	1.7	1.9	1.1	1.0	1.9	0.8	1.2	2.2	1.6	1.6	1.6	2.4

Table 2. Mnfcs – 2015 (percentage changes)

qxs[Mnfcs**]	1 Australia	2 NZ	3 China	4 Japan	5 Korea	6 SA	7 Canada	8 US	9 SAM	10 Austria	11 Denmark	12 Finland	13 France	14 UK	15 Ireland	16 Italy	17 Netherl	18 Portugal	19 Sweden	20 Eur	21 Turkey	22 ROW
1 Australia	3.5	3.6	6.5	6.0	5.5	5.3	1.6	1.9	1.8	4.4	3.9	4.5	3.1	3.4	5.2	3.2	3.5	4.3	4.2	3.9	4.8	4.3
2 NZ	3.4	3.5	6.4	6.0	5.3	5.2	1.5	1.8	1.8	4.3	4.0	4.8	3.2	3.4	5.1	3.1	3.6	4.4	4.2	3.9	4.9	4.2
3 China	4.7	4.9	8.0	7.5	6.9	6.7	2.9	3.2	3.2	5.6	5.3	6.1	4.5	4.7	6.6	4.5	5.0	5.7	5.6	5.3	6.3	5.6
4 Japan	-7.1	-7.0	-4.1	-5.0	-5.2	-5.4	-8.8	-8.5	-8.6	-6.4	-6.6	-6.0	-7.3	-7.2	-5.7	-7.4	-7.0	-6.3	-6.4	-6.7	-5.8	-6.4
5 Korea	2.6	2.6	5.6	5.2	4.6	4.4	0.7	1.0	1.0	3.4	3.1	3.9	2.4	2.5	4.2	2.3	2.7	3.5	3.4	3.1	4.0	3.4
6 SA	5.5	5.6	8.6	8.2	7.5	7.4	3.7	4.0	3.9	6.4	6.1	6.9	5.3	5.5	7.3	5.2	5.7	6.5	6.4	6.1	7.0	6.4
7 Canada	2.9	3.0	6.0	5.5	5.0	4.8	1.1	1.4	1.3	3.8	3.5	4.3	2.7	2.9	4.6	2.7	3.1	3.8	3.7	3.4	4.4	3.8
8 US	5.0	5.1	8.1	7.7	7.1	6.9	3.2	3.6	3.4	5.9	5.6	6.4	4.8	5.0	6.7	4.7	5.2	5.9	5.9	5.5	6.5	5.9
9 SAM	7.8	7.8	10.9	10.3	9.8	9.7	6.0	6.3	6.1	8.8	8.0	9.1	7.5	7.8	9.5	7.4	7.9	8.6	8.6	8.3	9.2	8.7
10 Austria	2.1	2.2	5.1	4.7	4.1	3.9	0.2	0.6	0.5	2.9	2.6	3.4	1.9	2.0	3.7	1.8	2.2	3.0	2.9	2.6	3.5	2.9
11 Denmark	1.1	1.2	4.1	3.7	3.1	2.9	-0.7	-0.4	-0.5	2.0	1.6	2.4	0.9	1.1	2.7	0.8	1.3	2.0	1.9	1.6	2.6	1.9
12 Finland	1.4	1.5	4.4	4.0	3.4	3.2	-0.5	-0.1	-0.2	2.2	1.9	2.7	1.2	1.3	3.0	1.1	1.5	2.3	2.2	1.9	2.8	2.2
13 France	2.8	2.9	5.9	5.5	4.9	4.7	1.0	1.3	1.3	3.7	3.4	4.2	2.7	2.8	4.5	2.6	3.0	3.8	3.7	3.4	4.3	3.7
14 UK	2.2	2.3	5.2	4.8	4.2	4.0	0.4	0.7	0.6	3.1	2.7	3.5	2.0	2.2	3.9	1.9	2.3	3.1	3.0	2.7	3.6	3.0
15 Ireland	4.8	4.8	7.9	7.5	6.9	6.6	2.9	3.2	3.2	5.7	5.4	6.2	4.6	4.8	6.5	4.5	5.0	5.8	5.7	5.3	6.2	5.6
16 Italy	4.4	4.5	7.5	7.1	6.5	6.3	2.6	2.9	2.8	5.4	5.1	5.8	4.3	4.5	6.2	4.3	4.7	5.5	5.5	5.0	5.9	5.3
17 Netherl	1.2	1.3	4.3	3.9	3.3	3.1	-0.6	-0.3	-0.3	2.1	1.8	2.5	1.0	1.2	2.9	1.0	1.4	2.2	2.0	1.7	2.7	2.1
18 Portugal	1.1	1.2	4.1	3.7	3.1	2.9	-0.7	-0.4	-0.5	1.9	1.6	2.4	0.9	1.0	2.7	0.8	1.2	2.0	1.9	1.6	2.5	1.9
19 Sweden	0.9	1.0	3.9	3.5	2.9	2.7	-0.9	-0.6	-0.6	1.8	1.5	2.2	0.7	0.9	2.6	0.7	1.1	1.8	1.7	1.4	2.4	1.7
20 Eur	1.1	1.2	4.1	3.7	3.1	2.9	-0.7	-0.4	-0.4	2.0	1.6	2.4	0.9	1.1	2.7	0.8	1.3	2.0	1.9	1.6	2.6	1.9
21 Turkey	7.0	7.2	10.0	9.8	9.1	8.8	5.1	5.4	5.4	7.9	7.6	8.4	6.8	7.0	8.7	6.7	7.2	8.0	7.9	7.6	8.8	7.9
22 ROW	2.5	2.6	5.6	5.2	4.6	4.4	0.7	1.0	1.0	3.4	3.1	3.9	2.3	2.5	4.2	2.3	2.7	3.5	3.4	3.1	4.0	3.4

Table 3. Svces – 2015 (percentage changes)

qxs[Svces**]	1 Australia	2 NZ	3 China	4 Japan	5 Korea	6 SA	7 Canada	8 US	9 SAM	10 Austria	11 Denmark	12 Finland	13 France	14 UK	15 Ireland	16 Italy	17 Netherl	18 Portugal	19 Sweden	20 Eur	21 Turkey	22 ROW
1 Australia	3.2	3.2	6.7	7.1	5.7	4.2	2.8	2.3	2.2	3.4	3.7	4.5	2.5	3.3	4.2	2.4	4.0	4.3	4.3	3.9	3.7	3.6
2 NZ	3.0	2.9	6.4	6.8	5.4	3.9	2.5	2.0	1.9	3.2	3.4	4.2	2.2	3.0	3.9	2.1	3.7	4.0	4.0	3.6	3.4	3.3
3 China	4.2	4.2	7.7	8.1	6.7	5.2	3.8	3.2	3.2	4.4	4.6	5.5	3.4	4.3	5.2	3.4	5.0	5.3	5.3	4.9	4.7	4.6
4 Japan	-5.1	-5.1	-1.9	-1.5	-2.8	-4.2	-5.4	-5.9	-6.0	-4.9	-4.7	-3.9	-5.8	-5.0	-4.2	-5.8	-4.3	-4.1	-4.0	-4.4	-4.6	-4.8
5 Korea	2.2	2.2	5.7	6.1	4.7	3.2	1.8	1.3	1.2	2.4	2.6	3.5	1.5	2.3	3.2	1.4	3.0	3.3	3.3	2.9	2.7	2.6
6 SA	5.4	5.3	8.9	9.3	7.9	6.4	4.9	4.4	4.3	5.6	5.8	6.6	4.6	5.4	6.3	4.5	6.2	6.5	6.5	6.1	5.8	5.7
7 Canada	2.4	2.4	5.8	6.2	4.8	3.3	2.0	1.4	1.4	2.6	2.8	3.6	1.6	2.4	3.3	1.5	3.2	3.4	3.5	3.0	2.8	2.7
8 US	4.2	4.2	7.7	8.1	6.7	5.2	3.8	3.2	3.2	4.4	4.6	5.5	3.4	4.3	5.2	3.4	5.0	5.3	5.3	4.9	4.7	4.5
9 SAM	6.3	6.3	9.9	10.3	8.9	7.3	5.9	5.3	5.3	6.5	6.8	7.6	5.5	6.4	7.3	5.5	7.1	7.4	7.5	7.0	6.8	6.7
10 Austria	2.1	2.1	5.5	6.0	4.6	3.1	1.7	1.2	1.1	2.3	2.5	3.3	1.3	2.2	3.0	1.3	2.9	3.2	3.2	2.8	2.5	2.4
11 Denmark	1.0	1.0	4.4	4.8	3.4	1.9	0.6	0.1	0.0	1.2	1.4	2.2	0.2	1.0	1.9	0.2	1.8	2.1	2.1	1.7	1.4	1.3
12 Finland	1.3	1.3	4.7	5.1	3.7	2.3	0.9	0.4	0.3	1.5	1.7	2.5	0.6	1.4	2.2	0.5	2.1	2.4	2.4	2.0	1.8	1.6
13 France	3.3	3.3	6.8	7.2	5.8	4.3	2.9	2.4	2.3	3.5	3.8	4.6	2.6	3.4	4.3	2.5	4.1	4.4	4.4	4.0	3.8	3.7
14 UK	2.0	2.0	5.4	5.8	4.4	3.0	1.6	1.0	1.0	2.2	2.4	3.2	1.2	2.0	2.9	1.2	2.8	3.1	3.1	2.7	2.4	2.3
15 Ireland	4.4	4.4	7.9	8.3	6.9	5.4	4.0	3.4	3.4	4.6	4.8	5.7	3.6	4.4	5.3	3.6	5.2	5.5	5.5	5.1	4.8	4.7
16 Italy	4.5	4.5	8.1	8.5	7.1	5.5	4.1	3.6	3.5	4.7	5.0	5.8	3.8	4.6	5.5	3.7	5.3	5.6	5.7	5.2	5.0	4.9
17 Netherl	0.9	0.9	4.3	4.7	3.3	1.9	0.5	0.0	-0.1	1.1	1.3	2.1	0.2	1.0	1.8	0.1	1.7	2.0	2.0	1.6	1.4	1.2
18 Portugal	1.1	1.1	4.5	4.9	3.5	2.0	0.7	0.1	0.1	1.3	1.5	2.3	0.3	1.1	2.0	0.3	1.9	2.1	2.1	1.7	1.5	1.4
19 Sweden	0.9	0.9	4.3	4.7	3.3	1.8	0.5	-0.1	-0.1	1.1	1.3	2.1	0.1	0.9	1.8	0.1	1.6	1.9	1.9	1.5	1.3	1.2
20 Eur	1.0	1.0	4.4	4.8	3.4	1.9	0.6	0.0	0.0	1.2	1.4	2.2	0.2	1.0	1.9	0.2	1.8	2.0	2.1	1.7	1.4	1.3
21 Turkey	6.4	6.3	9.9	10.4	8.9	7.4	5.9	5.4	5.3	6.6	6.8	7.7	5.6	6.4	7.3	5.5	7.2	7.5	7.5	7.1	6.8	6.7
22 ROW	3.0	3.0	6.5	6.9	5.5	4.0	2.6	2.1	2.0	3.2	3.4	4.3	2.3	3.1	4.0	2.2	3.8	4.1	4.1	3.7	3.5	3.4

Table 4. Food – 2016 (percentage changes)

qxs[Food**]	1 Australia	2 NZ	3 China	4 Japan	5 Korea	6 SA	7 Canada	8 US	9 SAM	10 Austria	11 Denmark	12 Finland	13 France	14 UK	15 Ireland	16 Italy	17 Netherl	18 Portugal	19 Sweden	20 Eur	21 Turkey	22 ROW
1 Australia	1.8	1.8	4.3	3.8	2.7	4.0	0.7	0.8	1.1	1.8	1.8	2.0	1.2	1.2	2.1	0.9	1.4	2.3	1.7	1.7	1.9	2.5
2 NZ	2.3	2.3	4.8	4.2	3.2	4.4	1.1	1.2	1.5	2.3	2.2	2.4	1.6	1.6	2.4	1.4	1.8	2.7	2.1	2.1	2.3	3.0
3 China	4.2	4.2	6.8	6.2	5.0	6.4	3.0	3.1	3.4	4.4	4.3	4.6	3.6	3.6	4.6	3.4	3.8	4.8	4.2	4.1	4.2	4.9
4 Japan	-5.0	-5.1	-2.6	-3.5	-4.1	-2.8	-6.0	-5.9	-5.9	-5.3	-5.3	-5.1	-5.8	-5.8	-4.9	-6.0	-5.5	-4.7	-5.3	-5.3	-5.0	-4.4
5 Korea	-1.0	-1.0	1.6	1.0	-0.2	1.2	-2.0	-2.0	-1.8	-1.1	-1.1	-0.8	-1.7	-1.7	-0.7	-1.9	-1.4	-0.5	-1.2	-1.1	-1.0	-0.3
6 SA	0.3	0.3	2.8	2.3	1.3	2.5	-0.7	-0.7	-0.4	0.3	0.3	0.5	-0.3	-0.3	0.6	-0.6	-0.1	0.8	0.2	0.2	0.4	1.0
7 Canada	1.8	1.8	4.3	3.8	2.7	4.0	0.8	0.8	1.1	1.8	1.8	2.0	1.2	1.2	2.1	0.9	1.4	2.3	1.7	1.7	1.9	2.5
8 US	3.2	3.3	5.8	5.2	4.1	5.4	2.3	2.3	2.5	3.3	3.2	3.5	2.6	2.5	3.4	2.4	2.8	3.7	3.2	3.1	3.3	4.0
9 SAM	4.0	4.0	6.5	6.0	4.8	6.1	2.9	2.9	3.2	4.1	3.8	4.2	3.2	3.2	4.1	3.0	3.4	4.3	3.8	3.8	3.9	4.7
10 Austria	1.5	1.5	4.0	3.5	2.4	3.7	0.4	0.5	0.7	1.5	1.5	1.7	0.9	0.8	1.8	0.6	1.1	2.0	1.4	1.4	1.5	2.2
11 Denmark	0.7	0.8	3.3	2.7	1.7	2.9	-0.3	-0.3	0.0	0.8	0.7	1.0	0.1	0.1	1.0	-0.1	0.3	1.2	0.6	0.6	0.8	1.5
12 Finland	1.0	1.0	3.5	3.0	1.9	3.2	0.0	0.0	0.3	1.0	1.0	1.2	0.4	0.4	1.3	0.2	0.6	1.5	0.9	0.9	1.1	1.8
13 France	2.1	2.2	4.7	4.1	3.1	4.3	1.1	1.1	1.4	2.2	2.1	2.4	1.6	1.5	2.4	1.3	1.7	2.7	2.1	2.1	2.2	2.9
14 UK	1.6	1.6	4.1	3.6	2.6	3.8	0.6	0.6	0.9	1.6	1.6	1.8	1.0	1.0	1.9	0.8	1.2	2.1	1.5	1.5	1.7	2.4
15 Ireland	3.7	3.7	6.1	5.6	4.6	5.8	2.5	2.6	2.8	3.7	3.7	3.9	3.0	3.0	4.0	2.8	3.3	4.1	3.6	3.5	3.7	4.3
16 Italy	2.8	2.8	5.3	4.8	3.7	5.0	1.7	1.7	2.0	2.8	2.8	3.0	2.2	2.2	3.1	2.0	2.4	3.3	2.7	2.7	2.8	3.5
17 Netherl	0.7	0.7	3.2	2.7	1.6	2.9	-0.4	-0.3	-0.1	0.7	0.6	0.9	0.0	0.0	0.9	-0.2	0.2	1.2	0.6	0.6	0.7	1.4
18 Portugal	0.1	0.1	2.6	2.0	1.0	2.3	-0.9	-0.9	-0.6	0.1	0.1	0.3	-0.5	-0.6	0.4	-0.8	-0.3	0.6	0.0	0.0	0.2	0.9
19 Sweden	0.5	0.5	3.0	2.5	1.4	2.7	-0.5	-0.5	-0.2	0.5	0.5	0.8	-0.1	-0.2	0.8	-0.4	0.1	1.0	0.4	0.4	0.6	1.3
20 Eur	1.0	1.0	3.5	2.9	1.9	3.2	-0.1	-0.1	0.2	1.0	0.9	1.2	0.4	0.3	1.2	0.1	0.5	1.5	0.9	0.9	1.0	1.7
21 Turkey	3.2	3.1	5.8	5.2	4.2	5.4	2.1	2.1	2.4	3.1	3.2	3.4	2.5	2.5	3.4	2.3	2.7	3.7	3.0	3.0	3.3	3.9
22 ROW	1.7	1.7	4.2	3.7	2.6	3.9	0.6	0.7	0.9	1.7	1.7	1.9	1.1	1.0	2.0	0.8	1.3	2.2	1.6	1.6	1.7	2.4

Table 5. Mnfcs – 2016 (percentage changes)

qxs[Mnfcs**]	1 Australia	2 NZ	3 China	4 Japan	5 Korea	6 SA	7 Canada	8 US	9 SAM	10 Austria	11 Denmark	12 Finland	13 France	14 UK	15 Ireland	16 Italy	17 Netherl	18 Portugal	19 Sweden	20 Eur	21 Turkey	22 ROW
1 Australia	3.4	3.5	6.4	5.9	5.3	5.1	1.7	1.9	1.8	4.2	3.7	4.3	3.0	3.3	5.0	3.0	3.4	4.2	4.1	3.8	4.7	4.1
2 NZ	3.4	3.5	6.4	5.9	5.2	5.1	1.6	1.9	1.9	4.1	3.8	4.6	3.1	3.3	5.0	3.0	3.5	4.2	4.1	3.8	4.7	4.1
3 China	4.7	4.8	7.9	7.4	6.8	6.6	3.0	3.2	3.2	5.4	5.1	5.9	4.4	4.6	6.5	4.3	4.8	5.5	5.4	5.1	6.1	5.5
4 Japan	-6.9	-6.9	-3.9	-4.8	-5.1	-5.3	-8.5	-8.3	-8.3	-6.3	-6.5	-5.9	-7.2	-7.0	-5.5	-7.3	-6.9	-6.2	-6.3	-6.6	-5.7	-6.3
5 Korea	2.6	2.7	5.6	5.2	4.6	4.4	0.9	1.2	1.1	3.4	3.1	3.8	2.3	2.5	4.2	2.3	2.7	3.4	3.3	3.0	4.0	3.3
6 SA	5.4	5.5	8.5	8.0	7.3	7.2	3.7	4.0	3.9	6.2	5.9	6.7	5.1	5.4	7.2	5.0	5.5	6.3	6.2	5.8	6.8	6.2
7 Canada	2.8	2.9	5.9	5.4	4.8	4.6	1.2	1.4	1.4	3.6	3.3	4.1	2.6	2.8	4.5	2.5	2.9	3.7	3.6	3.3	4.2	3.6
8 US	4.7	4.8	7.8	7.4	6.7	6.5	3.0	3.4	3.2	5.5	5.2	6.0	4.4	4.7	6.4	4.4	4.8	5.5	5.5	5.1	6.1	5.5
9 SAM	7.5	7.6	10.7	10.1	9.5	9.4	5.9	6.2	6.0	8.5	7.7	8.8	7.2	7.5	9.2	7.1	7.6	8.3	8.3	7.9	8.9	8.3
10 Austria	2.1	2.2	5.2	4.8	4.1	3.9	0.5	0.7	0.7	2.9	2.6	3.4	1.9	2.1	3.8	1.8	2.2	3.0	2.9	2.6	3.5	2.9
11 Denmark	1.2	1.3	4.2	3.8	3.1	2.9	-0.5	-0.2	-0.3	1.9	1.6	2.4	0.9	1.1	2.8	0.8	1.3	2.0	1.9	1.6	2.6	1.9
12 Finland	1.5	1.6	4.6	4.2	3.5	3.3	-0.1	0.1	0.1	2.3	2.0	2.7	1.2	1.4	3.1	1.2	1.6	2.4	2.3	1.9	2.9	2.3
13 France	2.9	3.0	6.0	5.6	4.9	4.7	1.2	1.5	1.5	3.7	3.4	4.2	2.7	2.9	4.6	2.6	3.0	3.8	3.7	3.4	4.3	3.7
14 UK	2.2	2.3	5.3	4.8	4.2	4.0	0.5	0.8	0.8	3.0	2.7	3.5	2.0	2.2	3.9	1.9	2.3	3.1	3.0	2.6	3.6	3.0
15 Ireland	4.8	4.8	7.9	7.5	6.8	6.6	3.0	3.3	3.3	5.6	5.3	6.1	4.5	4.7	6.5	4.5	4.9	5.7	5.6	5.2	6.2	5.5
16 Italy	4.4	4.5	7.5	7.1	6.5	6.3	2.7	3.0	3.0	5.3	5.0	5.8	4.2	4.4	6.2	4.2	4.6	5.4	5.3	4.9	5.9	5.2
17 Netherl	1.3	1.4	4.4	3.9	3.3	3.1	-0.3	-0.1	-0.1	2.1	1.8	2.5	1.1	1.3	2.9	1.0	1.4	2.2	2.1	1.7	2.7	2.1
18 Portugal	1.2	1.2	4.2	3.8	3.1	2.9	-0.5	-0.2	-0.3	1.9	1.6	2.4	0.9	1.1	2.8	0.8	1.2	2.0	1.9	1.6	2.6	1.9
19 Sweden	1.0	1.1	4.0	3.6	3.0	2.8	-0.6	-0.4	-0.4	1.8	1.5	2.2	0.8	0.9	2.6	0.7	1.1	1.8	1.7	1.4	2.4	1.8
20 Eur	1.2	1.3	4.2	3.8	3.2	3.0	-0.5	-0.2	-0.2	2.0	1.7	2.4	0.9	1.1	2.8	0.9	1.3	2.0	1.9	1.6	2.6	1.9
21 Turkey	6.8	7.0	9.9	9.6	8.9	8.6	5.1	5.4	5.4	7.7	7.4	8.1	6.6	6.8	8.5	6.5	7.0	7.7	7.7	7.3	8.6	7.7
22 ROW	2.7	2.7	5.7	5.3	4.6	4.4	1.0	1.2	1.2	3.4	3.1	3.9	2.4	2.6	4.3	2.3	2.8	3.5	3.4	3.1	4.0	3.4

Table 6. Svces – 2016 (percentage changes)

qxs[Svces**]	1 Australia	2 NZ	3 China	4 Japan	5 Korea	6 SA	7 Canada	8 US	9 SAM	10 Austria	11 Denmark	12 Finland	13 France	14 UK	15 Ireland	16 Italy	17 Netherl	18 Portugal	19 Sweden	20 Eur	21 Turkey	22 ROW
1 Australia	3.2	3.2	6.7	7.0	5.6	4.2	2.8	2.3	2.2	3.3	3.6	4.4	2.4	3.2	4.1	2.3	3.9	4.2	4.2	3.8	3.7	3.5
2 NZ	3.0	2.9	6.4	6.7	5.4	4.0	2.5	2.1	2.0	3.1	3.3	4.1	2.2	3.0	3.9	2.1	3.7	4.0	3.9	3.5	3.4	3.2
3 China	4.2	4.1	7.7	8.0	6.6	5.2	3.8	3.3	3.2	4.3	4.5	5.3	3.4	4.2	5.1	3.3	4.9	5.2	5.2	4.8	4.6	4.5
4 Japan	-4.9	-4.9	-1.7	-1.4	-2.7	-4.0	-5.3	-5.7	-5.8	-4.8	-4.6	-3.8	-5.6	-4.9	-4.1	-5.7	-4.2	-4.0	-4.0	-4.4	-4.5	-4.6
5 Korea	2.3	2.3	5.8	6.0	4.7	3.3	1.9	1.4	1.3	2.4	2.6	3.4	1.5	2.3	3.2	1.4	3.0	3.3	3.3	2.9	2.8	2.6
6 SA	5.3	5.2	8.8	9.1	7.7	6.3	4.8	4.3	4.3	5.4	5.6	6.4	4.4	5.3	6.2	4.4	6.0	6.3	6.3	5.8	5.7	5.5
7 Canada	2.4	2.3	5.8	6.1	4.8	3.4	2.0	1.5	1.4	2.5	2.7	3.5	1.6	2.4	3.3	1.5	3.1	3.4	3.4	2.9	2.8	2.6
8 US	4.0	4.0	7.5	7.8	6.5	5.0	3.6	3.1	3.0	4.2	4.4	5.2	3.2	4.0	4.9	3.1	4.7	5.0	5.0	4.6	4.5	4.3
9 SAM	6.2	6.2	9.8	10.1	8.7	7.3	5.8	5.3	5.2	6.3	6.6	7.4	5.4	6.2	7.1	5.3	6.9	7.2	7.2	6.8	6.7	6.5
10 Austria	2.2	2.2	5.7	5.9	4.6	3.2	1.8	1.3	1.3	2.3	2.6	3.3	1.4	2.2	3.1	1.4	2.9	3.2	3.2	2.8	2.7	2.5
11 Denmark	1.1	1.1	4.5	4.8	3.5	2.1	0.7	0.2	0.2	1.2	1.4	2.2	0.3	1.1	2.0	0.3	1.8	2.1	2.1	1.7	1.6	1.4
12 Finland	1.5	1.4	4.9	5.2	3.8	2.5	1.0	0.6	0.5	1.6	1.8	2.6	0.7	1.5	2.3	0.6	2.2	2.4	2.4	2.0	1.9	1.7
13 France	3.4	3.4	6.9	7.2	5.8	4.4	3.0	2.5	2.4	3.5	3.7	4.5	2.6	3.4	4.3	2.5	4.1	4.4	4.4	4.0	3.9	3.7
14 UK	2.1	2.0	5.5	5.8	4.5	3.1	1.7	1.2	1.1	2.2	2.4	3.2	1.3	2.1	3.0	1.2	2.8	3.1	3.1	2.6	2.5	2.3
15 Ireland	4.5	4.4	8.0	8.3	6.9	5.5	4.0	3.6	3.5	4.6	4.8	5.6	3.7	4.5	5.4	3.6	5.2	5.5	5.5	5.1	4.9	4.7
16 Italy	4.6	4.5	8.1	8.4	7.0	5.6	4.1	3.7	3.6	4.7	4.9	5.7	3.8	4.6	5.5	3.7	5.3	5.6	5.6	5.1	5.0	4.8
17 Netherl	1.0	1.0	4.5	4.7	3.4	2.0	0.6	0.2	0.1	1.2	1.4	2.2	0.3	1.0	1.9	0.2	1.7	2.0	2.0	1.6	1.5	1.3
18 Portugal	1.2	1.2	4.6	4.9	3.6	2.2	0.8	0.3	0.2	1.3	1.5	2.3	0.4	1.2	2.1	0.3	1.9	2.2	2.2	1.8	1.6	1.5
19 Sweden	1.0	1.0	4.4	4.7	3.4	2.0	0.6	0.1	0.1	1.1	1.3	2.1	0.2	1.0	1.9	0.2	1.7	2.0	2.0	1.6	1.5	1.3
20 Eur	1.2	1.1	4.6	4.8	3.5	2.2	0.7	0.3	0.2	1.3	1.5	2.3	0.4	1.2	2.0	0.3	1.9	2.1	2.1	1.7	1.6	1.4
21 Turkey	6.3	6.2	9.9	10.1	8.8	7.3	5.8	5.4	5.3	6.4	6.6	7.5	5.5	6.3	7.2	5.4	7.0	7.3	7.3	6.9	6.8	6.6
22 ROW	3.2	3.1	6.6	6.9	5.6	4.2	2.7	2.3	2.2	3.3	3.5	4.3	2.4	3.2	4.1	2.3	3.9	4.1	4.1	3.7	3.6	3.4

Table 7. Food – 2017 (percentage changes)

qxs[Food**]	1 Australia	2 NZ	3 China	4 Japan	5 Korea	6 SA	7 Canada	8 US	9 SAM	10 Austria	11 Denmark	12 Finland	13 France	14 UK	15 Ireland	16 Italy	17 Netherl	18 Portugal	19 Sweden	20 Eur	21 Turkey	22 ROW
1 Australia	1.8	1.8	4.3	3.7	2.8	3.9	0.8	0.8	1.2	1.7	1.7	2.0	1.1	1.1	2.0	0.9	1.3	2.2	1.6	1.6	1.9	2.5
2 NZ	2.3	2.3	4.8	4.1	3.2	4.3	1.2	1.3	1.6	2.2	2.2	2.4	1.5	1.5	2.4	1.3	1.7	2.6	2.0	2.0	2.4	2.9
3 China	4.2	4.2	6.8	6.1	5.0	6.2	3.1	3.2	3.5	4.3	4.2	4.5	3.5	3.5	4.5	3.3	3.7	4.7	4.1	4.0	4.2	4.9
4 Japan	-4.9	-4.9	-2.4	-3.4	-3.9	-2.7	-5.7	-5.7	-5.6	-5.2	-5.1	-4.9	-5.7	-5.6	-4.8	-5.9	-5.4	-4.7	-5.3	-5.2	-4.8	-4.3
5 Korea	-1.0	-1.0	1.6	0.9	-0.2	1.1	-1.9	-1.9	-1.7	-1.2	-1.1	-0.9	-1.7	-1.8	-0.8	-1.9	-1.5	-0.7	-1.3	-1.2	-1.0	-0.3
6 SA	0.6	0.7	3.2	2.5	1.6	2.7	-0.3	-0.3	0.0	0.5	0.6	0.8	0.0	-0.1	0.9	-0.3	0.2	1.1	0.5	0.5	0.7	1.4
7 Canada	1.8	1.8	4.3	3.7	2.7	3.8	0.8	0.8	1.1	1.7	1.7	1.9	1.1	1.1	2.0	0.9	1.3	2.2	1.6	1.6	1.9	2.5
8 US	3.0	3.1	5.6	4.9	4.0	5.1	2.1	2.2	2.4	3.0	2.9	3.2	2.3	2.3	3.2	2.1	2.6	3.4	2.9	2.9	3.1	3.7
9 SAM	3.9	3.9	6.4	5.8	4.7	5.9	2.8	2.9	3.2	3.9	3.7	4.0	3.1	3.1	4.0	2.8	3.3	4.1	3.6	3.6	3.9	4.6
10 Austria	1.6	1.6	4.1	3.5	2.5	3.6	0.6	0.6	0.9	1.5	1.5	1.7	0.9	0.9	1.8	0.7	1.1	2.0	1.4	1.4	1.7	2.3
11 Denmark	0.8	0.9	3.4	2.7	1.8	2.9	-0.1	-0.1	0.2	0.8	0.7	1.0	0.2	0.1	1.1	-0.1	0.4	1.2	0.6	0.6	0.9	1.6
12 Finland	1.1	1.2	3.7	3.0	2.1	3.2	0.2	0.2	0.6	1.1	1.1	1.3	0.5	0.4	1.4	0.2	0.7	1.6	1.0	1.0	1.2	1.9
13 France	2.2	2.2	4.7	4.1	3.2	4.3	1.2	1.3	1.6	2.1	2.1	2.4	1.6	1.5	2.5	1.3	1.7	2.6	2.0	2.0	2.3	2.9
14 UK	1.7	1.7	4.2	3.6	2.6	3.7	0.7	0.7	1.1	1.6	1.6	1.8	1.0	1.0	1.9	0.8	1.2	2.1	1.5	1.5	1.8	2.4
15 Ireland	3.7	3.7	6.2	5.6	4.7	5.8	2.7	2.7	3.0	3.6	3.7	3.9	3.0	3.0	4.0	2.8	3.3	4.1	3.5	3.5	3.8	4.4
16 Italy	2.8	2.9	5.4	4.8	3.8	4.9	1.9	1.9	2.2	2.8	2.8	3.0	2.2	2.2	3.1	2.0	2.4	3.3	2.7	2.7	2.9	3.6
17 Netherl	0.8	0.8	3.3	2.7	1.7	2.8	-0.2	-0.1	0.2	0.7	0.7	0.9	0.1	0.1	1.0	-0.2	0.3	1.2	0.6	0.6	0.9	1.5
18 Portugal	0.3	0.3	2.8	2.1	1.2	2.3	-0.7	-0.6	-0.3	0.2	0.2	0.4	-0.4	-0.4	0.5	-0.6	-0.2	0.6	0.1	0.1	0.4	1.0
19 Sweden	0.7	0.7	3.2	2.6	1.6	2.7	-0.3	-0.3	0.1	0.6	0.6	0.8	0.0	-0.1	0.9	-0.3	0.2	1.0	0.4	0.5	0.8	1.4
20 Eur	1.1	1.1	3.6	3.0	2.1	3.2	0.1	0.2	0.5	1.0	1.0	1.3	0.4	0.4	1.3	0.2	0.6	1.5	0.9	0.9	1.2	1.8
21 Turkey	3.2	3.2	5.9	5.2	4.3	5.3	2.2	2.3	2.6	3.1	3.2	3.4	2.5	2.5	3.5	2.3	2.7	3.6	3.0	3.0	3.4	3.9
22 ROW	1.8	1.8	4.3	3.7	2.7	3.8	0.8	0.8	1.1	1.7	1.7	1.9	1.1	1.1	2.0	0.9	1.3	2.2	1.6	1.6	1.9	2.5

Table 8. Mnfcs – 2017 (percentage changes)

qxs[Mnfcs**]	1 Australia	2 NZ	3 China	4 Japan	5 Korea	6 SA	7 Canada	8 US	9 SAM	10 Austria	11 Denmark	12 Finland	13 France	14 UK	15 Ireland	16 Italy	17 Netherl	18 Portugal	19 Sweden	20 Eur	21 Turkey	22 ROW
1 Australia	3.4	3.4	6.3	5.8	5.2	5.0	1.7	1.9	1.9	4.0	3.6	4.2	2.9	3.1	4.9	2.9	3.3	4.0	3.9	3.6	4.5	4.0
2 NZ	3.3	3.4	6.3	5.8	5.1	5.0	1.8	1.9	1.9	4.0	3.7	4.5	3.0	3.2	4.9	2.9	3.4	4.1	4.0	3.6	4.6	4.0
3 China	4.6	4.7	7.8	7.3	6.7	6.5	3.0	3.3	3.3	5.3	4.9	5.8	4.3	4.5	6.4	4.2	4.7	5.4	5.3	5.0	6.0	5.3
4 Japan	-6.7	-6.7	-3.7	-4.6	-4.9	-5.1	-8.2	-8.0	-8.0	-6.2	-6.4	-5.7	-7.1	-6.9	-5.4	-7.1	-6.7	-6.1	-6.2	-6.5	-5.5	-6.1
5 Korea	2.7	2.7	5.7	5.2	4.6	4.4	1.1	1.3	1.3	3.3	3.0	3.8	2.3	2.5	4.2	2.3	2.7	3.4	3.3	3.0	4.0	3.3
6 SA	5.3	5.4	8.3	7.9	7.2	7.1	3.7	3.9	3.9	6.0	5.7	6.5	5.0	5.2	7.0	4.9	5.4	6.0	6.0	5.6	6.6	6.0
7 Canada	2.8	2.8	5.8	5.3	4.7	4.5	1.2	1.4	1.4	3.4	3.1	3.9	2.4	2.6	4.4	2.4	2.8	3.5	3.4	3.1	4.1	3.4
8 US	4.5	4.5	7.5	7.1	6.4	6.2	2.9	3.2	3.1	5.1	4.9	5.6	4.1	4.3	6.1	4.0	4.5	5.2	5.1	4.8	5.8	5.1
9 SAM	7.3	7.3	10.4	9.8	9.2	9.1	5.8	6.0	5.9	8.1	7.4	8.5	6.9	7.2	8.9	6.8	7.2	7.9	8.0	7.6	8.5	8.0
10 Austria	2.2	2.3	5.2	4.8	4.1	3.9	0.7	0.9	0.9	2.9	2.6	3.3	1.9	2.1	3.8	1.8	2.2	2.9	2.9	2.5	3.5	2.9
11 Denmark	1.3	1.3	4.2	3.8	3.1	2.9	-0.3	-0.1	-0.1	1.9	1.6	2.4	0.9	1.1	2.8	0.8	1.3	2.0	1.9	1.6	2.6	1.9
12 Finland	1.7	1.8	4.7	4.3	3.6	3.4	0.2	0.4	0.4	2.3	2.1	2.8	1.3	1.6	3.2	1.3	1.7	2.4	2.3	2.0	3.0	2.3
13 France	3.0	3.1	6.0	5.6	4.9	4.7	1.4	1.6	1.6	3.7	3.4	4.1	2.7	2.9	4.6	2.6	3.0	3.7	3.7	3.3	4.3	3.7
14 UK	2.3	2.3	5.3	4.8	4.2	4.0	0.7	0.9	0.9	2.9	2.6	3.4	1.9	2.1	3.9	1.9	2.3	3.0	2.9	2.6	3.6	2.9
15 Ireland	4.8	4.8	7.8	7.4	6.8	6.5	3.2	3.4	3.4	5.5	5.2	6.0	4.4	4.7	6.4	4.4	4.8	5.5	5.5	5.1	6.1	5.4
16 Italy	4.4	4.5	7.5	7.1	6.4	6.2	2.9	3.1	3.1	5.2	4.9	5.7	4.2	4.4	6.1	4.1	4.5	5.3	5.2	4.8	5.8	5.1
17 Netherl	1.4	1.4	4.4	4.0	3.3	3.1	-0.1	0.1	0.1	2.1	1.8	2.5	1.1	1.3	3.0	1.0	1.4	2.1	2.1	1.8	2.7	2.1
18 Portugal	1.3	1.4	4.3	3.8	3.2	3.0	-0.3	-0.1	-0.1	1.9	1.6	2.4	0.9	1.1	2.8	0.9	1.3	2.0	1.9	1.6	2.6	1.9
19 Sweden	1.2	1.2	4.1	3.7	3.1	2.9	-0.4	-0.2	-0.2	1.8	1.5	2.3	0.8	1.0	2.7	0.8	1.2	1.9	1.8	1.5	2.5	1.8
20 Eur	1.3	1.4	4.3	3.9	3.2	3.0	-0.2	0.0	0.0	2.0	1.7	2.5	1.0	1.2	2.9	0.9	1.4	2.0	2.0	1.7	2.6	2.0
21 Turkey	6.7	6.9	9.7	9.4	8.6	8.4	5.1	5.3	5.4	7.4	7.2	7.9	6.4	6.6	8.3	6.3	6.8	7.5	7.5	7.1	8.4	7.5
22 ROW	2.8	2.8	5.8	5.3	4.7	4.5	1.2	1.4	1.4	3.4	3.1	3.9	2.4	2.6	4.3	2.4	2.8	3.5	3.4	3.1	4.1	3.4

Table 9. Svces – 2017 (percentage changes)

qxs[Svces**]	1 Australia	2 NZ	3 China	4 Japan	5 Korea	6 SA	7 Canada	8 US	9 SAM	10 Austria	11 Denmark	12 Finland	13 France	14 UK	15 Ireland	16 Italy	17 Netherl	18 Portugal	19 Sweden	20 Eur	21 Turkey	22 ROW
1 Australia	3.2	3.1	6.7	6.8	5.5	4.2	2.8	2.3	2.3	3.2	3.5	4.2	2.4	3.1	4.1	2.3	3.8	4.1	4.1	3.7	3.7	3.4
2 NZ	3.0	2.9	6.5	6.6	5.3	4.0	2.5	2.1	2.1	3.0	3.2	4.0	2.1	2.9	3.8	2.1	3.6	3.9	3.9	3.4	3.5	3.2
3 China	4.1	4.1	7.7	7.8	6.5	5.2	3.7	3.3	3.2	4.2	4.4	5.2	3.3	4.1	5.0	3.3	4.8	5.1	5.0	4.6	4.6	4.4
4 Japan	-4.7	-4.8	-1.5	-1.4	-2.5	-3.7	-5.1	-5.5	-5.6	-4.7	-4.5	-3.8	-5.5	-4.8	-3.9	-5.5	-4.1	-3.9	-3.9	-4.3	-4.3	-4.5
5 Korea	2.4	2.3	5.8	6.0	4.7	3.4	1.9	1.5	1.5	2.4	2.6	3.4	1.5	2.3	3.2	1.5	3.0	3.3	3.2	2.8	2.8	2.6
6 SA	5.1	5.1	8.7	8.8	7.5	6.2	4.7	4.3	4.2	5.2	5.4	6.2	4.3	5.1	6.0	4.2	5.8	6.1	6.1	5.6	5.6	5.4
7 Canada	2.4	2.3	5.8	6.0	4.7	3.4	1.9	1.5	1.5	2.4	2.6	3.4	1.6	2.3	3.2	1.5	3.0	3.3	3.3	2.8	2.9	2.6
8 US	3.8	3.8	7.4	7.5	6.2	4.9	3.4	3.0	2.9	3.9	4.1	4.9	3.0	3.8	4.7	3.0	4.5	4.8	4.7	4.3	4.3	4.1
9 SAM	6.1	6.0	9.7	9.8	8.5	7.2	5.6	5.2	5.2	6.2	6.4	7.2	5.2	6.0	7.0	5.2	6.7	7.0	7.0	6.6	6.6	6.3
10 Austria	2.3	2.3	5.8	5.9	4.7	3.4	1.9	1.5	1.4	2.4	2.6	3.4	1.5	2.3	3.2	1.5	3.0	3.2	3.2	2.8	2.8	2.6
11 Denmark	1.2	1.2	4.6	4.8	3.5	2.2	0.8	0.4	0.3	1.3	1.5	2.2	0.4	1.2	2.1	0.3	1.8	2.1	2.1	1.7	1.7	1.4
12 Finland	1.6	1.6	5.1	5.2	3.9	2.6	1.2	0.8	0.7	1.7	1.9	2.6	0.8	1.6	2.5	0.7	2.2	2.5	2.5	2.1	2.1	1.8
13 France	3.4	3.4	6.9	7.1	5.8	4.5	3.0	2.6	2.5	3.5	3.7	4.5	2.6	3.4	4.3	2.5	4.1	4.3	4.3	3.9	3.9	3.6
14 UK	2.1	2.1	5.6	5.7	4.5	3.2	1.7	1.3	1.3	2.2	2.4	3.2	1.3	2.1	3.0	1.3	2.8	3.0	3.0	2.6	2.6	2.4
15 Ireland	4.5	4.5	8.1	8.2	6.9	5.6	4.1	3.7	3.6	4.6	4.8	5.6	3.7	4.5	5.4	3.6	5.2	5.4	5.4	5.0	5.0	4.8
16 Italy	4.5	4.5	8.1	8.2	6.9	5.6	4.1	3.7	3.6	4.6	4.8	5.6	3.7	4.5	5.4	3.6	5.2	5.5	5.4	5.0	5.0	4.8
17 Netherl	1.2	1.1	4.6	4.7	3.5	2.2	0.7	0.3	0.3	1.2	1.4	2.2	0.4	1.1	2.0	0.3	1.8	2.0	2.0	1.6	1.6	1.4
18 Portugal	1.3	1.3	4.7	4.9	3.6	2.3	0.9	0.5	0.4	1.4	1.6	2.3	0.5	1.3	2.2	0.4	1.9	2.2	2.2	1.8	1.8	1.5
19 Sweden	1.1	1.1	4.6	4.7	3.5	2.2	0.7	0.3	0.3	1.2	1.4	2.2	0.3	1.1	2.0	0.3	1.8	2.0	2.0	1.6	1.6	1.4
20 Eur	1.3	1.3	4.8	4.9	3.6	2.4	0.9	0.5	0.4	1.4	1.6	2.3	0.5	1.3	2.2	0.5	1.9	2.2	2.2	1.8	1.8	1.5
21 Turkey	6.2	6.1	9.8	9.9	8.6	7.3	5.7	5.3	5.3	6.3	6.5	7.3	5.3	6.1	7.1	5.3	6.8	7.1	7.1	6.7	6.7	6.4
22 ROW	3.3	3.2	6.8	6.9	5.6	4.3	2.8	2.4	2.4	3.3	3.5	4.3	2.4	3.2	4.1	2.4	3.9	4.2	4.2	3.8	3.8	3.5

Commodity	Logistics Service
Sea transport services	Maritime Transport
Air transport services	Air Transport
Oil, gas, petroleum, coal products, chemicals, rubber, plastic products	Liquid Bulk Products
Motor vehicles and parts	Special
Live animals	Break Bulk Products
Paddy rice, vegetables, fruits, animal products, raw milk, wool, silk-worm cocoons, fishery, cattle meat, sheep meat, horse meat, meat products nec, vegetable oils and fats, dairy products, processed rice, food products nec.	Containerazable Agricultural Products
Beverages, tobacco products, textiles, wearing apparel, leather products, wood products, paper products, metals nec, metal products, transport equipment nec, electronic equipment, machinery and equipment nec, manufactures nec.	Containerazable Manufactured Products
Wheat, cereal grains nec, oil seeds, sugar cane, sugar beet, plant-based fibers, crops nec, forestry, coal, minerals nec, sugar, mineral products nec, ferrous metals.	Dry Bulk Products
Electricity, gas and water, construction, trade, transport nec, communication, financial services nec, insurance, business services nec, recreational services, public administration and defense, health, education and dwellings	Other Services

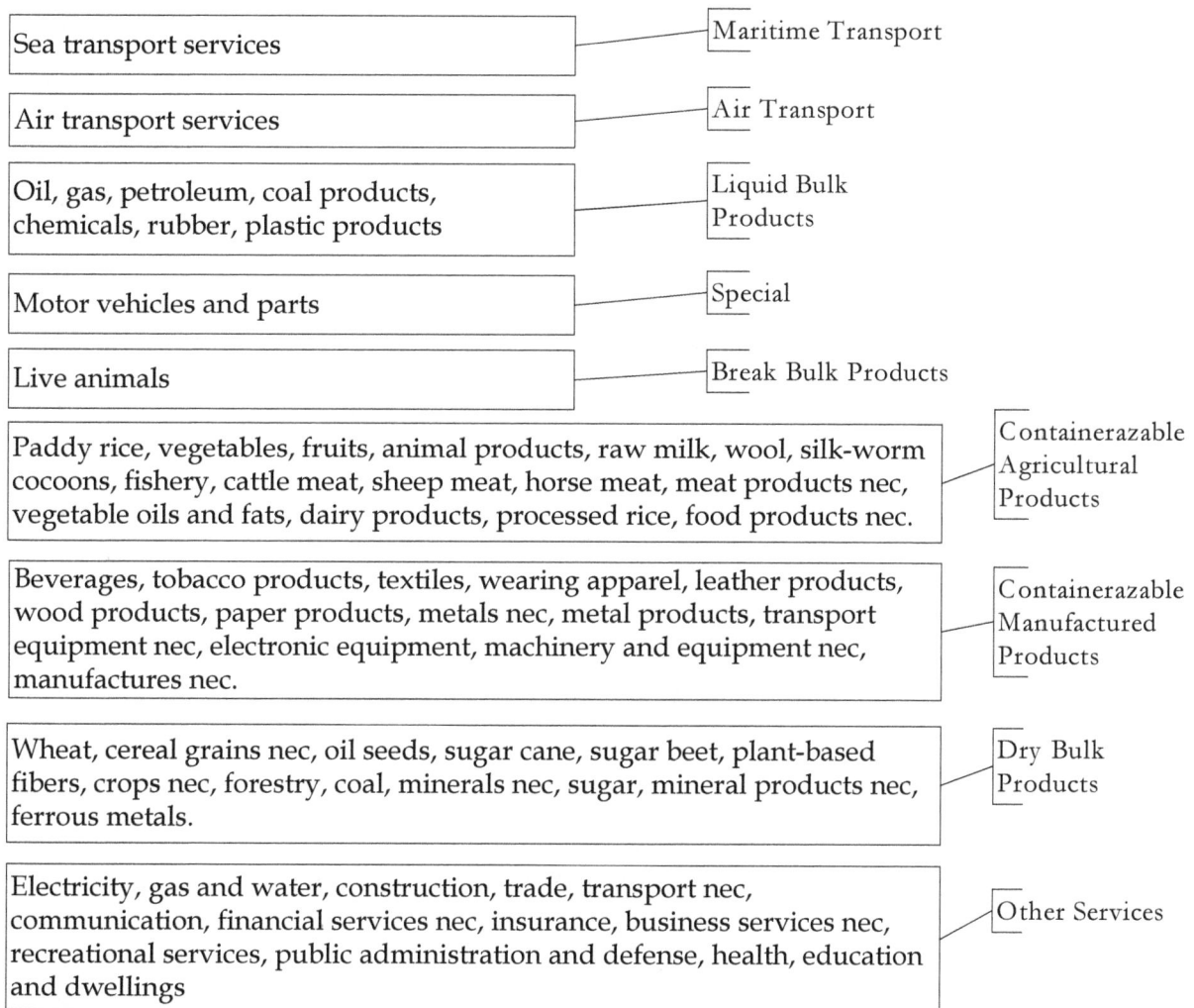

Figure 1. Commodities and sample aggregation (Source: GTAP Database)

Figure 1 depicts a sample aggregation of the commodities to represent their corresponding logistics services.

4. Conclusion

This study looked into the effects on logistics services based an economies in long-term. The results indicate that various logistics services from/to economies exhibit increases or decreases. These results are reported by using dynamic (CGE) simulations.

References

Akune, Y., Okiyama, M., & Tokunaga, S. (2015). Economic Evaluation of Dissemination of High Temperature-Tolerant Rice in Japan Using a Dynamic

Computable General Equilibrium Model. *Jarq-Japan Agricultural Research Quarterly, 49*(2), 127-133.

Arndt, C., Pauw, K., & Thurlow, J. (2012). Biofuels and economic development: A computable general equilibrium analysis for Tanzania. *Energy Economics, 34*(6), 1922-1930. doi: 10.1016/j.eneco.2012.07.020

Asafu-Adjaye, J., & Mahadevan, R. (2013). Implications of CO2 reduction policies for a high carbon emitting economy. *Energy Economics, 38*, 32-41. doi: 10.1016/j.eneco.2013.03.004

Bao, Q., Tang, L., Zhang, Z. X., & Wang, S. Y. (2013). Impacts of border carbon adjustments on China's sectoral emissions: Simulations with a dynamic computable general equilibrium model. *China Economic Review, 24*, 77-94. doi: 10.1016/j.chieco.2012.11.002

Barkhordar, Z. A., & Saboohi, Y. (2013). Assessing alternative options for allocating oil revenue in Iran. *Energy Policy, 63*, 1207-1216. doi: 10.1016/j.enpol.2013.08.099

Berrittella, M., & Zhang, J. (2015). Fiscal sustainability in the EU: From the short-term risk to the long-term challenge. *Journal of Policy Modeling, 37*(2), 261-280. doi: 10.1016/j.jpolmod.2015.02.004

Bhattarai, K. (2015). Financial deepening and economic growth. *Applied Economics, 47*(11), 1133-1150. doi: 10.1080/00036846.2014.993130

Bhattarai, K., & Dixon, H. (2014). Equilibrium Unemployment in a General Equilibrium Model with Taxes. *Manchester School, 82*, 90-128. doi: 10.1111/manc.12066

Boccanfuso, D., Joanis, M., Richard, P., & Savard, L. (2014). A comparative analysis of funding schemes for public infrastructure spending in Quebec. *Applied Economics, 46*(22), 2653-2664. doi: 10.1080/00036846.2014.909576

Bourne, M., Childs, J., & Philippidis, G. (2012). Reaping what others have sown: Measuring the Impact of the global financial crisis on Spanish agriculture. *Itea-Informacion Tecnica Economica Agraria, 108*(4), 405-425.

Breisinger, C., & Ecker, O. (2014). Simulating economic growth effects on food and nutrition security in Yemen: A new macro-micro modeling approach. *Economic Modelling, 43*, 100-113. doi: 10.1016/j.econmod.2014.07.029

362

Breisinger, C., Engelke, W., & Ecker, O. (2012). Leveraging Fuel Subsidy Reform for Transition in Yemen. *Sustainability, 4*(11), 2862-2887. doi: 10.3390/su4112862

Cheong, I., & Tongzon, J. (2013). Comparing the Economic Impact of the Trans-Pacific Partnership and the Regional Comprehensive Economic Partnership. *Asian Economic Papers, 12*(2), 144-164. doi: 10.1162/ASEP_a_00218

Chi, Y. Y., Guo, Z. Q., Zheng, Y. H., & Zhang, X. P. (2014). Scenarios Analysis of the Energies' Consumption and Carbon Emissions in China Based on a Dynamic CGE Model. *Sustainability, 6*(2), 487-512. doi: 10.3390/su6020487

Cockburn, J., Emini, A. C., & Tiberti, L. (2014). Impacts of the global economic crisis and national policy responses on children in Cameroon. *Canadian Journal of Development Studies-Revue Canadienne D Etudes Du Developpement, 35*(3), 396-418. doi: 10.1080/02255189.2014.934212

Dai, H. C., Masui, T., Matsuoka, Y., & Fujimori, S. (2012). The impacts of China's household consumption expenditure patterns on energy demand and carbon emissions towards 2050. *Energy Policy, 50*, 736-750. doi: 10.1016/j.enpol.2012.08.023

Decreux, Y., & Fontagne, L. (2015). What Next for Multilateral Trade Talks? Quantifying the Role of Negotiation Modalities. *World Trade Review, 14*(1), 29-43. doi: 10.1017/s1474745614000354

Doumax, V., Philip, J. M., & Sarasa, C. (2014). Biofuels, tax policies and oil prices in France: Insights from a dynamic CGE model. *Energy Policy, 66*, 603-614. doi: 10.1016/j.enpol.2013.11.027

Faehn, T., & Bruvoll, A. (2009). Richer and cleaner-At others' expense? *Resource and Energy Economics, 31*(2), 103-122. doi: 10.1016/j.reseneeco.2008.11.001

Femenia, F. (2010). Impacts of Stockholding Behaviour on Agricultural Market Volatility: A Dynamic Computable General Equilibrium Approach. *German Journal of Agricultural Economics, 59*(3), 187-201.

Femenia, F., & Gohin, A. (2013). On the optimal implementation of agricultural policy reforms. *Journal of Policy Modeling, 35*(1), 61-74. doi: 10.1016/j.jpolmod.2012.05.019

Furuya, J., Tokunaga, S., Okiyama, M., Akune, Y., Kunimitsu, Y., Aizaki, H., & Kobayashi, S. (2015). Economic Evaluation of Agricultural Mitigation and Adaptation Technologies for Climate Change: Model Development for Impact

Analysis and Technological Assessment. *Jarq-Japan Agricultural Research Quarterly, 49*(2), 119-125.

Gohin, A., & Rault, A. (2013). Assessing the economic costs of a foot and mouth disease outbreak on Brittany: A dynamic computable general equilibrium analysis. *Food Policy, 39*, 97-107. doi: 10.1016/j.foodpol.2013.01.003

He, J., Chen, X. K., Shi, Y., & Li, A. H. (2007). Dynamic computable general equilibrium model and sensitivity analysis for shadow price of water resource in China. *Water Resources Management, 21*(9), 1517-1533. doi: 10.1007/s11269-006-9102-7

Hosoe, N. (2014). Japanese manufacturing facing post-Fukushima power crisis: a dynamic computable general equilibrium analysis with foreign direct investment. *Applied Economics, 46*(17), 2010-2020. doi: 10.1080/00036846.2014.892198

Jiang, X. M., & Mai, Y. H. (2015). The social welfare housing project and its effects in China. *Journal of Systems Science & Complexity, 28*(2), 393-408. doi: 10.1007/s11424-014-3261-z

Lakatos, C., & Walmsley, T. (2012). Investment creation and diversion effects of the ASEAN-China free trade agreement. *Economic Modelling, 29*(3), 766-779. doi: 10.1016/j.econmod.2012.02.004

Lanzi, E., Chateau, J., & Dellink, R. (2012). Alternative approaches for levelling carbon prices in a world with fragmented carbon markets. *Energy Economics, 34*, S240-S250. doi: 10.1016/j.eneco.2012.08.016

Li, N., Wang, X. J., Shi, M. J., & Yang, H. (2015). Economic Impacts of Total Water Use Control in the Heihe River Basin in Northwestern China-An Integrated CGE-BEM Modeling Approach. *Sustainability, 7*(3), 3460-3478. doi: 10.3390/su7033460

Liang, Q. M., Wang, Q., & Wei, Y. M. (2013). Assessing the Distributional Impacts of Carbon Tax among Households across Different Income Groups: The Case of China. *Energy & Environment, 24*(7-8), 1323-1346.

Liang, Q. M., Yao, Y. F., Zhao, L. T., Wang, C., Yang, R. G., & Wei, Y. M. (2014). Platform for China Energy & Environmental Policy Analysis: A general design and its application. *Environmental Modelling & Software, 51*, 195-206. doi: 10.1016/j.envsoft.2013.09.032

Liu, Y., & Lu, Y. Y. (2015). The Economic impact of different carbon tax revenue recycling schemes in China: A model-based scenario analysis. *Applied Energy, 141,* 96-105. doi: 10.1016/j.apenergy.2014.12.032

Mabugu, R. E., Fofana, I., & Chitiga-Mabugu, M. R. (2015). Pro-Poor Tax Policy Changes in South Africa: Potential and Limitations. *Journal of African Economies, 24,* II73-II105. doi: 10.1093/jae/eju038

Mai, Y. H., Peng, X. J., Dixon, P., & Rimmer, M. (2014). The economic effects of facilitating the flow of rural workers to urban employment in China. *Papers in Regional Science, 93*(3), 619-642. doi: 10.1111/pirs.12004

Mariano, M. J. M., Giesecke, J. A., & Tran, N. H. (2015). The effects of domestic rice market interventions outside business-as-usual conditions for imported rice prices. *Applied Economics, 47*(8), 809-832. doi: 10.1080/00036846.2014.980576

Matovu, J. M. (2012). Trade Reforms and Horizontal Inequalities: The Case of Uganda. *European Journal of Development Research, 24*(5), 753-776. doi: 10.1057/ejdr.2012.35

Parrado, R., & De Cian, E. (2014). Technology spillovers embodied in international trade: Intertemporal, regional and sectoral effects in a global CGE framework. *Energy Economics, 41,* 76-89. doi: 10.1016/j.eneco.2013.10.016

Philip, J. M., Sanchez-Choliz, J., & Sarasa, C. (2014). Technological change in irrigated agriculture in a semiarid region of Spain. *Water Resources Research, 50*(12), 9221-9235. doi: 10.1002/2014wr015728

Philippidis, G., & Hubbard, L. (2005). A dynamic computable general equilibrium treatment of the ban on UK beef exports: A note. *Journal of Agricultural Economics, 56*(2), 307-312. doi: 10.1111/j.1477-9552.2005.00006.x

Qin, C. B., Bressers, H. T. A., Su, Z., Jia, Y. W., & Wang, H. (2011). Assessing economic impacts of China's water pollution mitigation measures through a dynamic computable general equilibrium analysis. *Environmental Research Letters, 6*(4), 15. doi: 10.1088/1748-9326/6/4/044026

Ricci, O. (2012). Providing adequate economic incentives for bioenergies with CO2 capture and geological storage. *Energy Policy, 44,* 362-373. doi: 10.1016/j.enpol.2012.01.066

Ruamsuke, K., Dhakar, S., & Marpaung, C. O. P. (2015). Energy and economic impacts of the global climate change policy on Southeast Asian countries: A general equilibrium analysis. *Energy, 81,* 446-461. doi: 10.1016/j.energy.2014.12.057

Saveyn, B., Paroussos, L., & Ciscar, J. C. (2012). Economic analysis of a low carbon path to 2050: A case for China, India and Japan. *Energy Economics, 34,* S451-S458. doi: 10.1016/j.eneco.2012.04.010

Schenker, O. (2013). Exchanging Goods and Damages: The Role of Trade on the Distribution of Climate Change Costs. *Environmental & Resource Economics, 54*(2), 261-282. doi: 10.1007/s10640-012-9593-z

Seung, C. K., & Kraybill, D. S. (2001). The effects of infrastructure investment: A two-sector dynamic computable general equilibrium analysis for Ohio. *International Regional Science Review, 24*(2), 261-281. doi: 10.1177/016001701761013150

Verikios, G., Dixon, P. B., Rimmer, M. T., & Harris, A. H. (2015). Improving health in an advanced economy: An economywide analysis for Australia. *Economic Modelling, 46,* 250-261. doi: 10.1016/j.econmod.2014.12.032

Wittwer, G., & Banerjee, O. (2015). Investing in irrigation development in North West Queensland, Australia. *Australian Journal of Agricultural and Resource Economics, 59*(2), 189-207. doi: 10.1111/1467-8489.12057

Wu, T., Zhang, M. B., & Ou, X. M. (2014). Analysis of Future Vehicle Energy Demand in China Based on a Gompertz Function Method and Computable General Equilibrium Model. *Energies, 7*(11), 7454-7482. doi: 10.3390/en7117454

Xie, W., Li, N., Wu, J. D., & Hao, X. L. (2014). Modeling the economic costs of disasters and recovery: analysis using a dynamic computable general equilibrium model. *Natural Hazards and Earth System Sciences, 14*(4), 757-772. doi: 10.5194/nhess-14-757-2014

Xie, W., Li, N., Wu, J. D., & Hao, X. L. (2015). Disaster Risk Decision: A Dynamic Computable General Equilibrium Analysis of Regional Mitigation Investment. *Human and Ecological Risk Assessment, 21*(1), 81-99. doi: 10.1080/10807039.2013.871997

ABOUT THE AUTHOR

Turkay Yildiz received his Ph.D. from the Institute of Marine Sciences and Technology, Dokuz Eylul University, Izmir, Turkey. He has degrees in Electronics, Management, and he received his Master's Degree in Logistics Management from Izmir University of Economics. He has a number of peer reviewed publications and conference presentations at various countries in such fields as transportation, logistics and supply chains. He also has various levels of expertise in the applications of IT.

www.ingramcontent.com/pod-product-compliance
Lightning Source LLC
Chambersburg PA
CBHW062017090426
42811CB00005B/882